Philosophical Studies Series

Volume 158

Philosophical Studies Series aims to provide a forum for the best current research in contemporary philosophy broadly conceived, its methodologies, and applications. Since Wilfrid Sellars and Keith Lehrer founded the series in 1974, the book series has welcomed a wide variety of different approaches, and every effort is made to maintain this pluralism, not for its own sake, but in order to represent the many fruitful and illuminating ways of addressing philosophical questions and investigating related applications and disciplines.

The book series is interested in classical topics of all branches of philosophy including, but not limited to:

- Ethics
- Epistemology
- Logic
- Metaphysics
- Philosophy of language
- Philosophy of logic
- Philosophy of mind
- Philosophy of religion
- Philosophy of science

Special attention is paid to studies that focus on:

- the interplay of empirical and philosophical viewpoints
- the implications and consequences of conceptual phenomena for research as well as for society
- philosophies of specific sciences, such as philosophy of biology, philosophy of chemistry, philosophy of computer science, philosophy of information, philosophy of neuroscience, philosophy of physics, or philosophy of technology; and
- contributions to the formal (logical, set-theoretical, mathematical, information-theoretical, decision-theoretical, etc.) methodology of sciences.

Likewise, the applications of conceptual and methodological investigations to applied sciences as well as social and technological phenomena are strongly encouraged.

Philosophical Studies Series welcomes historically informed research, but privileges philosophical theories and the discussion of contemporary issues rather than purely scholarly investigations into the history of ideas or authors. Besides monographs, *Philosophical Studies Series* publishes thematically unified anthologies, selected papers from relevant conferences, and edited volumes with a well-defined topical focus inside the aim and scope of the book series. The contributions in the volumes are expected to be focused and structurally organized in accordance with the central theme(s), and are tied together by an editorial introduction. Volumes are completed by extensive bibliographies.

The series discourages the submission of manuscripts that contain reprints of previous published material and/or manuscripts that are below 160 pages/88,000 words.

For inquiries and submission of proposals authors can contact the editor-in-chief Mariarosaria Taddeo via: massimo.durante@unito.it

Margherita Bianchi

Philosophical Implications of Plant Cognition

A Pluralistic Overview

Springer

Margherita Bianchi
Department of Philosophy
University of Rome "La Sapienza"
Rome, Italy

ISSN 0921-8599 ISSN 2542-8349 (electronic)
Philosophical Studies Series
ISBN 978-3-032-01327-9 ISBN 978-3-032-01328-6 (eBook)
https://doi.org/10.1007/978-3-032-01328-6

English translation — with modifications and additions — of the doctoral thesis "Le Implicazioni Filosofiche della Cognizione Vegetale" by Margherita Bianchi, University of Rome La Sapienza (2024).

This Springer imprint is published by the registered company Springer Nature Switzerland AG
The registered company address is: Gewerbestrasse 11, 6330 Cham, Switzerland

Acknowledgments

I would like to thank all those who have supported me, both directly and indirectly, and who have offered wise counsel throughout this diverse and complex research journey. I am grateful to my family, especially my parents, my friends, colleagues, and all the researchers I have had the pleasure to meet, who have combined passion with precision and originality in their work.

A special thanks goes to Professor E. Gagliasso, as well as to Professors S. Pollo, P. Donatelli, P. Calvo, and U. Castiello.

Contents

1 Introduction 1

2 Aims of Research 15
2.1 Main Goals and Prospects 16
2.1.1 The Challenge of Nonstandard Cognition: Are All Living Systems Cognitive Systems? 17
2.1.2 Toward a Cognitive Plant Ethology 17
2.1.3 Research Areas and Significance 20
2.2 Task of Scientific Research 23
2.2.1 Processes and Measurements 24
2.2.2 Theory and Practice in Scientific Practices 25
2.2.3 Science and Bias: Critical Reflection and Reduction 29
2.2.4 Progress in Science 32
2.3 Task of Philosophical Research 34
2.3.1 Terminological and Conceptual Analysis 36
2.3.2 Which Conception of "Intelligence"? 38
2.3.3 Why Plant "Cognition"? 40
2.4 New Focus on an Old Issue 43
2.5 Perceptual-Cognitive Challenges in Understanding Plant Behavior 48
2.5.1 Locomotion and Cognition 48
2.5.2 Theoretical and Technological Advances 50
2.5.3 The Invisible: Latent States of Life and "Silent Languages" of Plants 52
2.5.4 Plants and Sympathy: The Issue of Pain and Its Behavioral Manifestations 54
2.6 Contemporary Insights 56
2.7 Underlying Theoretical Framework 59
2.7.1 The Problem of Circularity and the Pluralistic Approach 60
2.7.2 Comparative Analysis of Cognition 63

2.7.3 Anthropocentrism, Anthropomorphism, and Anthropodenial ... 67
2.7.4 Toward an Inclusive Approach to Evolutionary Cognition ... 76
2.8 Ecology and Cognition: A Relational Approach to the Activities of Living Systems ... 83
2.8.1 From Systemic Theories to Different Levels of Organization in Nature ... 85
2.8.2 Beyond the Concept of Interaction: Conservation and Transformation ... 86
2.8.3 Relational Nature of Processes ... 90
2.8.4 Ecological Explanation of Relational Behaviors ... 93
2.8.5 Levels of Ecological Organization and Ways of Studying Ecosystems ... 94
2.8.6 Toward a Plant Cognitive Ecology: The Role of Context and Other Organisms ... 95
References ... 100

3 Plant Characteristics and Activities ... 115
3.1 Specificities in Plant Organization ... 116
3.1.1 Body Structure and Function: The Difference Between Sessile and Vagile Modes of Life ... 117
3.1.2 Phenotypic Plasticity ... 119
3.1.3 Levels of Complexity ... 120
3.2 Senses, Perception, Action ... 121
3.2.1 Signaling in Plants ... 122
3.2.2 Plant Perceptual Sensitivity ... 124
3.2.3 Plant Movements ... 128
3.2.4 Movements of Climbing Plants ... 129
3.2.5 Motor Intentions? ... 133
3.3 On Behavior ... 134
3.3.1 Some Peculiarities of Plant Behavior ... 136
3.3.2 Signaling and Communication in Plants ... 141
3.4 Processes of Memory, Learning, Decision-Making, Anticipation, and Attention ... 145
3.4.1 Types of Memory ... 145
3.4.2 Memory Processes in Plants ... 148
3.4.3 Learning Processes ... 150
3.4.4 Ability to Learn in Plants ... 152
3.4.5 Anticipatory Behavior ... 156
3.4.6 Decision-Making ... 159
3.4.7 Response Selection Processes in Plants ... 160
3.4.8 Attentional Processes ... 162
3.4.9 Attention in Plants? ... 163

3.5 Individuality and Consciousness 165
3.5.1 Individuality in Divisible Organisms: The Problem of Integration and Coordinated Responses 166
3.5.2 Conceptions of Consciousness 169
3.5.3 Consciousness and Self-Consciousness 174
3.5.4 Awareness in Plants? 175
3.6 Intelligence and Cognition in Plants 179
3.6.1 On Intelligence 180
3.6.2 Plant Intelligence? 184
3.6.3 Cognitive Behaviors in Plants 187
References 191

4 Philosophical Areas of Impact 205
4.1 Ontological Issues: Processes of Categorization of Reality 207
4.1.1 Habits of Thought: Utility and Simplification 209
4.1.2 Subdivisions Methods 210
4.1.3 Cognitive Ontologies: The Hallmarks of Cognition 215
4.1.4 Process Ontologies 217
4.2 Experiential Issues: Feeling and Acting 219
4.2.1 Perception-Action 219
4.2.2 The Concept of Agency 225
4.2.3 Are the Elements of Behavior Innate, Learned, or Both? . . 227
4.2.4 To Be a Plant 230
4.2.5 Observing Behavior: A Behaviorist Reintroduction? 232
4.3 Epistemological Issues: Situated and Natural Practices of Knowing 235
4.3.1 Epistemological Analyses 236
4.3.2 Conceptions and Types of Knowledge 237
4.3.3 Knowledge and Action: Relational Dimensions and Practical Forms of Knowing 240
4.3.4 Eco-Evolutionary Framework and the Biological Role of Knowledge 242
4.3.5 Knowing: Constraints and Possibilities 244
4.3.6 The Logic of Knowledge Processes 247
4.3.7 A Natural Logic? 249
4.3.8 Distinction of Quantity and Difference 250
4.3.9 Social Life and Game Theories 252
4.4 Plants in Cognitive Sciences 255
4.4.1 Historical Evolution of Approaches to Cognition 256
4.4.2 Bodily, Enactive, and Extended Turn of Cognition 260
4.4.3 4E Cognition and Systemic-Relational Approach to Plant Cognition 267
4.4.4 A Trivialization of the Cognitive? 280
4.4.5 Bio-Ecological Cognition: A Processual Perspective 281

4.5 Plants in the Philosophy of Mind. 284
4.5.1 Contemporary Philosophy of Mind and Historical Reconstruction of the Concept of Mind. 286
4.5.2 Embodied, Ecological and Processual Conception of the Mind. 290
4.5.3 Beyond the Human Mind. 292
4.5.4 Representational Cognition: An Open Question 297
4.5.5 Varieties of Conceptions of Representation. 298
4.5.6 Advantages and Disadvantages of Representations. 302
4.5.7 Representations in Plants? 306
4.5.8 Cognition and Temporality of Processes 312
References. 318

Index. 331

About the Author

Margherita Bianchi obtained her PhD in Philosophy from the Department of Philosophy at Sapienza University of Rome. Her research focused on the philosophical implications of plant behavior and cognition. During her doctoral studies, she also spent a visiting period at the MINT Lab, University of Murcia. She subsequently joined the Mind(the)Plant Lab at the University of Padua as a Postdoctoral Researcher. Her research interests include philosophy of cognitive science, philosophy of mind, epistemology, philosophy of biology, philosophy of science, and biosemiotics.

Chapter 1
Introduction

Abstract The structural and functional specificities of plants – sessile organisms that lack a nervous system – offer a valuable opportunity to critically reassess and broaden current research horizons regarding the type and extension of cognitive processes in the living world.
By identifying the most promising epistemic and practical approaches, this perspective lays the groundwork for a more comprehensive and integrated understanding of biological cognition, with particular emphasis on ecological interactions and the inherently relational development of organisms.

This work aims to initiate an interdisciplinary reflection on how ongoing studies of plant capabilities may clarify and enrich some questions in behavioral studies, cognitive science, and philosophy of mind. Additional connections are explored with research in the philosophy of science and biology, systems theory, complexity science, epistemology, and theories of knowledge.

This book investigates how the study of plant behavior can generate novel lines of inquiry by reframing, from an alternative perspective, traditional questions in the cognitive domain. It is structured around 1) a historical reconstruction of relevant debates, followed by 2) an analysis of recent scientific literature on plant characteristics and abilities, and 3) concludes with a discussion of the theoretical implications across various branches of philosophical research.

Such an approach may significantly contribute to the revision, or even transformation, of established conceptual frameworks, offering new insights into core notions such as perception, action, behavior, intelligence, cognition, and mind.

Keywords Plant behavior and cognition · Non-neural cognition · Ecological cognition · Philosophy of cognitive science · Philosophy of mind · Philosophy of science · Epistemology

Some years ago, when I decided to devote my research to the theme of plant cognition, plants – sessile organisms without a nervous system – did not appear in the textbooks of cognitive science and philosophy of mind. There was no section dedicated to the study of the abilities of these organisms, unlike the common references

M. Bianchi, *Philosophical Implications of Plant Cognition*, Philosophical Studies Series 158, https://doi.org/10.1007/978-3-032-01328-6_1

to the cognitive abilities of other animals and even to comparisons with the functioning of artificial devices. This absence was one of the main reasons why I decided to embark on this arduous and challenging path. Indeed, plant cognition is an emerging field of research that has gained much attention and aroused great interest only in recent years.

Regarding the structure of the book, I chose to integrate multiple levels of research into a single text: historical, scientific, and theoretical. The idea is both to make distinctions between different theoretical perspectives and to foster a dialogue throughout the book. The first part serves as an introduction, outlining the aims of the research and the general approach adopted to study these themes. This is followed by a section dedicated to the historical evolution of interest in and consideration of plants, tracing this trajectory from antiquity to the present, including the current state of research on plant cognition. The second part focuses on the scientific literature available on the characteristics and capacities of plants. Lastly, the third section explores the interdisciplinary implications of these studies in various areas of philosophical research, identifying potential connections with practical and experimental applications.

The main aim of this analysis is to initiate a reflection on the potential clarification and enrichment in the fields of cognition, behavior and intelligence, deriving from ongoing studies on plant abilities.

The study of plant cognition, considering the distinct body organization and ecological interactions of these organisms, presents a range of significant and often counterintuitive theoretical and practical challenges. It offers a valuable opportunity to explore important issues – primarily, but not exclusively, in cognitive science, behavioral studies, and philosophy of mind – from a fundamentally different starting point than traditional research on human and animal capacities. Indeed, this type of research allows for a critical rethinking and expansion of the scope of cognitive processes and abilities in the living world. It enables an examination of cognition in both neural and non-neural systems, aiming to provide a more comprehensive and integrated understanding of the evolution of cognitive processes. Particular attention is given to the ecological factors that shape and transform these processes throughout the inevitably relational development of organisms.

Thus, the study of plant behaviors – by integrating, reforming, and generating new questions – can significantly clarify, rethink, and potentially revolutionize some of our conventional categories of reasoning, such as reinterpreting our concepts of "perception," "action," "behavior," "memory," "learning," "cognition," and "mind." Hence, the title "*Philosophical Implications of Plant Cognition*" was chosen. Therefore, with a dynamic of mutual influence in the advancement of theoretical and experimental research, the direction of this research is twofold: 1) from theoretical reflection to clarify plant characteristics and behaviors and 2) from experimental observations of the peculiarities of plant abilities to deepen the issues subject to analysis in various fields of philosophical research.

From a methodological standpoint, I opted to raise more questions than provide (definitive) answers. My aim is to highlight the most promising approaches to the various issues explored throughout the book. Given the complexity and multidisciplinary nature of this ongoing research, I do not present a single thesis to defend. Instead, I offer numerous points and suggestions for reflection that can be developed in future studies. Therefore, this is a wide-ranging work in which much of the research mentioned is at an embryonic and hypothetical stage, and numerous challenges still lie ahead (one of the aspects that makes this research simultaneously stimulating and risky). These pages should be seen as an extended and multifaceted reflection on the impact of studies across various levels of analysis concerning plant activities in scientific and philosophical research. The manuscript serves as a foundational work plan, a theoretical basis for future in-depth studies. I adopt a pluralistic and interdisciplinary approach that remains open to future additions and modifications, aiming to explore the potential of the most promising approaches for studying plant capacities based on current evidence. Additionally, I indicate which paths are likely to be unproductive and should be excluded. Consequently, there will be no real conclusions. Instead, the various questions raised – many of which are closely interrelated – will receive not final answers, but rather provisional, renewed, or alternative theoretical perspectives throughout the text. The analysis is intended to be comprehensive, but not exhaustive, conclusive, or demonstrative. It highlights some of the most promising approaches and research directions currently available, which should be further explored and tested in the coming years.

Regarding the content of the next three chapters, it is helpful to briefly recall some of the main topics they address. In the chapter"*Aims of Research*," as can be guessed from the title itself, I outline the primary objectives and prospects of studying plant characteristics and capacities. This chapter highlights how such research can engage with various issues and expand the scope of consideration to include organisms traditionally not regarded as capable of cognition, thus raising several questions across multiple fields of study. Following a series of theoretical and methodological advancements in exploring the behavioral and cognitive aspects of plant activities, attention will be directed to the observation that, regardless of the type of organism and its species-specific behavioral expressions, there appear to be abilities common to all cognitive systems, whether neural or non-neural. These capabilities arise from multisensory perception processes and include abilities such as communication, memory, learning, and anticipation. When considered together, these aspects enable the initiation of behavioral responses that are not rigidly preprogrammed but can be adjusted and refined throughout the organism's lifespan.

Given this potential for variability and behavioral modulation, the question has begun to arise, as has been the case previously for other living things, whether it is possible and useful to treat plants as organisms with the ability to act and exhibit goal-directed behaviors, observable in their ability to solve survival problems, to affect the surrounding environment by leaving conditioning traces, and to distinguish multiple environmental factors, initiating different responses. Recent advances in the detection of these aspects have made it possible to begin framing these behaviors within the context of ethological-cognitive studies.

Constantly incorporating new findings from scientific-experimental analyses, the chapter goes on to emphasize the role of philosophical research on plant cognition. This research allows for the reformulation of critical questions, such as defining the concept of "cognition" and its boundaries. It prompts reflection on the key elements required to describe forms of cognition, mind, or intelligence, and investigates the processes that enable behaviors we recognize as goal-oriented actions. Philosophical research therefore plays a fundamental role in the field of cognitive science in clarifying the assumptions involved in the various approaches to the study of cognition. It also has the capacity to challenge commonly accepted beliefs and established certainties when necessary.

Another issue addressed in this work, central to the aims of a philosophical analysis, is the need to develop a more inclusive and conceptually adequate application of terms and notions traditionally employed in the study of human and animal cognition, one that can also account for the case of plants.

In the section dedicated to the transformation of the characteristics and capacities of plants over the centuries, I briefly recall the positions of some naturalists and philosophers who have contributed to the progressive understanding of plant activities, which generally tend to be neglected, misinterpreted or ignored.

I then proceed to examine the main perceptual-cognitive hurdles in understanding plant behavior. These challenges have hindered the possibility of adequately studying the capacities of these organisms, starting from the fact that cognition has generally been associated with organisms with a nervous system and capable of locomotion (plants are sessile and able to move but, generally, not to locomote from the place where they are rooted).

Owing to technological advances in the study of plant movements and the refinement of methods to detect the different forms of signaling inside and outside plant bodies, it has been possible to measure variations in responses depending on environmental stimuli and contexts of interaction. Some aspects have long been ignored, such as the "silent languages" of plants, which are related mainly to the ability to synthesize and recognize chemical substances but also to the communication of shapes and colors, which are closely linked to different interactions with animal organisms. In line with the ability to identify communication signals, I argue that another factor limiting our understanding of plant activities is their lack of overt expressions of pain, similar to our own, despite their high perceptual sensitivity. Unlike animals, whose reactions to wounds and damage are easily observed and elicit sympathy, plants do not display such visible signs, making it more difficult for us to relate to them.

In the following sections, I report on contemporary insights into the emerging field of plant cognition and behavior, retracing, in particular, the stages that led to the consolidation of this research in the last 20 years, indicating the main objectives of this young discipline.

I then focus on the basic theoretical framework and the approaches used to study plant behavior, specifying some of the most promising epistemic tools to carry out this branched research, within an empirically informed context of analysis by the biogenic approach to cognition.

In line with this purpose, I stress the need in the fields of comparative psychology and evolutionary cognition to increase research for the development of a branch of meta-analysis, i.e., for the thoughtful identification of behavioral traits (differently expressed at the phenotypic level across species) to be considered in the study of cognition among the living systems.

Within a comparative framework, I indicate the need for a critical comparison with the widespread tendencies that have both favored and hindered the understanding of the activities of non-human organisms. These interpretative models include forms of anthropocentrism (and anti-anthropocentrism), anthropomorphism (and anti-anthropomorphism). Then, once the risk of carrying out circular reasoning has been addressed, on the basis of comparisons that reduce to standard models of human (or animal) cognition, I invite the development of a bio-ecologically informed approach to the study of evolutionary cognition, such that it can also include the basic cognitive abilities of non-neural organisms such as plants.

Lastly, in the last section of the chapter, I express the need to develop a systemic-relational approach to the study of living activity to outline an ecology of cognitive difference, deepening the link between current studies on plant cognition and the assumptions that can be derived from ecological and ecosystem studies, which are typically multifactorial and multilevel. Making this connection explicit is particularly important in the case of plants, given the strong link of these sessile organisms to their living contexts, in which it is essential to coordinate behavioral activity in coexistence with neighboring species. Many plant activities can be carried out because plants interact with members of their species and other plant and animal species but also with an exorbitant number of organisms belonging to microbial communities. In highlighting the relational nature of processes, the importance of ecological relationships is at the basis of the possibility of the modulation of many behaviors.

This level of analysis also provides the opportunity to reflect on the evolution of how systemic studies have been conceptualized, particularly concerning the complexity and organization of information across different types of systems in various research fields. These range from studies in physics to processual approaches for examining living systems, which can also be further developed within eco-systemic studies. Additionally, this approach enhances our understanding of the diverse forms of plant behavioral relationships, emphasizing the role of context and other organisms in shaping and refining responses, all within the framework of a bioecological approach to cognition.

In the chapter "*Plant Characteristics and Activities*" I report the main characteristics and activities displayed by plants, covered in the most recent scientific literature. The aim of the chapter, which is rich in examples to substantiate the theoretical analysis, is to provide an overview of the peculiarities of plant systems, both from the perspective of their eco-evolutionary history and from a functional and interactional point of view. The in-depth examination of the specificities of the plant world, along with the characteristics and capacities of plants, serves as a crucial bridge between the research objectives outlined in the previous chapter and the implications of these studies for various areas of philosophical inquiry, which are explored in the last chapter.

In the first section, I revisit the unique aspects of plant organization, focusing on the structural and functional characteristics of plant bodies, the contrast between the sessile "lifestyle" of plants and the vagile lifestyle of animals, highlighting the phenotypic plasticity of plants. These differences are then explored in relation to various approaches to study the levels of complexity in living systems.

Next, I delve into the issue of plant perceptual sensitivity, noting that plants possess senses analogous to those of animals, though differently organized and arranged. I review the various types of signaling involved in communicating environmental information. Additionally, I outline the primary forms of movement characteristic of plant life, which are influenced by specific ecological factors. My focus then shifts to the movements of climbing plants, particularly in relation to potential aspects of motor intentions, understood as goal-directed behavior.

I then introduce different ways for studying behavior, which involve movement and the ability to produce, recognize, and emit chemical compounds. My focus then shifts to plant communication. I also explore the abilities underlying different forms of cognitive behavior, including memory, learning, anticipation, decision-making, and the potential presence of attentional capacities in plants.

Next, I address the concept of individuality in plants, emphasizing the problem of the different levels of integration of information and the possibility of coordinated responses in modular organisms. In this context, I also address the controversial issue of plant consciousness, understood as a form of awareness of the surrounding environment, of the plant's own conditions and ecological needs (such as resources to obtain and damage to avoid), and the fundamental ability to distinguish itself from other species and genetically related organisms.

I conclude the chapter with an in-depth look at forms of plant intelligence and cognition, two concepts that have been increasingly debated in recent years, which, after having raised sensible criticisms and perplexities, in their current and most appropriate meanings, are finding application in the analyses of the emerging area of research on plant interactions. Although the terms 'intelligence' and 'cognition' are often used interchangeably to describe plant activities, I broadly distinguish between their different uses. I refer to biological intelligence as the capacity of cognitive systems to regulate and modify their behavior to solve survival problems. This involves selecting various behavioral strategies depending on the circumstances and improving responses to specific stimuli over time. Instead, biological cognition can be understood as the set of species-specific capacities of a cognitive system, which include forms of memory, learning, and environmental communication, enabling an organism to interact meaningfully with its surroundings, respond to survival-related demands, and pursue goals. The set of cognitive processes and abilities can be examined individually and subsumed, for practical purposes, into a more integrated idea of cognition.

In the final chapter, "*Philosophical Areas of Impact*," I explain the potential advancement of current theoretical perspectives in relation to this growing attention to the interactional dynamics of plants. Regardless of the appropriateness of applying cognitive frameworks to plant activities, the diverse body of research in this area has the potential to expand our understanding across various fields. Such studies can enrich, or even transform, our perspectives on cognition by exploring the essential processes and conditions underlying different forms of cognitive abilities, their scope, and the levels of biological organization at which they emerge. A thorough investigation of cognitive behaviors in plants and other non-neural organisms can contribute to developing a processual, systemic, and context-dependent view of biological cognition, addressing various strategies for solving survival problems in accordance with the biological and ecological specifics of different organisms.

This type of analysis, in terms of reciprocal influence dynamics, holds dual relevance. On one hand, it can stimulate research on the peculiarities of plant abilities by examining both similarities and differences with animal organisms. On the other hand, it provides connections for exploring theoretical concepts across various fields, such as perception-action, cognition, and mind, refining our analysis by considering how these capacities manifest across different systems in nature.

In the section dedicated to ontological issues, and in line with analyses from the philosophy of biology, I focus on the importance of the processes of categorization of reality, understood as functional, nondefinitive or normative subdivisions of the elements and processes to be investigated to set up a given research. Considering the vast array of phenomena and the diverse methods of describing them, establishing these categories is crucial for structuring and conducting research effectively. This task is relevant because, depending on what is identified, what its boundaries are, its extent, and how certain elements and processes are considered, different theoretical and experimental analyses follow. Ontological considerations are deeply interconnected to epistemological reflections, with the latter serving as the starting point. Even in studies on cognition, efforts have been made, not always on the basis of solid evidence, to establish where (and why) to draw the boundary between what is considered cognitive and what is not. Therefore, I encourage the development of a flexible and empirically informed approach to be continuously updated. This approach should allow for accurate clarification of the subject matter, the focus of the study, and the criteria used (such as whether a system can be considered cognitive, and if so, based on which characteristics and capabilities). In the context of studies of cognitive ontologies, related to the identification of the distinctive features of cognition, a suggestion is to deepen both the potential and the critical aspects of process ontologies approaches.

In the section on experiential issues, which analyzes forms of perception-action, I focus on the implications of a detailed examination of current research on the sensory capacities of plants. This analysis broadens our understanding of how organisms without nervous systems perceive, act, and experience their environments. These capacities, which involve behavioral flexibility, sensory perception, and interaction, differ from those of animal organisms more closely related to our species. The comparison with a different range of sensors and perceptual-agentive possibilities suggests the presence of an experiential plurality, relating to the different ways of "being in the world" of the organisms, linked to their peculiar ways of perceiving reality, acting on it, and having some experience, not necessarily self-conscious or reflective. Underlying the diverse forms of perception-action, which are necessary to initiate behavioral responses appropriate to different environmental challenges, beyond the distinct and contingent eco-evolutionary histories that have shaped different structural and functional organizations, lies environmental uncertainty, stemming from the variability of life contexts, as well as the scarcity and competition for resources.

Subsequently, I focus on the concept of *agency* to evaluate whether a version of it could be applied to the description of the acting capacities of plant systems, similar to what has recently been indicated for some simple non-neural unicellular organisms capable of goal-directed behaviors. Without downsizing the complexity and efficiency of actions governed by nervous system-controlled movements in animals, understanding the various forms of behavior in nature involves examining whether a comparable form of action in animals can be identified in plants. In plants, this challenge is to determine whether their movement capacities, related to different survival goals and chemical abilities, such as synthesis, emission, and recognition of numerous chemical compounds, serve as effective strategies for action. Although these behaviors are based on different mechanisms and processes, they can still influence the surrounding environment, representing distinct ways of acting and interacting with the environment, addressing survival challenges, and affecting the behavioral responses of other life forms. It is therefore worth questioning whether, and in what sense, it is appropriate to refer to a form of "plant action."

Consistent with this perspective, I examine the distinction between more strongly programmed behaviors (inherited through evolutionary processes within a species) and responses learned during an organism's individual development. These acquired responses exhibit variability based on the individual organism's condition and needs, in specific contexts. This analysis aims to understand how to interpret plant behaviors with a degree of flexibility and self-regulation, moving beyond the notion of plants as merely responding to environmental stimuli in a wholly invariable way.

Moreover, given the obvious impossibility of feeling what another organism feels, I address the study of behavior and the associated difficulties in observational and experimental practices, involving a critical comparison with classical behaviorist approaches and with more recent reformulations. The goal is to highlight both the limitations of behaviorism and the aspects that can be effectively revised and integrated into contemporary studies, this time not to deny the presence of cognitive abilities but, on the contrary, to indicate how to observe and study them, considering both behavioral patterns and variations in physiological activity in response to various situations and environmental stimuli.

In the section on epistemological issues and natural practices of knowing, I first review the goals of contemporary epistemological analyses, which aim for a more realistic understanding of human reasoning practices and various conceptions of knowledge. I then explore the potential for developing a perspective on knowing that can address the cognitive modes and strategies of phylogenetically distant organisms. Rather than focusing on the intellectual and propositional aspects studied in human cognitive research, I shift the focus to the practical and agentive-relational aspects of knowing. This involves examining how processes of acquisition, transmission, and learning of information evolve throughout an organism's life and across generations.

The philosophical analysis of plant abilities could deepen (expand and problematize) our understanding of the diverse forms of knowledge, illuminating their differences, possibilities, and limitations in addressing various survival challenges. Different cognitive systems navigate their environments effectively, demonstrating the ability to discriminate and recognize beneficial and harmful elements. This suggests a need to explore cognitive modalities and strategies in relation to their practical side and the ability to act, within a framework that re-integrates cognitive processes into a bio-ecological context. Consequently, I emphasize the significance of bodily and ecological-relational constraints on the cognitive abilities of different organisms.

In the last part of this section, I focus on the ability of plants to behave, hypothesizing that, similar to animals, plants possess some ability to distinguish quantities and differences, to differentiate elements within their environment, to interact with them, and achieve goals. This is particularly evident in plant foraging strategies, which seem to involve the capacity to detect varying amounts of nutrients dispersed in the soil.

Given that plants can discern quantities and differences among relevant environmental elements, an area worth further exploration is the connection between game theory applications in social interactions and the behavioral strategies of plants in the presence of other organisms. This includes interactions with members of the same species, different plant species, or organisms from other kingdoms. It appears that plants do not act in isolation, but rather adjust their strategies in response to the presence and behavior of other organisms.

In the section focused on integrating plants into cognitive science research, I consider the most effective and appropriate approaches for studying cognition in these organisms. I explore how current research on plant capacities can shed light on aspects of behavior and cognition that are often overlooked or underexplored. This includes understanding cognition across different taxa, whether in neural or non-neural systems, and at different levels of complexity and organization. In this way, it becomes possible to gain insights into general cognitive processes and behaviors.

I review several theoretical frameworks, including the connectionist approach, ecological psychology, the broad 4E cognition research program (embodied, embedded, enacted, and extended), and systemic and self-organizing approaches to cognition. I suggest exploring potential connections between these theoretical perspectives and specific aspects of plant behavior that warrant further investigation. Specifically, I recommend continuing to explore the potential of embodied approaches to move beyond the strictly neurocentric analysis of cognitive phenomena. This includes examining cognition within a broader, multiprocessual, and multilevel framework. Cognition can be viewed as the outcome of a series of processes that produce self-regulating behaviors in living systems, considering the variability of their interaction contexts.

More specifically, in the context of ecological psychology, I suggest exploring the relationship between cognition and possibilities of action detectable in the environment, with a focus on the ecological relevance of the interactions of plants with the salient elements of their environments (the ability of plants to detect invariants). I then suggest deepening the explanatory potential of the concept of network to describe some aspects of plant cognition, drawing links between the assumptions of neo-connectionist theories, current network neuroscience and network theories that investigate the functioning of biological signaling networks within the bodies of organisms. Plant body systems can be analyzed as network systems both at the level of the physical structure of their modular bodies and at the functional level of exchange of different types of signals. In this regard, a proposal that deserves further investigation is related to the comparison between the functional characteristics of neural networks and other types of interacting cells. In contemporary terms, a comparison is made between the properties of connectomes and interactomes, to identify common characteristics and untranslatable differences. Even with different specializations, levels of complexity and degrees of integration, connectomes and interactomes, as intercellular signaling systems, share key characteristics. They possess the ability for contextually constrained self-organization, facilitate the exchange of information, modulate the frequency of transduction signals based on stimulus intensity, and undergo morpho-functional changes that lead to learning in response to environmental changes. An analysis of the variability of signaling networks within plant bodies could help to understand how these organisms form new information flows and develop new tissues and organs under different circumstances.

Another connection that I recommend deepening is between enactive approaches and the cognitive abilities of plants, which possess a different sensorimotor organization. This is evident in their branching behaviors, both above ground, where they expand to occupy space and maximize light exposure, and below ground, where their roots, in association with mycorrhizal fungi, extend into the soil to overcome obstacles and achieve survival objectives. Branching is a process of phenotypic adaptability that shapes in response to both endogenous and exogenous constraints. At the level of variable ecological interactions the organism plays an active role, owing to its ability to detect and respond flexibly to different signals. The "branching events", which can be interpreted as spatially multidirectional forms of growth, expand and extend the preexisting structure of the plant through the formation of new components that allow it to reach different types of targets (e.g., light and nutrients), establish chemical communication contacts (e.g., between roots of the same species or of different species and other organisms), to avoid obstacles (e.g., rocks in the ground) and potential hazards (e.g., pathogenic attacks or harmful substances). The ability to branch could represent in non-neural and rooted organisms incapable of locomotion but not of growth, movement or the possibility of influencing the surroundings, an equivalent strategy for solving survival problems in vagile organisms with a nervous system. Observation of branching dynamics could help understand some plant behaviors, such as competition for light or foraging dynamics, which in turn enable the organism to direct its growth toward specific directions.

The ability to discern differences, distinguishing between various elements, is a basic cognitive ability, not necessarily tied to the presence of metacognitive or reflective skills. In the case of plants, the specificities of this practical knowledge merit further exploration, particularly in relation to the development of studies on action-oriented sensorimotor cognition and other forms of practical intelligence.

Finally, in line with the analysis of the dynamic relational behaviors of plants, I suggest investigating the organisms' capacities through the lens of extended cognition to clarify how plants extend their abilities through interactions with neighboring organisms, which may, in some cases, enhance their activities or be essential for their execution. To more accurately explain certain plant behaviors – such as their communication with coexisting species, trophic dynamics, and especially their resistance and heightened responses to environmental stressors – it is essential to consider the critical role played by their interactions with other organisms. These include mycorrhizal fungi, certain bacterial species, and the broader soil microbiota, which influence plant activities in various ways. Such relationships, some of which are symbiotic, can significantly enhance the flexibility and effectiveness of plant behavioral responses and broaden the scope for ecological interactions, both quantitatively and qualitatively.

After clarifying why discussing plant cognition in these terms does not trivialize the concept of cognition but rather specify the conception of cognition at stake, I focus on a more inclusive understanding of bio-ecological cognition. This conception is not primarily tailored to address typically human capacities but allows for the inclusion of cognitive abilities of a wide range of life forms. I concentrate on a possible idea of biological cognition to develop a systemic-relational, process-oriented, and extended view of cognition, one that considers the unique bodily and functional characteristics of different organisms and their interactions with the environment. Therefore, cognition needs to be examined as a multifactorial, processual and contextual phenomenon throughout the evolutionary tree, paying attention both to the commonalities and to the species specificities between the different cognitive systems.

In the final section of the chapter, which is devoted to conceptions of the mind and specifically to an analysis of the concept of representation, I begin by tracing the history of the concept of mind, from its disembodied interpretations to the more recent embodied perspectives. These embodied approaches have been developed in contemporary philosophy of mind, in tandem with scientific advancements across the various disciplines that comprise the cognitive sciences.

An empirically informed reflection on the concept of mind aims to develop a bodily, relational and processual conception, going beyond the conceptions of the human mind, including studies on the specificities of other animal minds and possibly of other forms of life, to understand whether any conception of the mind can be enriched in certain aspects, following an analysis of the plants' capabilities. Although in the next pages I prefer to talk in terms of cognitive abilities, rather than in terms of "mind" or mental capacities, given the connotation of the concept of mind, which tends to be traced back to the analysis of the characteristics and capacities possessed by human beings and some other animals, facing the question cautiously and with the necessary critical sense, it is worth asking whether, and in what sense, the capacities of plants can be addressed in terms of the presence of proto-mind or mental capacities. Therefore, although there probably be no need to talk of mind when analyzing the capacities of plants or to conclude that plants possess a form of mind, this same reasoning on the theoretical-methodological transformation of the study of mental capacities, approached from an evolutionary-comparative perspective, broadens the horizons to include also the specificities of non-neural organizations. This approach could be instrumental in rethinking what the mind is in general, its role in nature, and the degrees and interspecific differences it encompasses. In doing so, it could help reposition the study of mental capacities within the complex array of bodily processes from which they emerge, or rather, that constitute them. The mind could be conceived as a set of capacities of the body, in relation to the environment, for the integration of relevant information and the coordination of behaviors to achieve different survival goals, not excluding, in principle, based on such a reading, plant organisms.

I subsequently detail the concept of representation within the different approaches adopted in cognitive science. I begin by examining the variety of conceptions of representation, along with the relative advantages (pro) and disadvantages (cons) of using these different concepts to explain cognitive abilities.

Then, I delve into this issue specifically in the context of plant cognition. Although further evidence is still needed, the question remains whether some form of representation is present in the cognitive behaviors of plants – if indeed it should be called that – or whether it is more appropriate to refer to different processes to explain plant behavior. Therefore, I reflect on whether to resort to some concept of representation or whether it is preferable to base the research on plant cognition on non-representational approaches, or even to initiate an analysis using different or new terms, concepts, and approaches. The issue is quite relevant because it is frequently argued that, without representations, there can be no genuine cognition – one that is tied to individual life experiences, shaped by specific interaction histories, and based on processes of memory, learning, and anticipation. Instead, it is suggested that what is being observed are merely simple reactions to environmental stimuli, which are unmodifiable and do not vary throughout the life course of organisms. Thus, one of the main controversies in considering plants as cognitive agents revolves around the presence or absence of mental representations and their role, which is often associated with the presence of a nervous system. However, this is not the only perspective on representations. There are conceptions of representations that differ from those traditionally conceived as "mental." I distance myself from some conceptions of representation that are most likely to be considered empirically inadequate or of little explanatory use in addressing the plant case. The attempt is to understand whether a significantly reformed concept of representation might be useful in explaining the ecological interactions of plants, or if it is ultimately unnecessary or even counterproductive and should be entirely discarded. The question remains open, requiring further investigation.

In the last part of the section, I highlight how the issue of representation provides a valuable opportunity to address several interconnected aspects, which I suggest deepening in future theoretical and experimental research. I refer to the connection between complexity and levels of emergence in cognitive systems, to the emphasis on predictive theories in the link with embodied and action-oriented approaches. All this is examined in the light of the relevance of the temporality of cognitive processes, in relation to the processses of memorization, learning, anticipation, and attention. In this context, I emphasize the development of processes across time and space. The study of plant activities provides a basis for further explorationof forms of cognition that are not hermeneutical or conceptual but pragmatic and action-directed, which relate more to how to do something rather than to the ability to know something or to know of knowing something.

Chapter 2
Aims of Research

Contents

2.1 Main Goals and Prospects 16
2.1.1 The Challenge of Nonstandard Cognition: Are All Living Systems Cognitive Systems? 17
2.1.2 Toward a Cognitive Plant Ethology 17
2.1.3 Research Areas and Significance 20
2.2 Task of Scientific Research 23
2.2.1 Processes and Measurements 24
2.2.2 Theory and Practice in Scientific Practices 25
2.2.3 Science and Bias: Critical Reflection and Reduction 29
2.2.4 Progress in Science 32
2.3 Task of Philosophical Research 34
2.3.1 Terminological and Conceptual Analysis 36
2.3.2 Which Conception of "Intelligence"? 38
2.3.3 Why Plant "Cognition"? 40
2.4 New Focus on an Old Issue 43
2.5 Perceptual-Cognitive Challenges in Understanding Plant Behavior 48
2.5.1 Locomotion and Cognition 48
2.5.2 Theoretical and Technological Advances 50
2.5.3 The Invisible: Latent States of Life and "Silent Languages" of Plants 52
2.5.4 Plants and Sympathy: The Issue of Pain and Its Behavioral Manifestations 54
2.6 Contemporary Insights 56
2.7 Underlying Theoretical Framework 59
2.7.1 The Problem of Circularity and the Pluralistic Approach 60
2.7.2 Comparative Analysis of Cognition 63
2.7.3 Anthropocentrism, Anthropomorphism, and Anthropodenial 67
2.7.4 Toward an Inclusive Approach to Evolutionary Cognition 76
2.8 Ecology and Cognition: A Relational Approach to the Activities of Living Systems 83
2.8.1 From Systemic Theories to Different Levels of Organization in Nature 85
2.8.2 Beyond the Concept of Interaction: Conservation and Transformation 86
2.8.3 Relational Nature of Processes 90
2.8.4 Ecological Explanation of Relational Behaviors 93
2.8.5 Levels of Ecological Organization and Ways of Studying Ecosystems 94
2.8.6 Toward a Plant Cognitive Ecology: The Role of Context and Other Organisms 95
References 100

M. Bianchi, *Philosophical Implications of Plant Cognition*, Philosophical Studies Series 158, https://doi.org/10.1007/978-3-032-01328-6_2

Abstract This chapter defines the research objectives and introduces the theoretical framework guiding the analysis. It is followed by a historical reconstruction of the interest in, and conceptualization of, plants from antiquity to the present, providing an overview of the current state of research on plant abilities. The chapter concludes with a focused examination of the relevance of systemic-relational approaches and ecological perspectives for the study of behavior and complex systems.

Keywords Anthropomorphism · Basal cognition · Bias in science · Cognitive ethology · Cognitive life sciences · Comparative cognition · Complex systems · Ecological interactions · Ecology and cognition · Interdisciplinarity · Philosophy of plant behavior and cognition · Pluralism · Scientific practices · Systemic approaches · Terminological and conceptual analysis

2.1 Main Goals and Prospects

In this first section, I summarize the theoretical aims, proposals, and clarifications that I intend to develop in this book, which focuses on the philosophical implications of ongoing research into the cognitive and behavioral capacities of plants. The work as a whole reflects the impact of scientific research across various fields on the different levels of analysis concerning the activities of plants.

The intent is not to provide definitive or conclusive answers, but rather to encourage a rethinking and broadening of research horizons regarding the cognitive processes and abilities present in the living world, an effort that cannot neglect to include the plant kingdom as well. A significant part of the interest lies in studying sessile organisms without nervous systems, which employ alternative strategies for interaction and problem-solving related to survival. These strategies involve processes of memorization, learning, and the anticipation of information through various types of signals and behaviors.

The emerging field of research on plant cognition – variously termed "minimal" (cf. Calvo & Keijzer, 2009), "basal" (cf. "basal cognition," Lyon et al., 2021), or simply a different form of cognition – is raising significant theoretical and practical challenges. It offers an opportunity to rethink important questions in the study of behavior, cognition, and intelligence, considering the distinct body organization and ecological interactions of plants compared to other (animal) organisms. One of the key objectives is to critically outline the relevant aspects that should be considered when interpreting plant behavior.

The study of cognitive abilities, addressed from both the animal and plant perspectives (neural and aneural), seeks to provide a more integrated view of the evolution of cognitive processes and the ecological interactions that shape and transform them throughout the inherently relational development of organisms.[1] At the conclusion of this chapter, I emphasize the importance of ecological studies – and the challenges they present – in fostering a more comprehensive, detailed, and realistic advancement of contemporary cognitive sciences.

2.1.1 *The Challenge of Nonstandard Cognition: Are All Living Systems Cognitive Systems?*

Broadening the horizons of consideration to include what Fred Keijzer (2021) refers to as "atypical cases" of cognition – organisms traditionally not considered to possess cognitive abilities, even at a basic level – presents cognitive sciences with an inevitable challenge. This challenge underscores the need to develop tests tailored to the diverse, species-specific abilities of various organisms, including plants (cf. Nick, 2021).

Regardless of the type of organism or the theoretical approaches, certain elements appear to be common to all forms of cognition. These elements extend beyond the animal kingdom and involve processes such as perception-action, memorization, learning, and non-random anticipation of responses based on signaling and communication (cf. Witzany, 2016).

Partially aligned with these considerations is the original conception by Maturana and Varela (1980) of the equivalence between living and cognitive systems, as well as the biogenic approach (Lyon, 2006), which will be discussed later. According to this perspective, the study of cognition should begin with the identification and understanding of basic biological aspects, to develop the most inclusive approach of the cognitive life sciences (cf. Keijzer, 2021).

2.1.2 *Toward a Cognitive Plant Ethology*

While earlier studies in behavioral and cognitive biology often excluded plants from their research or regarded them as relatively passive organisms, recent advancements – particularly in understanding their interactional and signaling behaviors – have led to a growing recognition of plants as possessing a wide range of abilities. Some of these abilities are commonly interpreted as cognitive within the animal kingdom (cf. Bianchi & Castiello, 2023a).

[1] "Evolution" refers to the genetic, morphological, and behavioral changes in organisms across generations, driven by variability among individuals within populations. These changes result from variation, inheritance, and the action of natural selection. In contrast, "development" refers to the process of change over an organism's lifespan. This process is characterized by flexibility, yet it is guided by conservative mechanisms that reliably preserve species-specific traits (cf. Bateson, 1976, p. 405; Pievani, 2017, p. 13).

For the development of a plant (cognitive) ethology, first, as Hiernaux (2021) noted, it is necessary to distinguish between simple behavior and cognitive behavior (or rather, between mere pre-programmed reaction and flexible, variable response), where in the second case, internally mediated "decision-making processes" would be involved, capable of varying according to the context of action and the most pressing needs of the plant for survival.

Some compelling examples of plant behavior, which will be explored in the following chapters, include the production and modulation of volatile organic compounds (VOCs) (cf. Dicke & Hilker, 2003; Baldwin et al., 2006; Karban, 2008, 2015) as part of defense systems against pathogens and herbivores. Through these mechanisms, plants appear capable of selecting *which* compounds to emit, *when* to release them, and *how* to modulate their quantity (cf. Kumar et al., 2020). Additional behaviors that appear to involve cognitive abilities include self-recognition (cf. Karban & Shiojiri, 2009), non-self discrimination (cf. Baluška & Mancuso, 2021), and kin recognition, alongside, more generally, the ability to distinguish among neighboring plants with whom to cooperate or compete (cf. Biedrzycki et al., 2010; Bilas et al., 2021). These capacities, emerging from multisensory integration, suggest an ability to recognize, differentiate, and retain significant information, particularly in relation to the biological dynamics of competition and cooperation for resources, especially at the level of the root system.[2]

Current research is delving into the mechanisms and processes that support plant abilities.

Studies on the mechanisms and processes underlying these capabilities are currently being conducted. This includes the molecular recognition of individuals of the same genetic origin, such as mechanisms preventing self-fertilization (cf. Nasrallah, 2002), and the physiological coordination among the roots or tendrils of the same plant. For instance, it has been observed that the tendrils of the same *Cayratia japonica* specimen (Fukano & Yamawo, 2015), when separated, react similarly to other tendrils, regardless of whether they are genetically identical, related or distant plants (Thompson, 2018, pp. 42–45).

In general, these and other behaviors considered cognitive in plants would therefore not seem to be rigidly predetermined but involve sophisticated forms of multisensory perception and perhaps multimodal signaling (cf. Bianchi et al., 2025a). They also encompass forms of memory (cf. Thellier, 2017; Galviz et al., 2020) and learning, such as habituation to repeated non-harmful stimuli (Martin et al., 2010; Gagliano et al., 2014), and the ability to anticipate or inhibit behaviors in response to environmental cues (Novoplansky, 2016; Calvo & Friston, 2017).[3]

---

[2] For instance, root behavior, mediated by root exudates, has been found to vary depending on whether roots encounter physical obstacles in the soil, non-self-roots that may represent a potential source of competition, or the plant's own roots or those of genetically related individuals (cf. Semchenko et al., 2014).

[3] For example, many plant species can anticipate future shading by initiating phenotypic changes in response to far-red and red light cues. Additionally, anticipatory behavior has been observed in the parasitic plant genus *Cuscuta* (dodder), which begins to grow toward and coil around the host

These capabilities highlight the processes that enable plants to enhance their responses in recurring situations.

Without delving into the various aspects that, when considered collectively, suggest the presence of basic cognitive abilities in plants – these will be explored in more detail in the following chapters – it is possible to begin considering the potential advantages of adopting a cognitive-behavioral and multilevel approach to understanding plant activities.

As a result of theoretical and experimental advances, it has become possible to consider – going beyond the mere physiological descriptions of processes – whether it is feasible and useful to treat plants as "agents" (in line with the ongoing expansion of the notion of *agency* [cf. Gilroy & Trewavas, 2023 for the plant case]). This perspective views plants as organisms that exhibit goal-directed behaviors, evident in their ability to affect their surroundings through various interactions, particularly chemical-tactile ones, which can alter soil composition and influence the development of neighboring species (cf. Parise et al., 2020). Recognizing this level of analysis appears to enrich our explanatory framework for observable and analyzable phenomena in the plant world. The future direction of this commitment lies in the increasingly detailed development of what has recently been termed "plant ethology" (or plant cognitive ethology). This should not come as a surprise to cognitive ethologists, as many of the current challenges faced by this emerging field are somewhat analogous to those that have historically characterized classical ethological research (e.g., Griffin, 1984; Allen & Bekoff, 1997) and continue to plague the field of animal studies. Indeed, the current biology of behavior has its roots in ethology, a field of research that emerged in the early twentieth century, particularly through the contributions of Karl von Frisch (1886–1982), Konrad Z. Lorenz (1903–1989), and Nikolaas Tinbergen (1907–1988), whose work primarily focused on understanding how animals behave in their natural environments.

When it comes to plants – despite the challenges posed by studying cognitive-behavioral phenomena in organisms that are structurally and functionally different from animals (with difficulties and counter-intuitiveness increasing as the distance from our own species grows) – the advantage is that many theoretical and methodological tools from traditional cognitive ethology are already available. The task now is to determine what can be effectively transferred from this more established field to the emerging study of plant cognitive behavior. This involves identifying what can be adapted and utilized, what needs significant modification, and what might need to be completely rethought or discarded.

Plant behaviors, like those of animals, can be understood as a series of continuous, context-dependent adjustments to the ecological conditions in which organisms live. All living beings, including plants, possess the capacity to distinguish between multiple environmental factors and respond to them through various strategies.

---

plant – investing energy to establish contact and access its vascular system – even before the actual uptake of nutrients has occurred (cf. Calvo & Friston, 2017).

2.1.3 Research Areas and Significance

The development of an interdisciplinary approach, while offering significant potential for the field of contemporary plant sciences, suggests a range of implications across various domains, especially in cognitive science and philosophy of mind. The impact on these distinct yet thematically interconnected disciplines will be explored in depth in Chap. 4.

Regarding the macro-areas that this analysis will address, I will now offer a brief introductory overview of some key topics.

A task that will be addressed more explicitly throughout the book – but which will also remain a recurring theme – is an ontological-epistemological exploration[4] of whether, where, and why to draw the boundary of cognition, both among different categories of organisms and in relation to the extension of cognitive processes beyond the "porous boundaries" of an organism's body (cf. Continenza et al., 2013). Living systems and their behavioral activities do not seem to be fully contained, and thus functionally exhausted, at the physical limits of their bodies, as they require an environment and external means to live and act. This, in turn, creates a loop of reciprocal influence, where the environment affects an organism's constitution and internal activities (metabolism in the broadest sense),[5] as well as its actions (cf. Thompson, 2007; Di Paolo et al., 2017). Therefore, limiting the investigation of cognition to the confines of the cell surface appears highly reductive.

In the study of cognition, one of the most challenging tasks is to analyze the theoretical and practical implications arising from the growing recognition of the importance of embodiment, the role of interaction contexts, and the extension of these processes within the broader framework of niche construction – in short, a rethinking of the constitutive and agentive "boundaries" of the organism.

---

[4] To avoid misunderstandings, it is important to clarify that the term 'ontology' in this context does not refer to traditional metaphysical conceptions of ontology, which are typically grounded in logical-dialectical arguments. Instead, it refers to contemporary philosophical analyses within the philosophy of biology, of physics, and cognitive sciences. These fields recognize the need to distinguish between different mechanisms and processes identified by the leading theories and approaches in the life sciences, given the vastness and complexity of the phenomena being studied. Identifying the processes to consider has both theoretical implications and practical consequences for subsequent experimental research. The goal of this type of analysis is not to impose a rigid categorization of reality based on speculative, non-empirical assumptions, but rather to develop a reflection grounded in available scientific evidence. This reflection should be flexible enough to adapt to the nonlinear and sometimes non-cumulative nature of scientific practices. Such an approach allows for a clearer understanding of what is being studied and in what terms, as research progresses.

[5] Living systems consist of complex, cyclically organized networks of biochemical processes, collectively known as metabolism, through which they produce and utilize energy. Living organisms actively produce and use energy, whereas nonliving entities merely undergo its effects (cf. LeDoux, 2020, pp. 51–52). The persistence of metabolic organization in living systems depends on contextual conditions that make these processes possible (cf. Keijzer, 2021).

When considering the means and strategies of knowledge acquisition for interaction and survival – strategies that involve the abilities to memorize, learn, communicate, transmit, organize, and use environmental information – especially in organisms whose capacities may differ significantly from our own, another objective is to shed light on the biological role of knowledge (cf. Cellucci, 2008; Bianchi, 2024) or on the process of knowing (in a anoetic sense). This intent is aimed at elaborating a perspective on knowledge, or rather on the different ways of knowing, rather than proposing a foundational theory. The study of cognitive phenomena in our species, including inquiries into both formal and practical forms of knowledge, is a specialized branch of research (cf. Piazza, 2017), especially within contemporary debates on theories of knowledge and epistemology. Nowadays, thanks to its many tools and the integration of other, more targeted approaches, this research can serve as a starting point for exploring the cognitive capacities of non-conceptual and non-introspective kinds in phylogenetically distant organisms – while also remaining aware of our own cognitive limitations. In analyzing the possibilities and limitations of our knowledge – knowledge that is culturally and biologically mediated by our physical constitution – when the topic of plant cognition is addressed, a broader question emerges regarding the terminological and conceptual distinction between cognition and knowledge. The concept of cognition, along with the processes and capacities it encompasses, in some of its contemporary specifications, appears to be more inclusive and applicable to a broader range, if not the entirety, of living beings compared to the concept of knowledge. Traditionally, the term 'knowledge' has been used to refer specifically to the mental (human) faculties involved in perceiving and learning through experience, facts, or information, often emphasizing its conceptual, representational, and propositional dimensions, as well as its connection to rationality and explicit awareness. However, much like the evolving understanding of "cognition," "knowing", when understood as a relational, dynamic, and adaptable capacity that enables learning and problem-solving throughout an organism's life, can also be extended to other species across the evolutionary tree. Obviously, here, I am not referring to intellectual forms of knowledge but rather to knowing, not as something an organism possesses, but as something that living beings actively engage in to orient themselves within their environment and interact with other organisms, with the possibility of identifying inter- and intraspecific differences. By observing behavioral success and flexibility, we can compare the extent and accuracy of an organism's "perceptual-agentive" capacity, and thus, to some extent, its "cognitive" capacity to communicate and solve survival-related problems within its environment. The focus is on understanding what aspects of an organism, what processes and behaviors, enable it to know and interact with its environment (cf. Michaels & Carello, 1981, p. 1; pp. 62–63).

Organisms perceive their surroundings with sufficient efficiency to navigate various obstacles and regulate their behavior, demonstrating the ability to discriminate and recognize different elements and communication signals. Something similar can be said about plants, particularly in relation to chemical synthesis and their ability to modulate movements across different contexts.[6] The structural and behavioral traits of organisms are the ongoing result of a non-finalistic, multilevel evolutionary process that has shaped diverse means and strategies. In these terms, the different forms of cognition are tied to the distinct bio-ecological and agentive possibilities of living systems. These can be studied from an evolutionary and ecologically informed perspective, viewing cognition as an activity of the entire organism in relation to its environment, in contrast with the more traditional view that focuses on the functioning of an intermediate system between sensory organs and effectors, such as the nervous system in animals or the genome in prokaryotes (cf. Cellucci, 2008, p. 198).

In the section dedicated to cognition within the broad multidisciplinary research program of contemporary cognitive sciences, various approaches can be adopted to study plant cognition and behavior (cf. Bianchi, 2024; Bianchi et al., 2025a). Some promising examples include the interdisciplinary research field of 4E Cognition (embodied, embedded, enacted, and extended cognition); predictive models that examine organisms' ability to anticipate and minimize prediction errors (cf. Friston, 2005; Calvo & Friston, 2017), (eco-)systemic and self-organizing approaches for studying chemical-electrical signaling processes in biological networks (cf. Baluška & Levin, 2016), and the approach of ecological psychology (cf. Gibson, 1979), which focuses on forms of direct, non-representational perception (cf. Heras-Escribano & Calvo, 2019).

From a pluralistic perspective, open to revisions and transformations, the most promising solutions are distinguished by the degree of empirical progress on a particular aspect, the type of problem being addressed, and the relevant level of explanation (cf. Bianchi, 2021).

Another closely related issue that warrants attention is the way, or rather, the various ways, in which the concept of 'mind' or 'mental capacities' is theorized and referred to. As a preliminary observation, it's important to note that when discussing the cognitive abilities of plants, it may not be necessary, and could even be counterproductive, to invoke the concept of mind (Bianchi, 2024). This is because the term is often associated with theoretically biased or reductive descriptions, frequently encephalocentric (centered on brain activity) or rooted in intuitive attributions of mental processes, in line with common sense psychology (cf. Keijzer, 2021). When discussing plant abilities, I do not rely on a concept of mind traditionally used to describe high-level cognitive abilities (i.e., abilities involving elevated degrees of abstraction from the immediate context of interaction) or aspects of introspective self-awareness. However, given the diverse interpretations of the concept, it might be beneficial to trace the historical evolution of the notion of mind and assess whether it is possible and advantageous to apply the concept of a "plant mind" to plants, or to refer to their "mental capacities". Even if we prefer to think in terms of

[6] Indeed, plants have a subtle perception of their surroundings, allowing them to distinguish different parts of the electromagnetic spectrum and detect gravity. They can move and react to various molecules, as well as acoustic and tactile signals, and are capable of synthesizing and emitting a wide range of chemical compounds (cf. Castiello & Guerra, 2020).

behavior or cognitive abilities, emphasizing the processual aspects of mental phenomena as multifactorial activities – ranging from the simplest and most basic to the most complex – could enable the development of a more inclusive conception, suited to the plant case: an ecological and relational conception of the "mental" (Bianchi, 2024).

Moreover, in the section dedicated to the concept of mind, I wonder if resorting to some conception of representation (cf. Souza et al., 2018; Debono & Souza, 2019) could help explain some variable behaviors throughout plant life. Although this issue remains controversial and further species-specific testing is needed, I build on the reflections initiated by Bianchi and Castiello (2023a, 2023b) to critically examine the potential relationship between representational aspects of behavior and the role of interactomes, as well as the broader functional biological networks of chemical-electrical signaling within plant bodies. Specifically, I investigate the connection between a potential molecular or metabolic conception of representation and the role played by various forms of plant memory, including epigenetic memory (cf. Ginsburg & Jablonka, 2021), immunological memory (Yakura, 2020), metabolic memory (cf. De la Fuente, 2015), and non-neural intercellular signaling (cf. Calvo et al., 2016). The central question, then, is whether it is possible to establish a link between a concept of representation and plants' abilities to detect, store, and learn environmental information, or whether it might be more appropriate to avoid using the concept of representation altogether when explaining plant behavior, and, consequently, to rely on entirely non-representational approaches. Rather than referring to representations (as discussed in the literature on the "representational demarcation challenge," cf. Lee's "piecemeal approach" [2023]), we might consider speaking of the capacity to train and structure networks of signaling processes. In this view, environmental information is organized in a constrained yet not rigidly determined manner, allowing for the flexibility of behavioral responses – a prerequisite for acting in both similar and changing contexts.

There are currently no unequivocal answers, and the question is open.

Beyond the ongoing debate on the presence or absence (and what types) of representations or representational aspects in plants, a heightened awareness of the implicit assumptions about what it means to possess a mind or mental (cognitive) capacities will enable us to more clearly and swiftly identify the most appropriate methods for observing behaviors, both in the field and in the laboratory, that indicate cognitive activity. This knowledge is crucial for understanding how various organisms, each with different capacities, initiate and modulate their responses in recurring action contexts according to their specific needs.

2.2 Task of Scientific Research

As noted by Gregory Bateson (1904–1980), we should not let our imperfect understanding fuel anxiety and increase the impulse to exert control. Instead, our studies should be guided by an older, though often neglected, principle: curiosity about the world of which we are a part. The reward for such an approach is not power, but beauty (cf. Bateson, 1979).

We know that the primary task of scientific research is to systematically explore, understand, and explain the complexities of the natural world. This involves not only the discovery of new knowledge but also the critical evaluation and refinement of existing theories and methodologies. Within the scientific community, the goal is to expand our understanding of how the world works by asking precise questions, formulating hypotheses, and conducting experiments that produce reliable data. This process contributes to building a body of knowledge that can inform practical applications, guide future research, and help solve problems.

This research – which seeks to explore the behavioral capabilities and modes of interaction of plants – is especially striking because it deviates from what is most intuitive and familiar to us. According to Jacques L. Monod (2017, p. 46), humanity's deep-rooted desire to be seen as necessary and inevitable stems from a long-standing effort to deny its own contingency, an effort reflected in all religions and many philosophical systems of the past. The original human cognitive enterprise, in its most abstract sense, could be viewed as an attempt to address these ancestral anxieties. We might experience something similar to what Carlo Rovelli (2014, pp. 9–11) described when he noted that, for much of our lives, we are "obsessed with ourselves." While it is natural to prioritize our immediate concerns, the beauty of scientific research lies in its ability to "open the window and look far ahead," broadening and deepening our understanding of what must be considered, even if it initially seems destabilizing.

Science is born from an act of humility, guiding us not to blindly trust our own intuitions (ibid., p. 227) or to uncritically accept what has been handed down by predecessors based on authority. Even today, it is common to naïvely believe that the purpose of science is to seek "truth" or absolute certainties. Scientific practice is heterogeneous, context-dependent, fallible, ever-evolving, and sensitive to changes in underlying theoretical frameworks and technologies used. While definitive certainties may never be achieved, the collaborative effort of the scientific community can produce sufficiently reliable answers, the best available at a given time. By remaining open to new perspectives and embracing methodological rigor, we recognize that scientific research is not static, but a continuous, iterative process that evolves over time and space with new insights and discoveries, helping us to address various challenges (cf. Bianchi et al., 2025b).

2.2.1 Processes and Measurements

Both a theoretical problem – regarding the types of processes identified and considered – and a practical one – concerning the methods and technologies employed – are connected to an epistemological awareness: that is, an understanding of the limits, conditions, and methods required to generate scientific knowledge, particularly in relation to the often blurred boundary between the knowing entity (the observer or experimenter) and the phenomenon under investigation (the process or network of processes being studied). This challenge is inherent in every form of analysis, including scientific inquiry, because the very mechanisms and processes of understanding are themselves subject to examination. This makes it arduous to delineate processes and ascertain the extent to which one is studying a natural phenomenon, and the extent to which one's underlying theoretical approach and methods of inquiry influence the study. How much of the phenomenon is being studied, and how much of oneself in relation to it? Certainly, there is no unmediated, non-relational approach to knowledge. The self is always entangled in the study of what lies beyond itself (the process or phenomenon under investigation), even when one adheres to the scientific method in pursuit of reliable knowledge of reality, knowledge that is verifiable or falsifiable. Every human research activity, whether explicitly or implicitly, defines the parameters of analysis and measurement, determines the terms of comparison, initiates comparisons, and interprets the data obtained based on the initial approach and questions.

In a similar, though not identical, vein, Werner Heisenberg's oft-quoted reflections come to mind. At different moments in *Physics and Philosophy* (1958), shaped in part by his long-standing dialogue with Niels Bohr, he reflects both on the nature of scientific inquiry and on the human condition, stating that what we observe is not nature herself, but nature exposed to our method of questioning, and later, in the final section, that one must never forget that in the drama of existence we are ourselves both players and spectators. The focus lies at the intersection of what nature is like and what we can meaningfully say about it through our methods of inquiry. Moreover, it is crucial not to conflate the influence of the observation process with the subjective or individual role of the observer, as this introduces a range of other epistemological and interpretive challenges (cf. Boncinelli, 2020).

In the study of plant activities, it is essential to discern what aspects of the phenomena we are truly investigating (and which ones are being overlooked) as well as how much of our own perspectives we project onto these phenomena. Therefore, avoiding projections and anthropomorphizations is of paramount importance.

2.2.2 Theory and Practice in Scientific Practices

Moreover, since the inner experiential side of the organism under examination is obviously not accessible, except to the organism itself (cf. Nagel, 1974; Calvo, 2017), all these aspects are somehow observed and interpreted by an experimenter. For such observations to be understandable and useful, they must count as evidence for the model on which the analysis is based. Observational and descriptive processes go beyond initial sensible experience; they are not limited to a passive recording of "factual data", conceived as immediately available, but are always mediated by theoretical conceptions.

The process of observation, in the field or in a controlled environment, which in turn includes measurement processes in its procedures, is not a trivial or superficial operation but rather the outcome of a long process in which knowledge is accumulated in variable contexts. Therefore, the observation of any behavior or interaction-communication system is usually the result of years of research with specific premises and implications.[7]

In scientific research, as Ernst W. Mayr (1982, pp. 833–834) observed, another aspect that should not be overlooked is that different scientists can draw different interpretative conclusions from observations of the same facts.[8] Each approach operates within its own foundational theoretical framework, which is often based on a set of assumptions that are implicit and not always explicitly stated during research practice.[9] In the study of living organisms, past examples of these unspoken assumptions include the belief in a fundamental, essentialist distinction between the organic and inorganic worlds, or the hierarchical view of life as progressing toward greater perfection.

The development of these pages emphasizes how ongoing research into the characteristics and abilities of plants is posing significant challenges to traditional conceptual and interpretative systems. One key contribution of this research is the rethinking of value assumptions associated with what was known in the pre-Darwinian era as the *Scala Naturae,*[10] by recognizing species-specific abilities in

---

[7] It is necessary to dispel the naïve idea of the observation process. For instance, LeDoux (2020, pp. 385–401), reflecting on how to detect consciousness in the animal world from a precautionary perspective, noted that if all we have to do to connect consciousness to behavior is to observe behavior, we would not need research. According to the author, consciousness cannot be inferred solely from observing activity in a particular brain area. While this cautionary approach and emphasis on rigorous experimental study are meaningful and certainly valid, it would be reductive to oversimplify or generalize the complex, theoretically guided, technologically supported, and methodologically refined nature of observational practices. Learning how to observe is a lengthy and delicate process of training. Moreover, there are various levels of observation, ranging from micro to macroscopic processes. In ethology, as Andrews (2020, p. 52) pointed out, an inexperienced student may struggle to identify individual animals during behavioral observation and must subsequently learn to distinguish the species' typical behaviors and individual variations. As far as plants are concerned, the challenge of detecting their interactions, often counterintuitive to our human perceptions, makes it even more difficult to discern their various behaviors without proper training and technological support.

[8] Even in the case of behavioral observation, whether you consider animals or plants, difficulties can arise when an activity is videotaped and "classified" by multiple observers, potentially introducing a source of bias. Generally, these types of disagreements are mitigated or resolved by examining which interpretation is most justified based on the type of question and objective of the research (cf. Andrews, 2020, p. 52).

[9] Despite the necessary interaction and even convergence at various points in the analysis, there are differences in the respective approaches to the issues. For example, while a biologist is trying to explain how an organ works, a theorist might be trying to clarify what it means to perform a function (cf. Michaels & Carello, 1981, pp. 86–87).

[10] The *Scala Naturae* represented a hierarchical scale of progressive perfection used to classify and organize the diversity of life. As Mayr (1982, pp. 200–201; 500–521) noted, Blumenbach (1782) viewed the *Scala Naturae* as the foundation of a natural system for ordering living organisms based on their various primary affinities, grouping similar entities together and separating the dissimilar ones. Lamarck, in his early writings, expressed a similar view. Linnaeus preferred to compare a classification to a map, in which each territory borders its neighbors. The original Platonic-

each organism, tailored to its own ecological context. The fact that these abilities manifest differently does not imply they are absent or less effective in achieving certain goals based on specific ecological needs. In nature, just as there are multiple ways (means and strategies) to achieve the same or similar results, it is possible to use the same (or similar) anatomical-structural and cognitive-behavioral solutions to respond to different problems (Bianchi, 2021, p. 166).[11]

A misinterpretation of these arguments could lead to the assumption that doing science simply involves operating within already structured and well-defined conceptual systems, where one must only become aware of, in order to better master their assumptions and implications. However, understanding theoretical approaches not only clarifies implicit assumptions but also provides the initial epistemic and methodological tools for subsequent empirical analyses. These tools are starting points, not endpoints, in the process of investigating a specific aspect of reality. The goal should not be to confirm a point within a particular theoretical framework, but rather to rigorously test a hypothesis. The result might lead to the integration or even abandonment of that hypothesis if it proves insufficient, in favor of one (or a set of hypotheses) that better explains the phenomenon being investigated. Scientific practice, through both theoretical and experimental developments, can significantly reshape initial conceptions, and it is both normal and beneficial for this change to occur.

In line with these considerations is the recognition of the value of error in scientific theories, whether as a corrective force or a potential heuristic tool. Errors can arise in various forms, such as those related to the methods of observation and measurement, cognitive or cultural biases, data collection and interpretation, or the design and preparation of experimental setups, including component selection and arrangement. As a result of mistakes, a positive effect is the possibility of stimulating research in (slightly or greatly) different directions and helping not to retrace paths that are certainly unsuccessful. Another beneficial aspect of error is the search for supporting evidence, which could have been ignored, to alternative hypotheses or theories to the wrong ones.

Mayr (1982, pp. 846–847) noted, for instance, how Geoffroyism[12] – the misconception that genotypes adapt directly in response to environmental stressors – nonetheless spurred valuable research into the relationships between the environment and adaptive traits, which subsequently proved useful for understanding evolution

---

Christian notion of a continuous succession from the inorganic world to the most perfect organism (humankind, followed by the angelic hierarchy and God) was gradually challenged through increasing and deeper engagement with biological diversity. This idea faced significant questioning, particularly by the second half of the nineteenth century, following the widespread acceptance of Darwin's anti-essentialist theory.

[11] An example of the first case involves the different nutritional strategies of autotrophs and heterotrophs, both aimed at the shared goal of obtaining energy from the environment. An example of the second case is the common use of photoreceptors for different purposes: in animals for "seeing," and in plants primarily for capturing energy to synthesize organic molecules.

[12] Specifically, the sideboard named after the French naturalist Hilaire Geoffroy (1772–1844).

by natural selection.[13] Although these are different cases, investigations into the relationships between organisms and contexts (or phenotypes and environmental factors), particularly the intracellular environment within the nucleus, have significantly contributed to the development of the now-thriving field of epigenetics. In the context of research on developmental processes, Conrad H. Waddington (1905–1975) introduced the term "epigenetics" in 1942 to describe the causal mechanisms by which genes bring about phenotypic development. He also proposed the influential metaphor of the "epigenetic landscape" to illustrate how gene regulation channels developmental trajectories, using a topological image of a ball rolling down a contoured surface to represent the progressive commitment of cells to specific fates. Today, the term "epigenetics" refers more specifically to heritable phenotypic changes that are not caused by alterations in the DNA nucleotide sequence (cf. Crisp et al., 2016), but rather by modifications to chromatin (such as methylation and acetylation), which can lead to changes in gene expression, activation, or silencing (cf. Pievani, 2017, pp. 51–52). Epigenetic markers that arise from environmental contexts can induce changes at various levels of biological organization, leading to phenotypic alterations that may persist and be passed on for several generations (cf. Ginsburg & Jablonka, 2009). In plant cells, compared to animal cells, epigenetic forms of memory are more widespread and play a critical role in their lives. For instance, in plants, epigenetic modifications are responsible for long-term developmental memory changes, such as those involved in vernalization (cf. Groszmann et al., 2011), or environment-induced behavioral changes, often triggered by stressors, that lead to heritable alterations in offspring (cf. Boyko & Kovalchuk, 2011). Some epigenetic changes can endure for extended periods, as seen in the five-spur variant with radial symmetry (Peloria) of the *Linaria vulgaris* flower (cf. Borghini & Casetta, 2013, p. 241; Trewavas, 2014, pp. 213–214).

Moreover, similar to "erring" is "neglecting." Aspects that were initially overlooked or deliberately omitted for the sake of simplification or due to a lack of knowledge may later prove to be relevant or even essential as research progresses.

A notable example from the plant world is the early work of systematizing and comparing species in the establishment of the first botanical gardens for plant study (cf. Cittadino, 2009, pp. 231–234). This activity, which initially followed morphological grouping criteria and abstracted species from their natural life contexts, resulted in the inevitable and gradual shift of attention to the relevance of ecological and developmental aspects.[14] These aspects have attracted increasing interest pre-

[13] A different case is the hypothesis of panspermia, dating back to the 1970s (although traces of the theory can already be found in the 1940s), proposed by the physicist, mathematician, and astronomer Fred Hoyle, together with his collaborator, the astrophysicist and mathematician Chandra Wickramasinghe. This hypothesis suggests that life originated in space and is therefore a global cosmic phenomenon (cf. Hoyle & Wickramasinghe, 1976; Hoyle & Wickramasinghe, 1981; Wickramasinghe, 2003). While this hypothesis currently lacks substantial evidence, it has nonetheless sparked valuable research into the search for organic forms in extraterrestrial environments, such as other planets, comets, and interstellar molecular clouds.

[14] E.g., by observing the lack of pollination in certain species within artificial habitats, researchers investigated the relationship between the presence of specific insects and plant reproduction.

cisely because they were initially poorly understood or underestimated (cf. Dröscher, 2008, pp. 147–153; Bianchi, 2021, pp. 29–30).

To make progress in research, it seems necessary to take risks, which inherently involves the possibility of making mistakes.

In addition to this, when identifying the relevant elements and processes to examine, one must also decide which factors (or set of factors) to prioritize. This decision depends heavily on the existing knowledge concerning the phenomenon, along with technological and experimental advances, the context of analysis, and the specific question being addressed. For instance, in research on plant behaviors that appear to be goal-directed – specifically studying the ability of climbing plants to plan and anticipate grasping movements based on the type of support – researchers have debated which aspect (or set of aspects) to focus on first (cf. Guerra et al., 2019; Ceccarini et al., 2020; Raja et al., 2020; Ceccarini et al., 2021). Since these abilities involve multisensory perception-action capacities (cf. Bianchi et al., 2025a), as shown in these studies, a thorough analysis should consider both circumnutation patterns, orienting and coiling movements, and chemical and electrical signaling detected during movement, as well as variation in gene expression during these processes.

2.2.3 Science and Bias: Critical Reflection and Reduction

Even the most meticulous research, which aims to consider multiple factors in detail, is not immune to biases, potentially affecting the ability to make accurate judgments.

While it is not realistically possible to move directly from the world to the concept without mediation, since research is generally guided by underlying ideas, whether explicit or implicit, the purpose of aligning theory with the world serves as an essential tool for countering unjustified intuitive judgments unsupported by evidence. This approach prevents the description of phenomena (including cognitive ones) from being disconnected from the actual functioning of a system, ensuring it is not merely reflective or speculative. In the case of cognition, the focus of interest here, both directions, from concept to world and from world to concept, interact and play a crucial role in cognitive science. Theories and concepts are needed to guide experimental work, which in turn, in an interplay of reciprocal influence, continuously challenges and can reshape the prevailing theoretical frameworks.

As Keijzer (2021) pointed out, the primary objective should not be to rigidly define cognition, but to clarify the material domain of this multi and inter-disciplinary field, identifying the relevant elements and processes to be considered across different species.

Similarly, upon noticing dwarfism or gigantism when certain species were transferred from their natural environments to the garden, attention was drawn to the connection between developmental aspects and trophic-ecological conditions (cf. Dröscher, 2008; Bianchi, 2021).

As Kristin Andrews noted in her book *How to Study Animal Minds* (2020),[15] a realistic and disillusioned position should recognize that biases characterize (to a considerable extent) the entire scientific enterprise.

As has emerged from the works of philosophy of science regarding the concept of objectivity, there is no "*view from nowhere*", no situated, disembodied, decontextualized perspective from which to answer a given question. At best, a "view from nowhere" serves as an ideal to strive for, helping to avoid extreme partiality. When taken literally, this notion risks oversimplifying complex realities, obscuring diverse modes of knowledge that arise from different perspectives, and ultimately leading to a loss of valuable information about reality and the specific systems being studied (cf. Andrews, 2020, pp. 30–34).

Before examining some examples of bias in cognition research, it is worth noting that certain aspects have long been overlooked or investigated to a lesser extent. This may reflect both human disinterest in certain topics and the fear of discovering that our species is not the only one to possess a certain capacity (even if it is expressed differently).[16]

In the study of behavior and (animal) cognition, various types of biases can be identified, both in laboratory and field research. Researchers engaged in fieldwork are typically more aware of the biases associated with laboratory studies, while laboratory experimenters tend to be more conscious of the biases in field research. Since fieldwork and laboratory work are complementary, identifying sources of bias in both contexts allows for more robust answers to research questions (ibid., p. 42). The same applies to methodologies used in analyzing plant capacities and behaviors.

In the field of animal cognition, Andrews identified at least four potential sources of bias related to the choice of measurement systems, theoretical terms, the types of arguments employed, and the context of analysis. In field research, potential biases related to measurement systems arise, for example, from the challenge of isolating and analyzing specific behavior (or an aspect of a specific behavior) within the complex web of ecological relationships. This complexity can make it difficult to determine which aspects should be considered, leading to the risk of either excluding relevant characteristics (whether causal or constitutive) of the process under investigation or being unable to account for all factors. This poses the challenge of deciding what to include and how to appropriately narrow or broaden the scope of the analysis.

---

[15] I will reference Andrews' text multiple times throughout this chapter, as many insights developed in the context of animal cognition studies can, with appropriate adaptation, be applied to research on plant cognition.

[16] A historical example of this can be seen in the disagreement between Japanese and Western scientists, particularly primatologists, regarding the way of conceiving the transmission of the baggage of behavioral information from one generation to the next, specifically, what organisms communicate and do. In many species, this process was only recognized much later in the West as a form of "animal culture," testifying the difficulty in acknowledging cultural forms in nonhuman animals (ibid., pp. 39–41).

In laboratory (or captive) research, the use of predefined data sheets can limit the scope of observation.[17] Additionally, it is common to significantly reduce or exclude explanations that involve the "motivational" aspects of an organism's (cognitive) activities. In both cases, biases can also arise from the operational terms used to describe behaviors, which may be poorly defined or misleading, either because they are too broad or overly specific.[18] Furthermore, the type of phenomenon under investigation may require different analytical approaches over time, including intergenerational analyses. For instance, cognitive abilities are challenging to study because they involve multilevel processes that cannot be reduced to a single factor. In the study of natural behaviors, data collection is influenced by human sensory and intellectual limitations, often confined to activities that are easily perceived through experimentation. For example, the location where an organism lives and acts may be hidden from view, or the process being studied might be difficult to detect without the aid of sensory-enhancing tools (in plants, observing the overall behavior of roots, which are typically hypogeal and photophobic, presents a particular challenge [cf. Mazzolai, 2021]).

A common criticism of laboratory or captive studies, which focus on accurate and fragmented examinations of the mechanisms that support cognitive functions, is their lack of ecological validity. These studies often fail to account for the dynamics of ecological (and, in some cases, social) interactions that play a crucial role in shaping relational behaviors. The tasks involved may not reflect an individual's "free" response capabilities, and the experimental setups may not adequately replicate the complexities of wild environments where these responses naturally occur.

---

[17] Colin Allen and Marc Bekoff (1997) revisited the issue of selecting which aspects to consider when constructing an ethogram. The options include describing behavior in formal terms, focusing on observable bodily movements, or in functional terms, considering the purpose of the behavior in question. Both approaches have their limitations. Formal descriptions may overlook key characteristics (e.g, in some species, play and fighting behaviors may appear identical rather than merely similar in an ethogram that only accounts for physical contact and proximity). On the other hand, functional descriptions risk misinterpreting similar behaviors (Andrews, 2020, pp. 48–49). Regardless of the specific limitations, ethogram generally narrow the scope of what behaviors are recorded as relevant. Moreover, as De Waal (1999) pointed out, the more inclusive the level of analysis, the more challenging it becomes to maintain a purely descriptive stance. Behavioral dynamics can be organized, for example, based on the type of interaction between two organisms (e.g., "A" addressing "B"), their social meaning (e.g., collaborative/competitive, or hybrid interactions), or the duration of relationships (e.g., short- or long-term allies/rivals). Regardless of how the information is organized, a common thread is the influence of a theoretical approach; behavioral observation is not a purely descriptive activity but involves interpretation and connotation, starting with the observer's selective perception and contextual sensitivity to salient elements of behavior. This is why the observer-experimenter is difficult to replace entirely by devices designed to objectively record movements and postures. Relying solely on such devices, without further work to connect and coordinate the data, risks accumulating a vast amount of unrelated, and therefore minimally useful, details.

[18] Andrews (2020, p. 49) provides the example of the term 'peering,' which, if defined as the close observation of another's actions from less than one meter away, would exclude all instances of close observation that occur just beyond that distance (for example, at 1.25 meters).

In field research, while the context may more closely resemble the organism's natural environment, there can be a lack of well-organized control, and researchers may encounter rare or partially observable events that are difficult to replicate. In contrast, laboratory or captive studies provide better control, but barriers, both physical, due to the experimental setup, and psychological, stemming from the use of communication and interaction systems different from those of the studied species, can limit the effectiveness of the research. This is especially true when working with non-human species, whether they are animals, plants, or other organisms (ibid., pp. 47–48; 60).

These are just a few examples of potential sources of bias. The adoption of a particular approach, methodology, and technology to investigate certain aspects inevitably excludes or overlooks others. Every procedure carries its own limitations and biases, some of which are not entirely avoidable. No scientific practice is entirely free from biases; all must strive to surface, explicitly articulate, and address these biases in order to reduce their impact. This is particularly true in multidisciplinary investigations, where the potential for bias, error, and misrepresentation increases due to the use of diverse tools, methods, and approaches. The complexity of interdisciplinary research underscores the importance of intersubjective collaboration, involving researchers with varied theoretical backgrounds.

The goal is not necessarily to identify a definitive experiment that proves whether an organism possesses a specific capacity (although developing appropriate tests is crucial). Instead, greater theoretical and methodological awareness is aimed at formulating hypotheses that provide more comprehensive, unifying explanations while respecting necessary distinctions. This involves coordinating the heterogeneous content of the scientific literature, synthesizing the varied results emerging from different research branches, which often address different aspects of a single, more complex and stratified phenomenon (ibid., p. 41; 61; 471).

2.2.4 *Progress in Science*

This is certainly not the place for addressing the complex and debated issue of scientific progress. As Kuhn (1962) and Feyerabend (1975) noted in different ways, the progress of research, shaped by highly heterogeneous practices and processes,[19] is often unplanned and unpredictable in its development in certain directions, which

[19]According to Poincaré (1914; ed. or. 1908), it is not possible to enclose individual disciplines within rigid boundaries. Doing so would be akin to making the whole thing fit into the part. Wondering what science was, he replied that it was first and foremost a work of classification, resulting from a grid of relationships. In this sense, scientific progress is driven by the discovery of unexpected connections between different fields (cf. Rispoli, 2012, p. 34).

usually accompany losses in other directions. Nonetheless, it is evident that the gradual acquisition of familiarity and accumulation of knowledge regarding certain phenomena allows for the transformation of what was once considered inexplicable, upsetting, improbable or worrying fact into something *not* inexplicable, upsetting, improbable or worrisome. Over time, with the application of specialized expertise, refined methodologies, and adequate resources, contradictions can be resolved, predictions sharpened, connections forged between previously unrelated phenomena, new frontiers of meaning explored, and the boundaries of both theoretical and practical inquiry expanded (cf. Bianchi, 2021, 2024).

In the context of our discussion, the study of plant cognition – which some critics have considered or incapable of adding significant novelties to traditional botanical studies (cf. Robinson et al., 2020) or completely counterproductive and misleading (cf. Alpi et al., 2007; Taiz et al., 2019) – raises important questions. Is it merely a descriptive extension of the processes carried out by a plant, or does it represent a genuine reconsideration of the boundaries and forms of behavior and cognition in nature, akin to a reform or even a conceptual revolution from which we still have much to learn? Surely, we need to be patient, continue this interdisciplinary, multilevel research, and wait for further results.

One thing can already be recognized, namely, that the level of ecological-relational and cognitive-behavioral analysis, which is maturing gradually (especially in the past 20 years [cf. Trewavas, 2003]), is helping to move beyond the earlier, overly enthusiastic and naïve phase, characterized by scientifically unfounded claims (cf. Tompkins & Bird, 1973). These claims included anthropomorphizing descriptions of plant consciousness or the notion that plants can "think" or respond as biologically related organisms would, ideas that were rooted in a mixture of extravagance and pseudoscientific speculation.

Although the respective ecological-evolutionary differences reveal that there are molecular and functional elements common to all living beings, reasoning in terms of forced comparison with the behavioral-cognitive peculiarities of other organisms would not make much sense (if not the certainly positive aspect of encouraging the exploration of new paths), given that in plants, the morphological structure, physiological organization and overall mode of life differ from those of humans or others animals (cf. Mancuso, 2017a, 2017b).

Moreover, as any effort to select relevant aspects, while it is an essential part of the process, it is never solely about the availability of data. Instead, it involves an ongoing comparison and evaluation of information (regarding knowledge acquisition processes, see Ippoliti & Cellucci, 2016, p. 16). It is necessary to remain open to future breakthroughs, new evidence, and the potential for significant re-evaluation, given the possibility of errors in interpretation, something that occurs in all types of research, especially in emerging fields.

This reminds us that theoretical-conceptual analyses, despite the risk of being reduced to mere definitional exercises and being insufficient on their own, are nonetheless indispensable when paired with experimental research (cf. Maher, 2020).

2.3 Task of Philosophical Research

If the aim of scientific research, through its most reliable methods, is to understand how the world works and to improve our capacity to predict phenomena, then it also requires the development of detailed conceptual frameworks, including theories and models.

By helping frame the themes from a historical perspective, what philosophical research can do for the sciences is to show that scientific inquiry is grounded in diverse theoretical-philosophical approaches and assumptions, thereby contributing to the elucidation of viable options and to the elimination, or at least the disentangling, of conceptual confusion (cf. Chemero, 2009; Morabito, 2020, pp. 132–133). In studies on plant abilities, both theoretical and experimental research are essential for developing a pluralistic approach, providing the most appropriate epistemic tools for understanding and reasoning about plant behavior and cognition.

Philosophical research plays a key role in rethinking the "boundaries" of cognition, fostering a more inclusive understanding of the diverse forms of cognition, from the simplest to the most complex or merely different, including those of organisms without a nervous system, such as plants. In cognitive science, this type of research contributes by clarifying the assumptions and principles underlying various approaches to the study of cognition, and by critically analyzing seemingly unquestionable results, challenging, when necessary, what is commonly taken for granted (cf. Bianchi, 2021, 2024). Philosophical analysis also contributes by mediating between the empirical sciences and the various disciplines that constitute contemporary cognitive sciences (cf. Calvo, 2016).[20]

The challenge lies in facilitating critical comparisons and fostering productive dialogue between fields that are often quite distant from one another, with the goal of integrating and expanding their respective results and horizons. In line with this objective, for a more cohesive and comprehensible framework, the analysis of specialized content is needed to overcome the terminological-conceptual technicalities of the various disciplines. By adopting a diversified approach and tools, the potential advantage of interdisciplinary collaboration among scholars from various fields (e.g., philosophers, psychologists, cognitive scientists, and plant scientists) lies in the ability to harmonize a wide range of contributions to develop new hypotheses, verification of predictions, and the design of experimental procedures, enabling quicker and more accurate evaluation of different interpretations of the results (cf. Bianchi, 2021, pp. 11–12; 165–172).

Within a "heuristic" conception of philosophy[21] (cf. Cellucci, 2008, pp. 34; p.10) – which sees one of its primary aims as the pursuit of new knowledge and

---

[20] E.g., cognitive neuroscience, cognitive psychology, philosophy of mind, computer science, artificial intelligence, linguistics, anthropology, etc.

[21] Rather than being "foundational," the premises or hypotheses used in cognitive processes are not arbitrary but plausible, meaning they are consistent with existing data. This is based on the understanding that no knowledge is "definitive," but rather provisional and subject to further investigation as new data emerge (Cellucci, 2008, p. 39).

discovery procedures, often exploring paths that are either underexplored or untouched by current sciences, and at times inspiring the emergence of new disciplines[22] – there is a heightened awareness of the risk of error that accompanies all new research. The research space is dynamic, where the starting point frequently builds on the conclusions of previous studies. Through a top-down and bottom-up process, flexible and context-sensitive, new scenarios can be constructed to test ideas, helping to identify reasoning errors (cf. Ippoliti & Cellucci, 2016, pp. 13–16).[23]

Philosophical research, in step with experimental scientific inquiry, can assist in reformulating existing questions or raising new ones.

Regarding the issues addressed in this work, some questions that may stimulate further research include reflecting on the fundamental and indispensable elements necessary to discuss basal forms of cognition, mind, or intelligence. In functional terms, this involves exploring the processes that enable information exchange, memory, and what we might describe, on a behavioral level, as goal-directed behavior or "actions." Another question is related to the way of conceiving cognition (and similarly, the mind) as a result of bodily organization in relation to the environment. Rather than being based on a single or exceptional ability, cognition is seen as emerging from a set of many capacities, such as perceiving and acting in a coordinated, goal-directed manner, learning, anticipating behavior, and prioritizing responses in recurring action contexts throughout individual development (cf. Bianchi, 2021). A thematically related question explores the extent to which cognitive dynamics extend beyond the physical limits of the body, forming a non-negligible – and, according to some views, even constitutive – relationship with the environment (see, for example, the extended mind hypothesis by Clark & Chalmers, 1998, and dynamical systems theory as developed by Thelen & Smith, 1994). In this view, organisms and their environments are considered a coupled system in constant interaction. The environment encompasses everything (a set of structures and processes) that contributes to maintaining the metabolic and functional organization of a living being, thereby meeting its needs. The environment is deeply intertwined with the organisms themselves. The unit of analysis becomes the agent-environment system, and the challenge lies in understanding to what extent these boundaries are distinguishable for practical purposes and how much can be differentiated by their distinct functions and roles in the organism's life (cf. Bianchi, 2022).

In the case of plants, questions remain about which aspects should be considered and how these aspects should be detected (cf. Bianchi et al., 2025a). The boundaries

---

[22] As Cellucci (2008, p. 30) noted, a hallmark of the philosophical enterprise is its ability to operate in undefined or unexplored territories, which, precisely due to this characteristic, can lead to fruitful developments, sometimes even giving rise to new sciences. For instance, the theoretical foundations of computer science emerged from Alan Turing's philosophical analysis of human computational behavior; cognitive science grew out of this type of analysis alongside traditional philosophy of mind, while Bayesian statistics largely stemmed from efforts to understand the nature of rational beliefs.

[23] Philosophical inquiry builds on previous investigations, with the ability to anticipate where certain ideas may lead, while also avoiding the repetition of paths that have proven difficult or unsuccessful in the past (cf. Cellucci, 2008, p. 30).

of individuality and their capacities for "action" are even more ambiguous. All these considerations must be rooted in empirical observations – whether already conducted or yet to be carried out – even if theoretically mediated, focusing on the physiology and relational dynamics of these organisms.

At a higher level of philosophical abstraction, one might also ask how best to articulate what can be observed. For instance, when considering what organisms do, and the possible awareness of what they do, one could explore the role of their cognitive modalities. This raises questions about whether these strategies are representational, and in what sense, whether they manifest similarly across taxa with or without a nervous system, whether they vary according to context, and whether they occur consistently or only under specific conditions (cf. Bianchi, 2021, pp. 165–172).

These are just a few examples of the many potential questions that could be explored.

Interestingly, each of these issues, when explored through the lens of plant behavior and signaling (cf. Van Volkenburgh & Brenner, 2024), becomes particularly challenging and stimulating due to the counterintuitive nature of the topics. A similar argument can be made for studies on the behavior and basal cognitive abilities of other organisms without a nervous system, such as bacteria and fungi, whose capabilities are increasingly being recognized and understood.

2.3.1 Terminological and Conceptual Analysis

Another arduous and often debated issue, which cannot be overlooked in this introductory phase, is the need to find, through conceptual analysis alongside growing empirical evidence,[24] a more inclusive and accurate application of many terms and expressions used in cognitive science. This will help determine the most effective way to conceptualize the behavioral and cognitive activities of plants in relation to ecological studies (cf. Cahill, 2015).

In general, theoretical terms can be operationalized, but the choice of *when* and *how* to do so has the potential to shape the micro-area of investigation. When a technical term is well-defined and clearly understood by the majority of those who use it in a certain field, it helps maintain clarity and consistency within that field.[25] When a term is implicitly transferred to other concepts, or the same term is used to express different ideas, ambiguity can arise. Interestingly, it is often from this plurality of meanings, common in scientific discourse, that new ideas can emerge

---

[24] The reference expressed here is to a critical commitment to an analysis of the conditions of application and use in different contexts of the concepts. It is a form of analysis different from the traditional one, not confined only to the theoretical side, which finds a point of contact with what has been referred to as "impure" conceptual analysis (cf. Glock, 2019), sensitive to empirical facts and experimental developments.

[25] The analysis of a concept is successful when it captures the general intuition of its users about its correct/adequate use.

through novel connections between a term or concept and a set of processes. Nonetheless, to avoid excessive confusion, it is essential to clarify the intended meaning of a term and the contexts in which it is used.

To give an idea of the vastness of the meanings available, there is no unanimous consensus even on how to refer to the field of research on the signaling, behavior, and plant "agency," relating to the ability to respond and learn from environmental stimuli to make decisions capable of ensuring survival. In fact, some prefer to refer to "cognition" (e.g., Calvo, 2007); others more frequently prefer "intelligence and behavior" (e.g., Trewavas, 2005, 2014); others prefer "plant gnosophysiology" (e.g., Michmizos & Hilioti, 2019), although the subject of debate is, in each of these cases, the interpretation of plants as systems that can detect information in the environment and respond to challenges in relatively flexible ways (cf. Brenner et al., 2006; Calvo, 2017; Trewavas, 2017). Thus, a considerable challenge in describing and explaining the capacities of non-neural organisms such as plants is finding new applications of the technical terms used in cognitive science, such as 'perception,' 'action,' 'memory,' 'learning,' 'behavior,' 'cognition,' 'intelligence,' 'mind,' etc., and the related models used to explain them (e.g., 'stimulus-response,' 'problem-solving skills,' 'flexible adjustment in changing contexts,' etc.).[26] When it is not possible to reuse or adapt some of these terms, even with necessary modifications, due to their being overly burdened with theoretical assumptions, it may be more practical to develop specific concepts tailored to the study of plant activities.

Most of the concerns about extending psychology or cognitive science terminology and reasoning to plants (e.g., Mallatt et al., 2021 and Taiz et al., 2019) stem from the claim to establish that plants are cognitive systems simply by relying on definitions of certain cognitive traits at a specific point in the research.[27] As Maher (2020) noted, in a discussion on the concept of consciousness, deriving what cognition is from a definition overstates the importance of definitions. A more sensible approach would be to let the productivity of research programs guide the extension of language to new contexts of analysis, rather than relying solely on definitions.

For each concept denoted by a certain term, it is necessary to assess both its strengths and weaknesses within a dynamic interplay between theoretical concepts and experimental observations. Theoretical concepts, on the one hand, reflect and preserve past research, while also guiding future inquiry. On the other hand, experimental observations, through new evidence and advancements, challenge and reshape previously accepted ideas, including what was previously accepted. Every

[26] In biology, a distinction is typically made between adjustment and adaptation to the environment. The former occurs within an organism during its individual development, while the latter is evolutionary, and involves multiple generations of individuals (cf. Oborny, 2019). In recent years, however, the term "individual adaptation" has been increasingly used in a way that closely resembles the concept of adjustment. For this reason, the term will be used in this sense throughout these pages.

[27] Definitions can also be interpreted not as "starting points" shaped by experiential problems, but rather as "end points" or possible hypotheses aimed at solving various problems (cf. Cellucci, 2008, p. 148).

transformation and integration of meaning is caused by and contributes to the evolution of the research process.

After the indispensable work of clarifying meanings, it is insufficient to merely examine various definitions; instead, it is necessary to understand how well these definitions apply in practice to describe the characteristics of plants and what modifications are needed to investigate specific behaviors. As Maher (2020) suggested, rather than defining, we should experiment, and, one might add, put the very definitions themselves to the test in this process, to see how far they can take us, with the awareness that they represent a starting point, not a destination. Indeed, while it is inevitable to begin with a certain level of knowledge and a corresponding conceptual approach, that initial point should not necessarily – and in many cases, should not – be the final stage of analysis. It is not a question of undisputed truths but of a process of research in the making, a theoretical and practical process (and theory is itself a form of practice; in the context of reflection of neopragmatism, see Margolis, 1996), which follows an evolution that is not perfectly linear and cumulative.

2.3.2 Which Conception of "Intelligence"?

Even when we frame the discussion in terms of "intelligence," these pages clarify in what sense the nervous system – traditionally associated with "intelligent" behaviors – is not indispensable to meet *all* the challenges posed by the environment.

Despite the emergence of a distancing from the most intransigent forms of "neurocentrism," the intent expressed here is certainly not to deny or to downplay the complexity and distinctiveness of animal nervous systems (which interact significantly with the rest of the body). Instead, it is to identify certain similarities between the functioning of nervous systems and biological organizations that lack them (e.g., unicellular organisms, fungi, plants) and then specify their differences. Every living system, indeed, organizes itself based on a set of structural and functional constraints that regulate the processes of synthesizing and recognizing chemical-electrical information within complex biological signaling networks.

The emerging field of research on plant cognitive abilities, along with related studies in the philosophy of plant cognition[28] (cf. Calvo, 2016; Heras-Escribano & Calvo, 2019; Bianchi, 2021, 2022, 2024) is beginning to shed light on how the physiological mechanisms of chemical-electrical signaling in non-nervous tissues of plants (and other non-neural organisms) can, through a chain of causal relationships from the microscale to the macroscale, form the basis for the development of behavioral processes linked to various species-specific needs.

[28] In this regard, see the research conducted at the *Minimal Intelligence Lab* (MINT Lab), a laboratory based at the University of Murcia.

The theoretical language of cognitive science, particularly cognitive neuroscience, is heavily influenced by assumptions related to neuronal bioelectrical signaling processes, which underpin animal behavior. In the absence of neurons in plants, scholars of plant cognition have identified forms of chemical-electrical signaling in plant non-neural tissues that exhibit functional similarities to those observed in animal nervous systems, albeit at different levels of integration and processing speed (cf. Calvo et al., 2017). From these simpler physiological mechanisms – responsible for bio-signaling within plant bodies (cf. Shemesh et al., 2010), between different parts of a single plant, and between different plants (cf. Heil & Karban, 2010) – emerge signal sequences that facilitate adaptive behaviors, including complex ones (cf. Castiello, 2019). These observations suggest that even non-neural living systems are capable of different levels of information processing. Although the integration and the speed of information processing differ, if we understand "information processing" as the non-random organization and use of information to perform communication tasks and support survival, plants – despite their distinct functional organization – also exhibit behavioral strategies based on forms of biological information processing. These strategies are ecologically relevant and efficient for their specific needs (cf. Bianchi & Castiello, 2023a, 2023b).[29]

In this work, I prefer focus on discussing "cognition" or "cognitive abilities" rather than concepts such as "mind," "mental capacities," or "intelligence." However, I also allocate space to these concepts to clarify their intended meanings. One goal, then, is to explicitly clarify the concepts used, while avoiding extremes in both the "exclusionist" approach (entirely avoiding terms or concepts from other contexts) and the "inclusionist" approach (forcing the use of terms in contexts outside their original or typical use). For example, when analyzing the concept of intelligence, the reference will be to the different manifestations of "biological intelligence" (cf. Stenhouse, 1974; Gilroy & Trewavas, 2023), rather than more sophisticated capacities like abstraction from the perceptual context and forms of imagination (cf. Bianchi, 2021, pp. 126–127).

The specific case of "plant intelligence" (as originally proposed by Trewavas, 2002, 2003, 2005, 2014) will be explored in relation to adaptability, the ability to regulate or adjust responses, during individual development (cf. Calvo et al., 2020). This discussion does not focus on adaptation from a phylogenetic perspective (cf. Pievani, 2017, pp. 80–81) – that is, the genetic inheritance of traits and variations from one generation to the next (though these processes are ultimately interconnected, as present possibilities are shaped by previous evolutionary pathways). Instead, the focus is on the metabolic and behavioral modifications that individual organisms make in response to environmental challenges throughout their life span (cf. Sultan, 2015). Moreover, whether we refer to forms of cognition or intelligence, the question of adaptive behavior to solve survival problems, which involves

[29] In this sense, we can observe a more biologically and ecologically grounded expansion and revision of the conception of classical cognitive psychology. In the traditional information processing paradigm, the "mark of the cognitive" is indicated by "*all the processes by which the sensory input is transformed, reduced, elaborated, stored, recovered, and used*" (Neisser, 1967, p. 4).

decision-making processes, is addressed from an ecological-relational and context-dependent perspective and is sensitive to the strategies of cooperation and competition of the various organisms involved, belonging to the same or different species (cf. Bianchi, 2022).

2.3.3 Why Plant "Cognition"?

Once the necessary precautions are established, while it is necessary to approach this subject with caution, it is important to acknowledge that reasoning about various processes, including cognitive ones, requires the use of specific terms and concepts. Even if many of the current ideas about what plants "do," and how and why they do it, eventually prove to be incorrect or in need of revision and refinement, it would seem overly reductive not to refer, in some way, to a form of plant "cognition" (or to cognitive abilities related to processes of perception-action, communication and problem-solving, memorization and learning of environmental information). In any case, in the study of cognition in general – as a complex, multilevel phenomenon involving multiple capacities that develop and change over time – one cannot expect a definitive answer or the same level of analytical precision as that achieved in more reductionist and less multifactorial fields of research (cf. Bianchi et al., 2025b).

More specifically, concerning plants, some scholars argued that we should not speak of true cognition in the case of plants. First, they suggested that cognitive phenomena are inherently tied to processes involving a nervous system, which limits cognition to the animal kingdom. Second, they pointed to the ability to form, store, and use internal representations of external events – even in the absence of the external stimulus itself – as a key feature of cognition (cf. LeDoux, 2020, pp. 260–262). This "offline" aspect of cognition enables a behavioral response to be detached from a perceptually present stimulus. It is precisely this offline capability that distinguishes cognitive processes from the non-cognitive biological processes of information processing (which are common to all living systems, from plants to fungi, unicellular eukaryotes, and bacteria) that allows organisms to respond to stimuli that are immediately present in their environment. In line with this view, Fred Adams (2018) proposed restricting the use of the term "cognition" to cases where organisms rely on a representational system. In this context, representations are understood as mental states grounded in propositional attitudes (e.g., beliefs, concepts, desires, etc.) and are subject to semantic evaluation (i.e., they can be true/false, accurate/inaccurate). According to Adams, cognition involves managing conceptual activities. In all other cases, behavior driven by information but lacking conceptual representations would not qualify as cognition. Thus, organisms without a nervous system, and which lack high-level representations directly tied to the possibility of "true knowledge," cannot engage in genuinely cognitive activities. However, as Miguel Segundo-Ortin and Paco Calvo (2019) argued in response to Adams' article, this approach to cognition, which defines conceptual representation as the hallmark of the cognitive, seems overly restrictive and limiting. It confines cognition to human cognitive abilities[30] while excluding other sophisticated forms of behavior (in animals and other living organisms) that do not involve mental representations. Equating cognition solely with the conceptual level overlooks non-representational theories of cognition, which are currently being explored for their explanatory power and broader scope in accounting for diverse processes and entities. According to Joseph LeDoux (2020, p. 261), if cognition is not defined by the possession of mental representations, it becomes synonymous with the mere ability to generate responses to various environmental stimuli, thus diluting the term to the point of losing its meaning. While LeDoux acknowledges a certain "intelligence" in plant behavior, understood as the ability to solve problems through interactions with the environment, and despite recognizing the broader and more inclusive conception of "cognition," he cautions against extending the term too far. Even if this broader definition gains acceptance, LeDoux suggests reserving the term "cognition" for processes that involve internal representations guiding behavior. He further invites us to find a new term to describe behaviors, even complex ones, that are not driven by representations (ibid., pp. 64, 66).

Therefore, why should be relevant to discuss in terms of "cognition" in plants?

The position expressed here (at least for now, in the absence of evidence to the contrary) does not agree that using the term "cognition" to encompass all possible forms of cognitive abilities, where organisms can learn, solve problems, and improve their responses over time, empties the term "cognition" of meaning. Instead, this approach could offer a more inclusive way of addressing the topic, allowing for a broad framework, before the need arises to specify and explore individual cognitive phenomena in more detail, from the most sophisticated forms involving higher abstraction to the simpler, more context-dependent interactions. Therefore, it seems necessary to continue discussing cognition, while specifying the sense and level at which it is being applied in each context (cf. Bianchi & Castiello, 2023a, 2023b).

The presumed absence of (mental) representations in the classical sense, in organisms such as plants, does not seem to rule out the possibility of discussing forms of cognitive behavior. Additionally, as will be explored in detail later, there is no single, universally clear way of defining the concept of representation. For instance, representation could be understood as a map or model – an internal configuration of molecular processes – that guides behavior, even in organisms without a nervous system, extending the reasoning to the involvement/importance of the role of information memorization and learning mechanisms (cf. Bianchi & Castiello, 2023a, 2023b). A precautionary perspective suggests being patient and awaiting the results of studies on plant behaviors that appear to involve offline forms of behavior (cf. Hiernaux, 2021), related to the ability to inhibit immediate responses to present

30 On the other hand, even those who are convinced that the study of human cognition is constitutively different from that of non-human cognition to support such a strong and far-reaching claim must do careful work to establish what (and what does not) count as cognition.

stimuli in favor of greater or more certain future benefits, such as in resource acquisition (cf. Novoplansky, 2016).

In any case, the valuable dialogue between different perspectives, some of which are highly critical of each other, has helped to raise the fundamental question of what is meant by "cognition." Cognition is, indeed, an umbrella term (cf. Schlicht, 2018), and throughout the history of cognitive science, various approaches have offered different interpretations of what it encompasses, how it should be understood or defined, and which biological entities and processes should fall within its scope of analysis. Much of this depends on the specific problems being investigated and the technological tools available at a given time in a particular field of research (cf. Bianchi et al., 2025a). There are several ways to characterize cognition. Some broader, more inclusive definitions attribute "minimal," "basic" or "basal" forms of cognition to all self-organizing living systems, or even equate living systems with cognitive systems (e.g., Varela et al., 1991), capable in different ways of "manipulating" their niche to maintain and improve their conditions. On the other hand, narrower, more exclusive definitions restrict cognition to organisms with nervous systems, or, even more selectively, to those capable of forming conceptual representations.

Furthermore, it could be argued that discussing "cognition" or using behavioral terms in the context of plants is misleading, unnecessary, or even incorrect, as descriptions at the physiological level may suffice. However, the same could be said for describing any animal (including human) behavior or cognitive abilities. For instance, one could limit themselves to explaining what happens inside a mammal's brain, detailing the chemical, electrical, and configurational activities triggered in response to external stimuli by specific groups of neurons. However, a cognitive-behavioral description, which includes the underlying mechanisms but goes beyond them by integrating multiple levels of explanation, enriches our understanding and can open new paths of inquiry, even if some of these ideas may later prove incorrect (cf. Colaço, 2022). Lastly, the cognitive-behavioral level of explanation has already proven advantageous for our species, as it allows us to concisely refer, at once, to a range of phenomena, before specifying individual cases, that involve memory, learning, anticipation, response improvement, and problem-solving, in both similar and novel contexts of action.

What makes research on plant behavior and cognition – as well as that of other non-neural organisms – particularly compelling is its long-term potential to reveal both shared traits and species-specific peculiarities. Such findings could significantly deepen and expand our general understanding of how different biological systems function and interact within their ecological niches, according to their specific needs. Achieving these goals, however, requires considerable patience, a willingness to take the risk of being wrong, and an openness to revising our initial hypotheses, which are often shaped by arguments and theoretical models developed to explain the characteristics and capacities of humans or other animals. It is not appropriate to draw rushed conclusions, that present themselves as definitive, regarding a specific aspect. There are still many challenges and interesting questions to be addressed; on the other hand, this is precisely the essence of "research."

2.4 New Focus on an Old Issue

This work primarily focuses on contemporary scientific and philosophical literature concerning the characteristics and capacities of plants. However, entirely omitting some turning points in the evolving understanding of plant behavior would be overly simplistic. Therefore, in this section, I briefly mention some pivotal passages, from ancient times to the present, related to attempts (not widespread but still present) by certain naturalists and philosophers to examine plant traits and abilities, which have otherwise often been overlooked (cf. Bianchi, 2021).

As Luciana Repici (2015, p. 7) noted, plants do not represent the "absolutely other," but the other by difference. As early as the fifth century B.C., prominent philosophers began laying the groundwork for interpretative models that notably focused on the complexities and comparisons between plants and other forms of life. These early inquiries addressed fundamental questions, such as the specific nature of plants and whether they possess a "soul," thereby raising the broader question of whether plants can be considered "animate entities."

As Gianfranco Pellegrino and Marcello Di Paola (2019, pp. 27–38) mentioned, Plato regarded plants as capable of perception (in the sense of having appetites, or basic forms of "desire"). In *Timaeus* (c. 360 B.C.), he compared humans to "celestial plants," with their heads as roots, reaching toward the *Hyperuranium* in their quest to connect with the realm of Ideas.

Aristotle, on the other hand, made a significant contribution to understanding the organic world through his pioneering efforts to compare and classify living beings, which also raised important questions about the differences between animal and plant characteristics. He expanded and refined Plato's concept of the soul, distinguishing between the nutritive, sensitive, and intellectual souls. In this view, the soul is not ontologically separate from the body, as later Christian traditions would suggest, but rather the form of the body, what enables living matter to exhibit the traits that define it as living: the soul is the form of a natural body that has life in potential (cf. *De Anima*, II, I, 412a, 20–21). Thus, for an organism to possess a soul means to be capable of carrying out its vital functions (cf. Di Francesco et al., 2017, pp. 37–38). Plants possess only the most basic form of life, the nutritive soul (*tò threptikón*), a quality shared by all living beings (*De Anima*, 413b; Aristotele, 2008, p. 457; Pellegrino & Di Paola, 2019, pp. 32–35). While the tripartite division of the soul was not explicitly intended to create a hierarchy of living forms, the act of distinguishing between them helped lay the groundwork for later hierarchical models of the natural order, based on varying degrees of perfection. Aristotle's approach has profoundly influenced the history of biological studies, including botany, from Pliny to Linnaeus (ibid., p. 38).

Special attention should also be given to Theophrastus, a disciple of Aristotle, often regarded as the "father of botany." In his works *Historia Plantarum* and *De Causis Plantarum* (late 4th to early 3rd century B.C.), Theophrastus provided detailed descriptions and an informal classification of plants, based on their various morphologies – what is now referred to as biological or growth forms – distinguishing between trees, shrubs, and herbaceous plants. He also used criteria related to the cultivability of different species. Notably, Theophrastus recognized plants' ability to perceive, differentiate, and respond to elements in their environment (cf. Pellegrino & Di Paola, 2019, pp. 68–70), which could today be described as "proto-ecological" observations. In the development of natural philosophy, particularly through the continuation and refinement of Theophrastus' approach, what has been termed a form of "philosophical botany" emerged. This perspective involved a broad inquiry into phenomena such as plant development, fruiting, species-specific diseases, and the causes of plant death.

As Repici (2015, pp. 8–9) pointed out, despite this progress, Hellenistic philosophies did not develop botany as a specialized discipline. This was likely due to the disappearance of key biological writings on plants from the Aristotelian tradition – the content of which remained known to a restricted audience, probably within the school – and because the dominant philosophical focus of the time shifted more toward human-centered reflection.

In Rome, philosophical botany did not fully develop; instead, a more "moralizing" one emerged, particularly in the works of Seneca, Epictetus, and Marcus Aurelius (ibid., p. 132).[31] However, in the case of some thinkers, such as Lucretius and Seneca, plants were studied within the broader framework of natural philosophy, treated as natural entities.

In the centuries that followed, it was medical botany that carried forward Aristotle's legacy (ibid., p. 42). Until the sixteenth century, plant studies were largely centered on medicinal herbs, or *Simples*, valued for their therapeutic properties and practical applications. The most widely consulted text by ancient apothecaries was Dioscorides' *De Materia Medica Libri Quinque* (c. 60 A.D.), written by Pedanius Dioscorides (c. 40–90 A.D.).

A new phase for botany, often referred to as modern or proto-scientific, began with the contributions of Luca Ghini (1490–1556), Pietro Andrea Mattioli (1501–1578), and the so-called "German fathers of botany," Otto Brunfels (1488–1534), Hieronymus Bock (1489–1554), and Leonhart Fuchs (1501–1566). Their work progressively shifted plant observations and descriptions towards a more scientific and realistic approach, moving away from allegorical or symbolic interpretations and those primarily linked to human utility. The increasing focus on accurate plant identification led to the establishment of the first botany chairs in Europe, such as in Padua around 1533, which preceded the creation of the Botanical Garden in 1545 (cf. Dröscher, 2008, pp. 133–153; Bianchi, 2021, pp. 23–30).

Over the centuries, attention to the plant world has explored various aspects, though rarely focusing on their behavioral, and, even less so, cognitive characteristics.

For example, Carl Nilsson Linnaeus (1707–1778), who adopted and further developed the binomial naming system, still adhering to an essentialist view, focused

[31] For example, in Epictetus the condition of the ears of wheat offers the opportunity to reflect on some choices that human beings are required to make (*ibid.*).

predominantly on classification methods that facilitated the quick naming and memorization of plants. Linnaeus used what has been termed the "sexual system,"[32] based on characteristics considered well-defined and relatively stable, such as the number and shape of stamens and pistils (cf. Mayr, 1982, pp. 177–178). Another noteworthy figure is Gustav T. Fechner (1801–1887), who, drawing from earlier thinkers, redirected attention toward questions that went beyond mere classification and taxonomy. Likely in reaction to the mechanistic and reductionist explanations of his time, Fechner, from a panpsychist and romantic perspective, often provocatively, argued that plants were not mere machines. He suggested that, as part of a natural unity, plants possessed characteristics akin to those of animals, along with an animating force (cf. Fechner, 1848).

We cannot fail to remember the significant interest in the plant kingdom of Charles R. Darwin (1809–1882) and several members of his family, particularly his son, Francis Darwin (1848–1925).[33]

Modern plant sciences, at many levels of analysis, owe much to the pioneering insights and observations of the Darwin family (cf. Bianchi, 2025). These early contributions laid the groundwork for an approach that has been revisited, expanded, and integrated into contemporary studies on plant behavior and intelligence. Charles Darwin himself published several influential monographs primarily focused on botanical issues. Notably, in 1862, he published *On the Various Contrivances by which British and Foreign Orchids are Fertilized by Insects and on the Good Effects of Intercrossing*, where he examined the mechanisms behind the diverse structures enabling orchids to be pollinated by insects, as well as the detrimental effects of self-fertilization (Darwin, 1862, pp. 1, 359–360). The theoretical aim was twofold: to elevate the status of the plant kingdom and to provide empirical evidence supporting his theories, in response to accusations of speculation following the publication of *On the Origin of Species by Means of Natural Selection* (1859) (cf. Bellon, 2011).

The second specific botanical monograph, *The Movements and Habits of Climbing Plants*, was published in 1875, following an earlier version that had

---

[32] For Linnaeus, reproduction reflected the divine order's design.

[33] Charles Darwin's uncle, Josiah Wedgwood II, was one of the founders of the Royal Horticultural Society, while his grandfather, Erasmus Darwin (1731–1802), had written on botanical themes in *The Botanic Garden* (1791), a work comprising two poems: *The Economy of Vegetation* and *The Loves of the Plants*. Charles Darwin's son, Francis Darwin, a professional botanist, played an essential role in conducting observations and "home experiments" with Charles on various plant species (cf. Bianchi, 2025). Francis had the opportunity to learn the most advanced techniques in plant physiology at Julius von Sachs' laboratory (1832–1897) at the University of Würzburg, which enabled him to study plant movement and responses to stimuli (cf. Citizen, 2009, pp. 235–236). At that time, interest in plant sensitivity was growing, with various perspectives emerging. As Maher (2017, p. 15) notes, Jacques Loeb (1859–1924) argued that plant sensitivity could be understood entirely in mechanical terms, without invoking the presence of a mind or sense organs. In contrast, in 1888, Alfred Binet (1857–1911) proposed that microorganisms also possessed a "psychic life." Later, in 1908, Francis Darwin considered the possibility of a specific form of intelligence and "plant consciousness," suggesting that plants exhibited a basic equivalent of animal consciousness (cf. Bianchi, 2021).

appeared as an article in 1865 in the *Journal of the Linnean Society*. This essential work, still a point of reference for studying certain forms of plant behavior, saw Charles and Francis Darwin explore the movement characteristics and specialized organs that allow climbing plants to attach to various types of support. The starting point for their observations was the potential modifications of circumnutation, i.e., a subtle spiraling motion observed in growing plant parts. This was also one of the earliest texts to mention plants' ability to selectively discriminate between supports (which could be unsuitable if, for instance, they were too thick or smooth).

In the same year, Darwin also published *Insectivorous Plants*, a work that brought attention to a unique class of plants, what we now call "carnivorous" plants, a concept previously unimaginable. Some of the observations in this pivotal text helped to spark a series of important studies on plant electrical signaling, which later became a cornerstone of the modern field of plant electrophysiology (cf. Kutschera & Briggs, 2009).[34]

In 1876, Darwin published *The Effects of Cross and Self Fertilisation in the Vegetable Kingdom*, in which he revisited some of the topics covered in his earlier work on orchids. His primary aim was to demonstrate that the offspring of cross-pollinated plants are more numerous and vigorous than those of self-fertilized plants, even though genetic studies were not yet developed at the time.

In 1877, Darwin published *The Different Forms of Flowers on Plants of the Same Species*, based on articles previously presented to the *Linnean Society*. This volume focused on the differently shaped flowers produced by certain plants, either on the same or distinct varieties (Darwin, 1877, p. 1), and on heterostyly, a characteristic involving plants with styles of varying lengths.

In 1880, he published another monograph entirely devoted to botanical studies. *The Power of Movement in Plants* was a collaborative effort between Charles Darwin and his son Francis. The main objective of this work was to describe the various classes of movement common to most plants. It can be interpreted as a thematic continuation of his earlier work on climbing plants and is now regarded as a classic study on tropisms (cf. Ayres, 2008). Darwins focused particularly on the phototropic responses of the seedling of the grass *Phalaris canariensis L.*, observing that only the apical part was sensitive to light. They hypothesized that some substance was transmitted from the light-exposed upper part to the lower part, enabling it to bend accordingly (Darwin & Darwin, 1880, p. 4).[35] Moreover, the

[34] Darwin hypothesized the presence of a form of "nervous matter" in certain plant species, particularly after observing the transmission of a "motor impulse" to the marginal tentacles of the Sundew. He noted that the time required for this transmission could vary depending on factors such as the leaf's age, living conditions, and the type and size of the object being stimulated (Darwin, 1875a). This idea was later confirmed through the experiments of John Burdon-Sanderson (1828–1905) and Sir Jagadish Chandra Bose (1858–1937). Bose, in particular, measured action potentials first in *Mimosa pudica* (a plant known for its rapid movements) and subsequently in other plant species. In his work *The Nervous Mechanism of Plants* (1926), Bose described a nerve-like impulse transmitted along the vascular tissues of plants.

[35] Darwinian hypotheses were later confirmed by the experiments of Peter Boysen-Jensen (1883–1960) and Frits W. Went (1903–1990) who identified the involvement of auxin, one of the main plant hormones in phototropic responses.

Darwins noted that climbing plants not only had the ability to locate potential supports in their environment, but could also distinguish between suitable and unsuitable characteristics of those supports, such as material and size, before attempting to reach and attach to them. *The Power of Movement in Plants* also presents significant research on gravitropism, by root apex, which plays a key role in determining the direction of root growth (ibid., pp. 72–73). However, the work is perhaps best known for its famous concluding passage: "*It is hardly an exaggeration to say that the tip of the radicle, thus endowed, and having the power of directing the movements of the adjoining parts, acts like the brain of one of the lower animals; the brain being seated within the anterior end of the body, receiving impressions from the sense-organs, and directing the several movements.*" (ibid., p. 573). This analogy, summarized in what is now referred to as the "Root-Brain Hypothesis" (cf. Baluška et al., 2009), has been explored in recent years. It draws a comparison between the functioning of the brains of simpler animals and root behavior, specifically the ability of root tips to transmit information (via chemical and electrical signals), thereby directing movements in adjacent parts.[36]

Overall, Darwin's research approach has fostered deeper and broader reflections on the different forms of intelligence in nature, including in plants, highlighting both their unique characteristics and the continuities across kingdoms (cf. Giaculli, 2019). The "Darwinian revolution," scientific, conceptual, and ethical (cf. Pollo, 2019, pp. 187–188), triggered a series of consequences and set a precedent for the future, transforming how we perceive the relationships between living beings and, particularly, our place in the world. Among its key implications were the shift from viewing the biological world as static to recognizing it as evolving (in fact, this was already the case with the work of Jean-Baptiste de Lamarck [1744–1829]), and the replacement of the creationist explanation with a materialist one, centering on species evolution by natural selection, based on a unfinalized relationship between variation and reproductive or survival success. This led to the critique of finalistic explanations, opening the door to reflections on the role of chance in nature, and to challenging anthropocentric justifications (cf. Mayr, 1982, pp. 500–501). This is on a general level.

In this specific area, the botanical studies initiated by the Darwins have been productively carried forward through subsequent biological, ecological, and more recently, ethological research (cf. Fay & Chase, 2009; Isnard & Silk, 2009; Thompson, 2018, p. 107). These developments have laid the basis for contemporary philosophical analyses, particularly in the field of cognitive science and philosophy of mind (cf. Bianchi, 2025).

After Darwinian interest, the story of consideration of the plant world continued. As Umberto Castiello (2019, p. 12) pointed out, studies on plant behavior, especially movement, were initially explored in Anglo-Saxon treatises on comparative psychology (e.g., Warden, 1928; Fuller, 1934; Warden et al., 1935; Appelwhite, 1975), before merging into the field of plant physiology.

[36] For a critique of this hypothesis, cf. Taiz et al. (2019).

This brief (and incomplete) overview of the scholars who have focused on plants aims to illustrate the development and transformation of our understanding of certain plant characteristics over time. However, despite these efforts and a few other notable exceptions, the behavioral-relational, communicative, or what have more recently been termed "intelligence" or "cognition" aspects of plants have generally remained underexplored or neglected. Why is that?

2.5 Perceptual-Cognitive Challenges in Understanding Plant Behavior

In this section, I briefly address why the cognitive abilities of plants have generally been neglected, due to various types of practical difficulties and theoretical hurdles. One reason is the counterintuitive nature of perception and cognition systems (or forms of intelligence) that differ from our own.

I emphasize how the increasingly widespread use of techniques for detecting plant movements and signaling mechanisms and processes, along with the development of theoretical approaches more attuned to the biology and ecology of different species, has expanded the experimental framework for analyzing these phenomena.

Building on this, I discuss how the practical and theoretical intersection of these advances has led to the development and consolidation of the field of plant behavior and cognition. This discipline goes beyond traditional plant biology and ecology, integrating findings from multiple levels of analysis with behavioral-cognitive perspectives, highlighting the ability of plants to organize and use environmental information in a non-random way.

2.5.1 Locomotion and Cognition

To begin with, there is a common human tendency to overlook plants and their activities, a phenomenon known as "plant blindness" (cf. Wandersee & Schussler, 1999; Mancuso, 2017a, 2017b). This form of "blindness" (or "plant awareness disparity" [see Parsley, 2020]), which has evolutionary roots (cf. Balding & Williams, 2016), is linked to the perceived stasis and lack of locomotion in plants, as well as their relative harmlessness compared to animals, since plants lack the ability to "attack" us. Although plants exhibit a wide repertoire of movements (cf. Castiello, 2019, pp. 59–79), these occur on a much slower timescale than those of most animals, rendering them nearly imperceptible to the human eye without technological assistance to enhance motion detection. As a result, one of the primary reasons plants have been excluded from the cognitive domain is the extensive tendency to associate cognitive abilities with movement, particularly with the ability for free locomotion, the basis of self-realization (cf. Jonas, 1966).

An example can help illustrate this point. From a human perspective, even when observing the behavior of a microscopic organism like the gram-negative bacterium *Escherichia coli*, which moves away from danger (e.g., moving away from toxic or harmful substances) and swims toward nourishment, we get the impression that something is "happening" (cf. Trewavas, 2014, p. 11). Despite its simplicity, this behavior appears to be goal-directed, hinting at a form of "proto-intentionality."

In general, beings that move appear to have intentions, and those with intentions seem to possess a mind (cf. Kahneman, 2012). From a psychological and perceptual perspective, humans tend to associate movement, especially locomotion, with cognition or intentional behavior. In this context, Pietro Perconti (2017, pp. 48–53) highlighted the similarity between "motor triggers"– mechanical stimuli that activate a bodily reflex, causing a body part (e.g., a limb) to move beyond conscious control – and "mental triggers,"[37] which involve automatic processes prompted by specific environmental stimuli that lead to a mentalistic interpretation of certain behaviors or movements. In both cases, mechanisms are activated and are not the result of conscious deliberation.

In line with this, Gunnar Johansson (1911–1998) demonstrated the connection between kinematic information and certain human characteristics in the automatic and selective process of perception. Perconti (2017) also referred to the phenomenon of attributive dualism, a type of ontologically neutral dualism tied to the evolutionary tendency to perceive organisms, particularly humans, as bodies guided by intentional minds. This predisposition inclines us to interpret the behavior of others "as if" it were driven by rationality or intentionality.

However, beyond these tendencies, locomotion alone does not seem to be a reliable criterion for distinguishing between cognitive and non-cognitive entities, nor for separating animals from plants (cf. Linson & Calvo, 2020; Bianchi, 2021, p. 156). For the subjects of interest here, movement is not a definitive mark of the cognitive abilities. (In general, it is problematic to identify a single, definitive "hallmark of cognition," as cognition, and the abilities it involves, is a complex, multifactorial, and multilayered phenomenon).

Additionally, it's important to distinguish between locomotion and movement. While plants do move,[38] they are rooted and generally unable to autonomously relocate from one point to another (i.e., lack the ability for locomotion from point "A" to "B") to achieve their goals and solve survival problems (cf. Wang et al., 2023). However, as Calvo and Keijzer (2011) pointed out, and as is increasingly recognized, being rooted does not equate to a lack of behavioral capacity.

Despite having different energy requirements and metabolic needs (cf. Segundo-Ortin & Calvo, 2022) and a distinct sensorimotor organization compared to animals,

[37] Examples of "mental triggers" include the ability to perceive faces with different expressions in certain patterns, and more relevant to our discussion, the predisposition to distinguish between nonliving and living beings – or between different types of living beings – based on their presence, absence, or the type of movement they exhibit.

[38] This holds true not only for plants with rapid movements, such as *Mimosa pudica* or *Desmodium gyrans*, but also for certain carnivorous plants like *Dionaea muscipula* or some species of *Drosera*.

plants can adapt to their environments and solve survival problems through the flexibility of their responses over the course of their life (i.e., throughout their individual development). The potential for phenotypic modification in plants encompasses metabolic, morphological, and behavioral changes, providing them with diverse strategies for inter-action. These strategies, primarily linked to chemical-electrical signaling, tactile perception, and growth dynamics, enable plants to adaptively respond to environmental challenges.

Such flexible and context-dependent variations are what we refer to when discussing forms of (cognitive) behavior in plants.

2.5.2 Theoretical and Technological Advances

A question that remains in the background is the extent to which recent (and ongoing) developments in these areas stem from shifts in theoretical frameworks, as opposed to advancements in technologies and techniques for detecting plant behavior and signaling dynamics. This raises a broader issue: to what extent, and in what ways, do theoretical approaches shape research outcomes, and how significantly does the availability of technology influence the investigation of a specific problem? It is most likely that progress results from a confluence of factors: conceptual advancement, technological development, and methodological-experimental refinement (cf. Bianchi et al., 2025a).

This is well illustrated by the technical and technological innovations[39] that have accompanied nearly all major advancements in various branches of scientific research. For example, in fields like cytology, the invention of the microscope (ca. 1590), the development of various types of lenses and objectives, the phase-contrast microscope (1934), and later the scanning electron microscope (1938), have greatly refined our understanding of cellular processes, both in terms of morphology and function (cf. Mayr, 1982, p. 853). Similarly, in astronomy, when Galileo Galilei (1609) pointed his telescope at the sky, he discovered previously unimaginable celestial entities, such as the lunar mountains, Jupiter's moons, and Saturn's rings.

In the case of plants, beyond the challenges posed by their different life rhythms and movements, additional obstacles arise from their size, whether too large or too small. Just as many microorganisms in the rhizosphere are undetectable without a microscope – an aspect that complicates our understanding of soil ecosystems and their crucial role in supporting above-ground systems (cf. Mazzolai, 2021, p. 61) – similarly, difficulties arise in studying large-scale plant systems, such as clonal colonies. These systems, composed of interconnected components (macro-modules), experience vastly different living conditions, including variations in microclimate, resource availability, and soil composition.

[39] The terms "technique" and "technology" are not synonymous. Generally, "technique" refers to the methods or ways of performing an activity efficiently, while "technology" relates to the application of scientific knowledge in the creation of tools and procedures.

In terms of practical advances, the development of precise techniques for detecting and recording movement has made it possible to observe various types of plant movement across different contexts. This progress dates back to the first observations by Pfeffer (1896), who used time-lapse techniques (cf. Stolarz et al., 2014).

Later, electrodes were developed for electrophysiological studies (cf. Volkov, 2012, 2017), enabling the monitoring of electrical activity in plants, which proved valuable for identifying physiological responses and subsequent morphological changes (cf. Karban, 2008) that might otherwise go unnoticed (cf. Calvo & Trewavas, 2020). These advancements have allowed researchers to observe the context-dependent flexibility of plant responses, challenging the previously dominant view that plants reacted to their environment in a rigid, predetermined manner (cf. Trewavas, 2017).

Regarding the use of time-lapse techniques, some authors have criticized videos that show the accelerated growth of stems or roots, arguing that they make plant movements appear more similar to those of animals, thereby attempting to suggest the presence of consciousness or intentionality (cf. Taiz et al., 2019). While it's essential to approach species-specific behaviors with care, avoiding projections and anthropomorphisms, the goal of such techniques is not to "animalize" plant behavior but to make it visible and more comprehensible to humans, who perceive time at different scales (cf. Bianchi et al. 2025a).[40] Similarly, slowing down animal movements is a common practice used to better observe behaviors in the wild.

Beyond technological advances, there are additional practical challenges in observing spontaneous plant behaviors in nature, which are generally not recorded (cf. Trewavas, 2003). This has been a significant limitation for those involved in the debate on plant cognition, as the theoretical discussion requires tools capable of revealing or clarifying more or less complex behavioral patterns. As Segundo-Ortin and Calvo (2019) pointed out, it would be mistaken to conclude that plants lack the range of behaviors seen in animals simply because the tools and methodologies used were (or are) inappropriate for the organisms being studied (cf. Calvo, 2017). Thus, the techniques and devices employed are crucial for exploring the factors and processes under investigation (cf. Bianchi, 2021, p. 35).[41]

For instance, despite undoubted advances, there remain objective challenges in studying the subterranean behaviors of roots, given the nontransparency of the soil and the vast extent of root systems, which are intricately connected in symbiotic

[40] E.g., even in the case of the comparatively very fast growth of plants such as bamboos (about one centimeter per hour), without some recording device it would be difficult for the human senses to observe this phenomenon (cf. Trewavas, 2003).

[41] It's important to remember that the modular growth of plants, which is highly sensitive to variations in local microenvironmental conditions, exposes their bodies to multiple stimuli and potential changes. Due to this variability, even in controlled laboratory environments, particularly during the early stages of seedling development, when plants are more flexible and adaptable, it is extremely challenging to design experimental setups and plan tests without influencing their growth and structure.

relationships with mycorrhizal fungi and a variety of soil microorganisms (cf. Bonfante, 2021).

As Barbara Mazzolai (2021, pp. 78–79) noted, few technologies currently exist that can observe belowground behaviors effectively. One noninvasive technique is nuclear magnetic resonance imaging (MRI), which is used to capture both two- and three-dimensional images in soils of varying compaction. Another method adapted for the study of underground activities is X-ray and neutron computed tomography (CT), which are employed to investigate root growth and interactions, particularly their architecture (shape and branching patterns) in relation to water and nutrient uptake processes. Technological and applied research will be able to integrate and further develop the most suitable techniques for observing increasingly broader portions of root behavior, with greater resolution detail.

2.5.3 The Invisible: Latent States of Life and "Silent Languages" of Plants

Another significant consideration is that not every process or behavior manifests in an observable, continuous manner across time and space. Phenomena may be described at different stages and through various levels of analysis (cf. Bianchi et al., 2025b).

As Paola Bonfante (2021, p. 15) noted, a researcher does not "see" solely with the eyes of the body but also with "the eyes of the mind," recognizing what is being studied. For example, the activity of root systems remains hidden at first glance, as they lie beneath the soil, just like the complex interactions between plants and various microorganisms, whose microscopic scale renders them invisible to the naked eye.

In the animal kingdom, there are instances of cryptobiosis or anabiosis – states of metabolic suspension – into which ultra-resistant organisms, such as tardigrades, can enter. These states can persist for extended periods under conditions of environmental desiccation, extreme temperatures, or even in outer space. In such extreme conditions, the metabolism of these tiny invertebrates halts, only to reactivate when more favorable conditions return (cf. Mazzolai, 2021, pp. 66–68).

A similar example of latent life can be found in lichens. These organisms are formed by the symbiotic association between a cyanobacterium or alga (the autotrophic organism) and an ascomycete or basidiomycete fungus (the heterotrophic). Lichens, as poikilohydric organisms, possess the remarkable ability to survive without water. Under desiccation, their metabolism slows to basal levels, resuming normal activity when humidity levels rise (cf. Bonfante, 2021, p. 82).

Another example of latent metabolism is seed dormancy, where seeds, in a not-so-metaphorical sense, can "move through time and space." A study on dormant Arabidopsis seeds (Topham et al., 2017) identified a mechanism that integrates environmental information – specifically temperature fluctuations – by regulating the balance of the antagonistic hormones abscisic acid (ABA) and gibberellin (GA).

These hormones, located in distinct cell types, engage in signaling cross-talk to create a "bistable switch" that determines the seed's developmental fate. This system allows seeds to break dormancy when optimal conditions are detected, enhancing their chances of germination across diverse ecosystems.

Viewing seeds as a type of "spacecraft" capable of traveling through time and space not only familiarizes us with the concept of noncontinuous metabolic processes but also deepens our understanding of how eco-evolutionary dynamics enable some seeds to germinate far from the parent plant, reducing future competition among seedlings (cf. Mancuso, 2018).

Despite the lack of locomotor ability, many plant species effectively disperse their seeds via wind, water currents, or specialized mechanisms, like catapult-like structures. Some even rely on animals, including humans, for seed transport. For instance, certain fruits, such as those of *Xanthium* or wild carrot (*Daucus carota*), have hooked surfaces that attach to the fur of animals, facilitating dispersal. Others, like mistletoe, produce sticky berries that are transported over long distances by birds.

Returning to the central theme of these sections (the perceptual and cognitive challenges related to understanding the characteristics and, particularly, the behaviors of plants) there are limitations to what we can grasp directly or easily. Even with dedicated effort, such as studying processes across different levels of organization, phases, conditions, and environments, much remains elusive.

As Lorenz (1952) observed – the ethologist and zoologist widely regarded as one of the founders of modern ethology, known for his studies on animal behavior and imprinting – many animals can perceive a wide range of signals that are imperceptible to humans. For example, one of his dogs was able to anticipate his reactions and actions with remarkable accuracy, testifying to its specialized sensory and behavioral abilities. For instance, take the olfactory capabilities of dogs: compared to humans, their olfactory receptors can detect an astonishing range of odors, even at extremely low concentrations. It's no exaggeration to say that dogs primarily "experience" and "understand" their surroundings through their highly developed sense of smell, which is thousands of times more sensitive than ours. The reality they perceive and navigate is likely different from that of humans and other species, as each organism's view of its environment is always mediated by its biological and ecological specializations.

Every organism operates within certain constraints, possessing both capacities and limits that shape its species-specific traits,[42] and this holds true for plants as well. Plants, relying on distinct perceptual systems, detect a range of environmental signals that many animals cannot, such as recognizing various chemicals in the soil,

[42] It is not a question of superiority and in many cases not even of greater complexity: it is a question of different perceptual-cognitive specializations. From an eco-evolutionary point of view, each organism, including plants, has the characteristics and abilities necessary for its survival and does not "lack" the abilities of other living beings (cf. Baker, 2017, p. 147), such as, for example, in our case, the ability to breathe underwater or to fly, or in the case of a hamster, to carry out photosynthesis.

sensing magnetic fields, and distinguishing between different types of light radiation, etc. Imagining what it might be like to experience the world as a plant – perceiving its environment through these unique channels of exchange and interaction – requires a considerable leap of imagination (cf. Calvo, 2017). Yet, this perspective is destined to remain partially beyond our understanding, as it is not something we can directly "sense" or fully grasp.

There are, in fact, aspects of plant life that remain beyond full comprehension or even detection, despite advances in technology. According to Gagliano et al. (2017, pp. XVII–XVIII), plant communication and interaction occur on two distinct levels: "extrinsic language" and "intrinsic language." The former refers to the various ways in which scientists, philosophers, and scholars conceptualize and articulate the specific characteristics of plants, encompassing the scientific and philosophical lexicon used to describe them. The latter refers to the concrete mechanisms of multisensory perception, action, and communication, such as the chemical-hormonal, electrical, or tactile signals exchanged between plants and other organisms (cf. Baluška et al., 2005; Bianchi et al., 2025a).

However, it is difficult to fully separate these two levels, as any behavior or form of communication is inevitably tied to our interpretive frameworks, which are always partial to some degree. As Marder (2017, p. 105) reminded us, research into the "silent languages" of plant forms, colors, and scents (cf. Gagliano, 2017) also requires us to listen to the gaps and silences within language, allowing space for the untranslatable and, therefore, for the unspeakable.

2.5.4 Plants and Sympathy: The Issue of Pain and Its Behavioral Manifestations

Another longstanding obstacle to the investigation of plant activities from a cognitive-behavioral perspective lies in the difficulty of extending sympathy to plants. In this context, "sympathy" denotes the capacity to comprehend the emotions of others, especially their suffering. Typically, sympathy is reserved for beings traditionally considered "sentient," those capable of experiencing pleasure and pain (cf. Pollo, 2019, pp. 196–197). Unlike many animals, for example, plants do not recoil or cry when struck with a fist or axe (when generally being damaged). However, as Khait et al. (2023) reported, stressed or injured plants can emit airborne ultrasounds, as detected in an acoustic chamber, signals that can convey information to other organisms. Research into bioacoustic phenomena in plants is still in its early stages (cf. Calvo & Segundo-Ortin, 2023). In the coming years, many aspects of this field may be further clarified. For instance, it has been hypothesized that plants might emit sounds to acoustically inhibit the growth of unrelated (non-kin) plants (cf. Demey et al., 2023).

However, there appears to be a distinction between experiencing pain and merely detecting physical harm. As Godfrey-Smith (2021, pp. 107; 244–247) noted, the term "nociception" refers to the sensory process of detecting damage and the corresponding behavioral response. An organism's reaction to a harmful stimulus may be driven by "insentient" nociception, where damage is detected and a response is triggered, without the experience of pain.

Many biologists, indeed, do not consider the mere ability to detect and respond to injury as sufficient evidence of pain or a similar sensation. Instead, they look for additional signs that are more closely linked to the presence of pain or something analogous. Examples of such signals include wound protection and healing, the production of analgesic substances, the avoidance of potentially harmful behaviors or situations, and the ability to weigh certain behaviors by balancing costs and benefits.

It's worth noting that even in the animal kingdom, such as in certain insects, where there is a functional equivalence between the cephalic ganglion and the mammalian midbrain (cf. Klein & Barron, 2016), the issue of pain is more complex than it might initially appear. Increasingly, various forms of sentience are being recognized and studied in a wide range of species, from mammals, birds, reptiles, and fish to invertebrates (cf. Mikhalevich & Powell, 2020; Crump et al., 2022).

Research has explored "nociceptive sensitization" in insects, a state characterized by heightened sensitivity and a protective response following injury, which can last hours, or even days. This leads to changes in decision-making and responses to other stimuli. On a broader level, interesting findings have emerged from studies on honey bees. For example, Melissa Bateson et al. (2011) found that when honey bees were subjected to shaking, they were more likely to interpret ambiguous stimuli in a "pessimistic" way, anticipating negative outcomes.

When it comes to plants, despite their high sensitivity to environmental stimuli, their aneural and modular nature, and their ability to lose large portions of their bodies (such as through herbivory), it is likely that they do not experience anything equivalent to animal pain. However, there has been limited scientific research and philosophical discussion on the potential counterpart of pain in plants (cf. Viola, 2019, p. 216). In the absence of clear evidence, the most prudent approach is to remain open to future findings, without entirely ruling out the possibility of a form of aneural plant sentience (cf. Segundo-Ortin & Calvo, 2023), especially given the lack of consensus on the biological traits required for sentience (cf. Godfrey-Smith, 2016).

Although experiencing pain is not evolutionarily advantageous for plants, it is interesting to note that even in the absence of a nervous system, plants contain signaling molecules that are also involved in pain perception and transmission in animals. Additionally, plants possess mechanisms for rapidly transmitting and interpreting damage-related signals, such as those triggered by a predator attack, which help coordinate responses aimed at halting or repairing the injury (cf. Toyota et al., 2018). Beyond the aspects of sentience related to the ability to experience pleasure and pain in animals (some authors have linked sentience to consciousness), certain chemical substances, true signaling molecules, have been discovered in plants. These include glutamate, acetylcholine, dopamine, norepinephrine, histamine, serotonin, and gamma-aminobutyric acid (GABA), well-known neurotransmitters in animals (cf. Bouché et al., 2003; Žárský, 2015; Volkov, 2017). In plants, for instance, glutamate released upon wounding triggers calcium ion influx, enabling the systemic propagation of electrical signals throughout the plant body (cf. Toyota et al., 2018).

From an evolutionary perspective, it is not surprising that many molecular signaling mechanisms existed before the emergence of animal nervous systems (cf. Baluška & Levin, 2016). These chemical, electrical, and hydraulic signaling processes are currently being studied, and they present a challenge for understanding how they relate to the integrated behavioral responses of plant systems in dynamic environments (cf. Huber & Bauerle, 2016). Electrochemical excitability, particularly in the phloem vascular cells, enables plants to respond in a coherent and adaptive manner to environmental changes and physical damage.

On a practical level, one way to study plant sensitivity and response is through the use of anesthetic substances (cf. Grémiaux et al., 2014; Baluška & Yokawa, 2021), which can disrupt their behavior and environmental responsiveness (the plants' sensitivity is temporarily halted by anesthetics). While much remains to be understood about the biomolecular mechanisms by which anesthetics alter plant reactivity (cf. Baluška et al., 2016), these substances offer a valuable tool for research. Anesthetics like ethanol or ethylene, which are naturally produced by plant tissues under stress, can also be externally administered to observe their effects.

In plants with rapid movement, such as *Dionea muscipula* (Venus flytrap) or *Mimosa pudica* (Sensitive plant), the application of anesthetics inhibits their typical responses. In the case of *Dionea muscipula*, the predatory response (closure of the trap leaves) is blocked, while in *Mimosa pudica*, the anesthetic prevents the anti-predatory defense of folding its leaves, which normally exposes its thorns (cf. Yokawa et al., 2018).

2.6 Contemporary Insights

Despite the challenges mentioned earlier and the likelihood of further impediments or counterintuitive findings, particularly over the past two decades, significant theoretical and technological advances have been made. Alongside advances in philosophy, cognitive science, and the biological and ecological sciences, these developments have prompted a rethinking and conceptual revision of how plant behaviors are described, explained, and theorized. These are behaviors that, in organisms with nervous systems (whether centralized or distributed), are typically understood as expressions of cognitive capacities (cf. Castiello, 2021).

Several labels have been proposed to define this area of research on plant cognitive behavior, including "plant neurobiology," "plant ecophysiology," "plant gnoseology", and "plant cognitive ecology." Some terms, like *plant ecophysiology*, tend to be more neutral and less conceptually ambiguous, though they may reinforce the misconception that plants are passive components tightly bound to their environment. The neologism *plant gnosophysiology*, proposed by Michmizos and Hilioti

(2019), seems apt for describing plant information-processing systems in a dynamic and interactive manner (cf. Cahill et al., 2010; Souza et al., 2017). This term helps free the field from the biases and expectations associated with models rooted in the functioning of animal nervous systems.

In this field, while the phenomena and processes under investigation are broadly similar across various approaches and interpretations, for the sake of clarity and to help consolidate the research area, I refer here to the general framework of plant behavior and cognition. The specific subfield may be specified subsequently. Compared to the early days of this discipline, when its innovative ideas sparked considerable debate, many of the initial controversies, particularly those related to terminology, have since been mitigated or partly resolved. These controversies were primarily due to the use of terminology borrowed from animal research, which some critics viewed as zoomorphic (not merely anthropomorphic) or simply misleading (cf. Alpi et al., 2007; Struik et al., 2008). The term "plant neurobiology" has been especially criticized for referring to neurons, which plants do not possess. However, the original intent behind the term was to employ "neuro" in plant biology as a heuristic tool, meant to establish connections and guide future research (cf. Brenner et al., 2006; Calvo, 2016; Bianchi, 2021, pp. 128–129). As noted earlier, it is through the convergence of theoretical and practical advancements – particularly in the detection of electrochemical signaling processes within and outside plant bodies, as well as a deeper understanding of plant movements – that this field of research has been able to grow and solidify.

Historically, the Society for Plant Neurobiology was established in 2005, followed by the launch of its journal, *Plant Signaling & Behavior*, in 2006. The formal foundation of plant neurobiology as a field was laid out in the paper *Plant Neurobiology: An Integrated View of Plant Signaling* (cf. Brenner et al., 2006), published in the journal *Trends in Plant Science*. As the titles suggest, the primary goal of scholars studying plant abilities is to develop a unified approach to understanding plant signaling and behavior. This approach has been largely influenced by revisiting observations made by Charles Darwin and his son Francis, particularly their brain-root hypothesis, presented in *The Power of Movement in Plants* (cf. Darwin & Darwin, 1880).

As Adrian Frazier (2021) pointed out, the early debates in plant biology largely centered on two opposing perspectives: the strictly physiological approach of plant physiologists, and the cognitive perspective of plant "neurobiologists," who sought to create a more integrated and comprehensive view of plant abilities from a cognitive-behavioral standpoint. The more conservative view holds that plant science already operates within a sufficiently clear and coherent explanatory framework, and that a cognitive-behavioral approach adds little of significance (cf. Robinson et al., 2020). Plant behavior can be seen as the outcome of minimal need for active modulation by the organism itself (cf. Mallatt et al., 2021). Although this debate has not been fully resolved, much of the tension has lessened, as researchers increasingly recognize the value of a common framework for analyzing plant behavior and interactions. This approach allows for more effective examination of hypotheses, identification and correction of errors, overcoming biases, and the design of more targeted, species-specific tests, all with greater critical insight and speed, thanks to the broader, interdisciplinary perspective (cf. Bianchi, 2021, p. 172).

The study of plant behavior sheds light on various plant characteristics that, in the animal equivalent, are considered broadly "cognitive." Examples include plants' ability to recognize and communicate more intensively and accurately with genetically related organisms (cf. Murphy & Dudley, 2009; Biedrzycki et al., 2010), their sensitivity to risk (cf. Dener et al., 2016), and their capacity for nonassociative learning processes such as habituation and sensitization. Other traits observed in plants include conditioning, salient information retention (cf. Bose & Karmakar, 2003; Thellier & Lüttge, 2013; Galviz et al., 2020), anticipatory behaviors (cf. Garcia Rodriguez & Calvo Garzon, 2010; Novoplansky, 2016; Calvo, 2018), and decision-making abilities (cf. Hodge et al., 2009; Reid et al., 2015; van Loon, 2016; Trewavas, 2017; Segundo-Ortin & Calvo, 2018, 2023). Thanks to their ability to process different types of information, plants can select and initiate responses, allowing them to modulate their behavior, such as altering the rate and direction of growth, and the synthesis and emission of chemicals. These aspects reflect current research, but they do not rule out the possibility of exploring other, less understood or unknown areas.

From a process-oriented perspective, studies on the functional parallels between plant and animal anatomy and physiology – particularly between animal nervous systems and the biological networks that organize and transmit information in plants – are becoming increasingly detailed. In essence, the original idea was that, even though plants lack nerve cells, they still engage in chemical and electrical signaling activities within and between cells in various tissues (cf. Maher, 2017, p. 61). Thus, in alignment with this perspective, plants possess complex biological signaling networks that serve integrative and coordinative functions within the organism (cf. Baluška et al., 2006; Barlow, 2008; Calvo, 2007; Trewavas, 2014; Bianchi, 2022). Through empirically and evolutionarily informed studies of how information is generated, transmitted, and processed – that is, how biologically and ecologically meaningful signals are organized in multicellular organisms (cf. Miguel-Tomé & Llinás, 2021) – research is shedding light on the complex dynamics of plant signaling. These findings are prompting a broader reconsideration of how we define and understand systems of internal communication and coordination across living organisms, highlighting functional parallels in molecular and electrical signaling mechanisms (cf. Baluška & Mancuso, 2009; Baluška, 2010; Segundo-Ortin & Calvo, 2023).[43]

[43] One example of plant signaling is the ability of plant cells to generate and sustain action potentials (APs), which play a crucial role in essential functions such as photosynthesis, respiration, movement, and responses to beneficial or harmful stimuli (cf. Favre & Agosti, 2007; Felle & Zimmermann, 2007). The range of plant signaling varies depending on the molecular components involved, as well as the speed and distance of signal propagation, whether over short or long distances. In addition to action potentials, plants produce local electrical potentials (LEPs), variation potentials (VPs), and system potentials (SPs) (cf. Vodeneev et al., 2016; Gilroy et al., 2016; Debono & Souza, 2019).

Signal propagation occurs via plasmodesmata, channels that traverse cell walls. These electrical signals can then travel greater distances through the network of phloem vessels. The vascular system facilitates systemic communication between different plant organs, ensuring coordinated responses (cf. Lucas et al., 2013; Calvo et al., 2020; Segundo-Ortin & Calvo, 2023).

Given the advances in plant physiology and ecology, the aim of studies on plant abilities – along with their approaches to data collection, organization, and interpretation – has been to deepen our understanding of how plants perceive and "act" within their environment. These studies have gone beyond the boundaries of traditional research, integrating insights from various levels of analysis to offer a signal-mediated, behavioral, and ecological interpretation of plant interactions (cf. Bennett, 2021). The primary focus is on plant phenotypic plasticity and adaptability in response to environmental influences, and on the capacity of different systems to modify their surrounding environment.

Recently, the contributions of philosophy to the broader field of cognitive science have grown, fostering the development of a more inclusive and critical theoretical approach. As Heras-Escribano and Calvo (2019) pointed out, the shared theoretical aim in studying plant signaling and behavior from a philosophy of cognitive science perspective is to move beyond merely identifying signaling processes and organizing information at the molecular level. Instead, the focus is on understanding the flexibility of plant responses, often described as "soft-wired," in contrast to the rigid "hard-wired" behaviors.

This growing recognition of plants' flexible behavioral repertoire, especially in dynamic interactional contexts, has paved the way for considering plant behavior in terms of cognition or "intelligence" (cf. Trewavas, 2014; Bianchi, 2022, 2024; Bonato et al., 2023).

2.7 Underlying Theoretical Framework

In this section, adopting a pluralistic approach that remains open to future developments and revisions, I summarize significant ongoing work to clarify the theoretical framework that enables us to study plant cognition. I highlight the most useful and promising approaches for advancing this vast and ramified research.[44]

Based on recent experimental advances and the selection of theoretical approaches, we are witnessing shifts in the boundaries of cognition and a gradual inclusion of forms of cognition identified across *taxa*, evaluated according to species-specific characteristics (cf. Clark, 1997; Wheeler, 2005). This process of selection and differentiation ultimately depends on the specific methods we use to detect and conceptualize systems and processes, that is, in a broad sense, on how we interpret "cognition" (cf. Schlicht, 2018).[45]

---

[44] In this context, the disparate epistemic tools, among which, as we will see, models for the study of cognition, are conceived as frameworks to facilitate scientific investigation (cf. Knuuttila, 2011). Even concepts, which gain coherence within networks of shared meaning, can be regarded as epistemic tools in their role of guiding research.

[45] From the integration of various approaches, a picture is emerging in which "cognition" encompasses a wider range of phenomena, including different forms of perception, motor control, memory, language, and, in some species, even the motivational aspects of behavior and the emotional sphere (cf. Akagi, 2018).

The current debates surrounding the various interpretations of cognition, which highlight the lack of consensus, primarily focus on how to conceptualize cognition itself (i.e., the set of processes that define it) and, on a practical level, how to establish clear criteria for distinguishing between "cognitive" and "non-cognitive" phenomena (cf. Keijzer, 2021). A classic example that spurred the search for a defining criterion of cognition (its nature and scope) arises from the conception of the extended mind (cf. Clark & Chalmers, 1998; Rowlands, 2010). Over the years, this targeted inquiry has evolved into a broader interdisciplinary concern that has garnered significant attention (e.g., Kaplan, 2012; Ramsey, 2007; Bayne et al., 2019).

One point that many cognitive scientists and philosophers (those aligned with naturalism) agree upon is that cognition, although its definition may evolve with theoretical and practical advancements, must be understood in scientific terms and anchored to a material domain of investigation, as is the case in other scientific disciplines (cf. Ramsey, 2007, p. 4207; Keijzer, 2021).

Subsequently, moving away from traditional neurocentric or zoocentric approaches to the study of behavior and cognition, I emphasize the need within comparative psychology and evolutionary cognition to expand research in the development of a branch of meta-analysis. This would involve studying which traits, functional equivalences, or differences should be considered when investigating the cognitive behaviors of various organisms, taking into account their structural, functional, and ecological differences, which are shaped by distinct evolutionary and developmental paths. This initiates a reflection, one that requires further exploration, on what to prioritize when studying different organisms, including which aspects (genetic, epigenetic, phenotypic, and thus behavioral) should be considered in order to analyze their specialized abilities in the most biologically and ecologically appropriate way.

I also address several biases, typical of our species, that have commonly skewed studies in comparative animal cognition, and that could similarly impact research into other forms of cognition. Specifically, I highlight the issues of anthropocentrism, anthropomorphism, and zoomorphism.

Finally, following the core principles of the biogenic approach to cognition, I further develop an empirically informed framework for studying biological cognition, which includes the cognitive capacities of aneural organisms, such as plants.

2.7.1 The Problem of Circularity and the Pluralistic Approach

A primary aim expressed here is to clarify the approaches adopted and the flexible methods used to address the issues discussed throughout the book. First, regarding the methodological approach, I highlight two key challenges, "two sides of the same coin," that could undermine studies of plant cognition if not approached through a comparative and empirically informed analysis. These challenges include what might be termed the problem of "explanatory circularity" and the "disregard" or "overlooking" or "denial" of cognitive-behavioral traits in organisms that are structurally and functionally different from the more familiar (generally animal) systems (cf. Bianchi, 2024).

Regarding the first problem, circularity, there is often a common, and perhaps inevitable, tendency to explain plant abilities by relying on existing approaches and models that were originally developed and tested to account for animal cognition, which is more familiar and well understood. This initial reliance should not be surprising, as comparisons can only be made using known elements or well-characterized phenomena. However, the problem arises when the analysis fails to identify and account for the relevant differences between the cases at hand, leading to an unconscious reduction or forced alignment with the paradigmatic models taken as the point of departure. The risk, then, is that by beginning with an established framework, one ends up returning to it without sufficiently exploring or explaining any differences. In short, a theoretical-experimental starting point should not necessarily coincide with the end point, quite the contrary. Circular reasoning, in relation to the specific characteristics that enable cognition, occurs when research begins with a specific explanatory model and continues to rely exclusively on that model (or similar ones), excluding everything that does not conform to its criteria.

This gives rise to the second problem, which is closely related to the first. There is a common tendency to exclude or outright deny cognitive abilities that do not fit within the framework of more emblematic or widely recognized capacities, simply because they cannot be aligned with the model used as a benchmark. For example, the belief that cognition must necessarily involve the use of internal representations ("mental" in the classical sense) to respond flexibly to changing environmental situations (cf. LeDoux, 2020, p. 269), may restrict the analysis of cognitive abilities to a narrower range of organisms, thereby excluding many others that utilize different mechanisms to interact with and adapt to environmental challenges. Instead, it is necessary to remain open to diversity, even if at first this approach might seem counterintuitive (cf. Bianchi, 2024).

The intent here is twofold. On the one hand, the aim is to avoid overly restrictive comparisons that could distort the case studies. However, with the careful use of metaphors and analogies, cautious comparisons can guide the analysis in new and fruitful directions. On the other hand, in the study of "atypical" forms of cognition, particularly non-neural ones such as plant cognition, the goal is to avoid reductionist conclusions, when relying exclusively on the findings of disciplines like molecular biology, without incorporating insights from fields such as psychology, philosophy of mind, or cognitive science. These disciplines build upon the same core data but also address phenomena at a behavioral and relational level, a macroscopic level that, while inclusive of lower-level processes, cannot be entirely reduced to them (cf. Bianchi, 2021, p. 170).

At a general level, I am not referring here to a specific, rigid, or exclusive "paradigm."[46] Instead, the concept guiding this discussion aligns with the

[46] The Greek etymology of the term refers to a model or example, and in its verb form, it means "to show, present, or compare."

twentieth-century epistemological understanding of a scientific paradigm: a theoretical framework (comprising assumptions and perspectives on reality) and a set of experimental practices shared by the scientific community – or, more realistically, by a portion of it – within a particular historical period for the study of certain phenomena (cf. Morabito, 2020, p. 3). Within the broader macro-paradigm (or research program in the widest sense) shared by scholars of plant behavior, the proposal presented here is rooted in a form of explanatory pluralism. This approach aligns with the traditionally accepted view in the philosophy of science that theories at different levels of description, like psychology and neuroscience, can co-evolve, and mutually influence each other, without the higher-level theory being replaced by, or reduced to, the lower-level one (de Jong, 2002, p. 441). In this context, "higher" and "lower" levels are not understood hierarchically but as distinct, though interconnected, levels of description for analyzing phenomena.

The greatest advantage of a flexible pluralistic approach lies in its ability to connect different disciplines, each of which, due to varying specializations, addresses different aspects of a phenomenon using diverse tools. This results in findings expressed at different levels, sometimes integrable, always comparable. Once the challenge of managing these diverse methods is resolved on a case-by-case basis, the overall richness that emerges offers potential benefits that would be difficult to achieve by other means. In pluralistic approaches, it is generally believed that the pragmatic side of explanations should be emphasized while the reductionist aspect is minimized (though not eliminated), without sacrificing the precision of the analysis (cf. McCauley & Bechtel, 2001; Di Francesco et al., 2017, p. 174). A pluralistic approach does not conflict with the possibility of achieving a unified vision of science or, when narrowing the focus, a coherent understanding of a specific field of investigation.

In the case under consideration, this approach implies, for example, that the study of plant cognitive abilities is not merely a terminological or semantic issue but involves a plurality of approaches to inquiry, each best suited to examine a specific aspect of the subject. This also means recognizing that the process of understanding is always in progress, constantly subject to refinement and revision, continuously evolving (cf. Bianchi, 2021; Bianchi et al., 2025a).

Adopting a fundamentally pluralistic explanatory approach, there are multiple frameworks for the scientific-philosophical investigation of cognition, which also seem well-suited for studying plant cognition from a biological-adaptive and ecological-relational perspective.

In developing this book, particularly in the section on the most promising perspectives in cognitive science for exploring plant abilities, I discuss the heterogeneous 4E Cognition research program (cf. Bianchi, 2022, 2024), predictive processing models (cf. Calvo & Friston, 2017), systemic studies on the self-organization of plant signaling via chemical-electrical analysis of functional networks (cf. Baluška & Levin, 2016), dynamical systems theory (cf. Van Gelder & Port, 1995), enactivism (cf. Thompson, 2007; Hutto & Myin, 2013; Di Paolo et al., 2017), pragmatic and action-oriented approaches (cf. Dominey et al., 2016; Chemero, 2009), and ecological psychology (cf. Gibson, 1979).

In this analysis, within the fruitful dialogue between plant biology's concepts and experimental findings, several positions are provisionally incorporated, which align with the post-cognitivist macro-area of cognitive research, typically understood as embodied, situated, and extended (cf. Varela et al., 1991; Gomila & Calvo, 2008). These positions contribute to a growing body of literature that challenges traditional understandings of concepts like "intelligence, "cognition," and "consciousness." Instead of focusing on conceptual competence, these approaches emphasize notions such as agency, goal-directed behavior, and adaptivity during the development of organisms, potentially extending the analysis to other forms of life, including plants (cf. Margulis, 2001; Calvo Garzón & Keijzer, 2009; Baluška & Levin, 2016; Baluška & Reber, 2019).

Post-cognitivism, therefore, views cognition as a set of multilevel processes that arise from the reciprocal interaction between bodily and environmental factors. This perspective integrates well with both original and contemporary formulations of the life–mind continuity thesis (cf. Godfrey-Smith, 1994; Wheeler, 1997; Thompson, 2007), which, in its various forms, has helped reconsider the conventional boundaries between life, adaptation, and cognition in a more integrated and nuanced way. This has led to increased hybridization between thematically related areas, recognizing the close connection between these processes.

With regard to plants, these considerations become even more pertinent. "Plant biological cognition" refers to the continuous and dynamic interaction between modular plant structures and environmental factors. From this perspective, the ability of plants to dynamically adapt to their niches through ongoing processes is studied by examining how they detect and respond to environmental information (cf. Linson & Calvo, 2020).

It is therefore evident how the fields of cognitive science, biology, plant sciences, and plant behavior, can mutually influence and enrich one another in meaningful ways (cf. Heras-Escribano & Calvo, 2019).

2.7.2 *Comparative Analysis of Cognition*

One of the objectives for future research is to move away from the mindset that has characterized certain approaches in comparative psychology, which relied on subjecting the analysis of the characteristics and behaviors of organisms to a single test, often without sufficient consideration of the biological and ecological specificities of the species in question. Such methods overlook the necessary nuances needed to accurately investigate their capacities.

Therefore, I stress the need to establish what could be referred to as a branch of "meta-analysis" within the philosophy of biological cognition. This would involve developing a flexible, non-reductionist, approach capable of integrating findings from multiple specialized studies on the same broad topic, even if conducted at different levels of analysis. Particular attention should be paid to the phenotypic and behavioral diversity across species, enabling researchers to select and accurately assess the salient traits relevant to the capacities being examined in each organism.

A comparative analysis that avoids the primary pitfalls and biases in the search for similarities and differences between species can contribute to the development of more precise scientific methodologies for highly interdisciplinary research, such as this one (cf. Andrews, 2020, p. 38).

Dwelling for a moment on the original meaning of "comparative," this approach shifts away from making judgments based on a single characteristic (or a limited set of traits) in the study of behavior or cognition. Instead, it emphasizes the careful selection of which traits to examine, offering a progressively more reliable method to select, contextualize, and compare information for studying the diverse behavioral systems across different species (cf. Timberlake, 2007).

From this perspective, cognition is understood as the capacity to orient, form relationships, and acquire information about the world in accordance with each organism's unique capabilities. In this regard, "cognition," in a broad sense, is the capacity to use information to interact effectively with the environment. In the evolutionary *continuum* of life, and specifically regarding the various forms of behavior and cognition, we should speak not of "degrees" of cognition but of distinct cognitive specializations: different ways of perceiving, orienting, acting, and communicating within one's environment.[47]

Comparative psychology historically traces its roots to Darwin's approach, which laid the foundation for developing cognitive theories based on the experimental observation of behavioral traits and mental faculties across different forms of life, from the simplest to the most complex. Darwin himself (1859, p. 488), in line with his view of the organic world, famously wrote, "*In the distant future I see open fields for far more important researches. Psychology will be based on a new foundation, that of the necessary acquirement of each mental power and capacity by gradation. Light will be thrown on the origin of man and his history*." In this way, Darwin distanced himself from "psycho-physical dualism" (cf. Gazzaniga, 2019, p. 71).

The work that significantly contributed to the development of comparative psychology is *The Descent of Man, and Selection in Relation to Sex* (1871). Initially adopted by psychology departments, it was later used by fields such as anthropology, biology, primatology, zoology, and philosophy (cf. Andrews, 2020, p. 1; cf. La Vergata, 1988, p. 231). In a now-famous and provocative passage, Darwin (1871,

[47] While not contradicting the evolutionary perspective, the concept of "degree" has often been misinterpreted and sometimes used to support hierarchical thinking. However, by adopting a continuity-based view – where human capacities and those of other living beings exist on a spectrum without abrupt leaps – and, in a broad sense, a gradualist view, where differences in capacities reflect varying levels of complexity, it becomes clear that organisms cannot be meaningfully compared on a single scale or evaluated along a singular path. E.g., an organism with a greater capacity for abstraction and judgment but simpler perceptual sensitivity is not inherently more "conscious" than another with the opposite traits. The most empirically sound approach acknowledges multiple dimensions corresponding to different specializations, each of which may exhibit varying degrees of complexity when compared (cf. Godfrey-Smith, 2021, pp. 304–305).

p. 105) stated, "*Nevertheless the difference in mind between man and the higher animals, great as it is, is certainly one of degree and not of kind.*"

Sometime later, early behavioral psychologists displayed limited interest in studying animal behavior for its own sake. Instead, they focused on discovering the general laws of behavior, particularly through stimulus-response contingencies, with the primary goal of predicting and controlling human behavior.

As De Waal (1999) observed, the original "comparative" aspect of comparative psychology was largely anthropocentric, structured around a progressive hierarchy from "lower" to "higher" forms of life. Even with the shift from behaviorism to the so-called "cognitive revolution" in the latter half of the twentieth century, Darwinian biological insights were gradually accepted. The prevailing view that cognitive behavior requires the presence of a nervous system remains the standard perspective today (cf. Lyon et al., 2021). Presently, the diverse and interdisciplinary field of comparative psychology addresses a wide range of methods and questions employed by scientists from various disciplines and backgrounds[48] to study animal mental capacities (cf. Andrews, 2020, p. 1).

From a practical point of view, testing the abilities of different organisms requires designing experiments that avoid unintentional guidance by experimenters, which might influence the organisms' responses (ibid., pp. 13–14). A classic example is the case of Clever Hans (a horse believed to be able to count), which highlights the importance of body language in transmitting information within and between species. This example also underscores the fact that not all cues can be controlled or detected. While visual cues are most obvious and relevant to humans, other species may rely on different, more developed sensory modalities (cf. Bianchi et al., 2025a).

The perception-action mechanisms, and ultimately the cognition an organism has of its environment, are based on its specific behavioral capacities, which are linked to different structural and functional specializations. Some organisms are more sensitive to visual stimuli, others to sound, touch, or, in the case of plants, chemical cues in the soil or air. The preference for one sensory-cognitive channel over another does not exclude the possibility of multisensory integration when solving survival challenges, as these adaptations are shaped by the distinct eco-evolutionary pathways of each species (cf. Bianchi et al., 2025a). These considerations highlight the importance of understanding the biology and ecology of organisms in order to select the most appropriate tests to examine their capacities (cf. De Waal, 2016, pp. 25–28; Bianchi, 2021, pp. 123, 155–156).

One of the most intriguing aspects of comparative evolutionary research is the discovery that certain cognitive traits – once believed to be exclusive to organisms with nervous systems – have existed in simpler forms; the brain enhanced and expressed them in more sophisticated ways (cf. Gazzaniga, 2019, p. 285).

To study cognitive behavior in non-neural organisms, such as plants, and make meaningful comparisons while identifying commonalities and unique traits, it is

---

[48] For example, cognitive scientists, biologists, ethologists, and animal welfare researchers each approach the study of animal behavior and cognition differently, from slightly different perspectives and levels of analysis.

essential to have a clear understanding of what a nervous system is and how it functions. The nervous system is a morpho-functional unit specialized in receiving, transmitting, and processing internal and external stimuli through bioelectrical signals. Its primary role is sensorimotor integration, enabling organisms to interact with their environment in a coordinated and timely manner, essential for survival. This exchange system, involving receptors and effectors, underpins the sensory, motor, and cognitive functions of animals.

Like any biological system, the nervous system has its own evolutionary history (for details, see LeDoux, 2020; cf. Godfrey-Smith, 2021, pp. 70–73). It evolved from early physiological and genetic-molecular processes (the so-called presynaptic building blocks) into a complex network of specialized cells, neurons, capable of forming synapses.[49] A significant advancement in our understanding of the nervous system has been the recognition of its role as an "ecological control device," achieving its objectives in alignment with the body's capabilities and influenced by the surrounding environment (cf. Clark, 2008, pp. 5–6; Di Francesco & Piredda, 2012, pp. 145–146). Furthermore, in animal brains, it is not absolute brain size that matters most, but rather the ratio between brain and body mass (known as the encephalization quotient), along with factors such as neuronal density and the functional organization of neural networks (cf. LeDoux, 2020, p. 318; Preuss & Wise, 2022).

Consistent with the aims of this study, it is important to move beyond the constraints of a neurocentric framework by exploring more distributed and modular nervous systems, as seen in some invertebrates. At the same time, plant behavior, despite the absence of a nervous system, demonstrates highly efficient mechanisms of communication, including chemical signaling via airborne volatiles and root-based interactions (cf. Yoneyama & Bennett, 2023), as well as the coordination of electrical signals in response to various stimuli.

This approach enables the study of cognitive capacities in non-neural organisms within the broader framework of comparative behavior and cognition, aligning them with other forms of life. The time seems ripe to include plants in the analysis of comparative cognitive behaviors. Extending this consideration to the plant kingdom presents a significant challenge, though a crucial one, especially at the observational

[49] In choanoflagellates, flagellated protists believed to be the closest unicellular relatives of metazoans (with animal ancestors likely resembling these protozoa), and later in sponges, there are elements and processes that, through molecular signaling and coordinated gene expression, paved the way for the emergence of nervous systems. The first neural networks developed in the common ancestor of Eumetazoa, marking an evolutionary step toward more complex forms of coordination and communication in multicellular organisms. These early systems eventually gave rise to the nerve nets observed in Cnidaria. This form of organization provided faster sensorimotor integration than chemical diffusion alone (cf. LeDoux, 2020, pp. 143; 165), though it was still unable to produce localized responses based on the specific body part that encountered a stimulus. A precursor to the specialized neuronal groups typically concentrated in the head region of most animals can be found in a few clusters of neurons located in the umbrellas of jellyfish. A fundamental evolutionary step was the concentration of sensory and motor cells within the same body region and the use of rapid electrical communication to overcome the limitations imposed by spatial distance (ibid., pp. 172–177).

level of behavioral variations. This is due to the multilevel phenotypic plasticity of plants, which encompasses numerous processes, including growth, often involving morphologically irreversible transformations (cf. Bianchi et al., 2025b).

In the field of comparative biology, attention is given to the morpho-functional characteristics of organs and bodily systems that vary across changing contexts, both ecologically and in terms of specific physiological processes. For example, animal respiration can be studied as both a fundamental component of an organism's overall functioning – situated within a larger, densely interconnected ecological system – and as a more basic physiological process of gas exchange, primarily involving the intake of oxygen and the release of carbon dioxide and water.[50]

Similarly, in the context of evolutionary studies of comparative cognition – ranging from macroecosystem interactions to micromolecular and bioelectrical processes – plant cellular specializations offer a unique set of constraints that, in some respects, are functionally equivalent to those found in animals (cf. Calvo & Trewavas, 2020; Linson & Calvo, 2020, p. 15).

In summary, as evidence of the richness of eco-evolutionary and developmental outcomes, cognitive-behavioral strategies reveal multiple pathways to achieve similar goals, despite the complexity of multifactorial influences, including various genetic traits and phenotypic expressions. The same anatomical and behavioral solutions can be adapted to meet different challenges (cf. Bianchi, 2021, pp. 166–167). In evolutionary biology, a distinction is made between homologies – traits that share the same evolutionary origin but perform different functions, providing evidence of common descent along with the universality of the genetic code (cf. Pievani, 2017, p. 35)[51] – and, on the other hand, analogies – traits that are functionally similar but with different embryonic and evolutionary origins. The former depends on the history of the species, while the latter is determined by the ecological niche in which the organisms live (ibid., pp. 27–28).

2.7.3 *Anthropocentrism, Anthropomorphism, and Anthropodenial*

In a comparative analysis, the closely intertwined issues of anthropocentrism (and anti-anthropocentrism) and anthropomorphism (and anti-anthropomorphism) inevitably arise. These widespread tendencies, if not critically examined and mitigated, can prevent the correct understanding of observed behaviors.

For the purposes of the present discussion, it is essential to address these tendencies, commonly debated in animal studies, as they also apply to research on plant

---

[50] For example, both animal lungs and plant leaves contribute to the cellular respiration processes of their respective organisms, functioning as integral components of the systemic mechanisms that maintain homeostasis within their ecological niches.

[51] An example of homology from the world of plants is between chlorophyllian green leaves, flower leaves, tendrils and leaf thorns.

cognition. Indeed, these issues may be even more pronounced and challenging in plant studies due to the counterintuitive nature of this research, complicated by the considerable phylogenetic distance and the structural and organizational differences between plants and animals.

The most problematic tendency identified in research on cognitive abilities in non-human entities is anthropocentrism, which considers human beings, and all that pertains to them, as central to the universe, prioritizing them as of greater importance and value than the rest. This perspective can manifest across various levels, including personal opinions, religious beliefs, and theoretical approaches of different kinds.

In biological contexts, there is often confusion between the concept of a species' "uniqueness" (the idea that each species, and each individual within it, is unique) and the notion of "specialness," which implies the qualitative superiority of human characteristics and abilities over those of other living beings (for a discussion of these issues, particularly regarding language, see Cosentino & Ferretti, 2011, p. 135).[52]

In comparative analyses, the principle of anti-anthropocentrism serves as a reminder that non-human organisms are not incomplete versions of humans. Rather, they possess species-specific perspectives, goals, and behavioral strategies.

It may seem obvious to acknowledge that thinking in mildly anthropocentric terms is understandable, within certain limits, given the emotional bonds and shared interests among members of our species. The issue arises when anthropocentric views are treated as objective truths, leading to the disregard of the importance and needs of other organisms. This narrow viewpoint undermines a comprehensive understanding of the living world's interconnectedness, which ultimately affects human interests, whether directly or indirectly.

In the study of plant activities, an additional challenge beyond anthropocentrism is confronting the broader framework of zoocentrism ("animals at the center"), which subsumes anthropocentric tendencies.

A tendency that should not be confused with anthropocentrism, although the two share an original connection, is the inclination to anthropomorphize the characteristics and behavior of other living beings.[53] In the field of animal cognition studies,

[52] There is certainly no intent here to downplay human (cognitive) abilities. As Stephen Hawking reminded us, the greatest enemy of knowledge is not ignorance, but the illusion of knowledge. *Homo sapiens*, a species living on a small planet of an average star, has nonetheless managed to grasp a great deal about the workings of the universe, though much remains to be discovered (*work in progress*). The purpose of these reflections is to highlight the distinctions between different kingdoms and species.

[53] The term "anthropomorphism" derives from Greek and literally means "in the form of a human" or "attributing human form or characteristics to non-human entities." As De Waal (1999) noted in a key article on this topic, anthropomorphism has carried negative connotations since the time of Xenophanes (ca. 570 BC), who criticized Homeric poetry for depicting the gods as human-like, arrogantly presuming that they must resemble us. In a provocative remark, the philosopher suggested that if horses could represent their gods, they would depict them in their own likeness.

this refers to the habit of attributing human faculties – whether truly unique to humans or mistakenly believed to be so – to other organisms.

Despite a shared core, this tendency should not be conflated with common-sense psychology (or folk psychology), which is the everyday understanding of psychological phenomena. In fact, during the initial stages of comparative research, folk psychology can be useful in distinguishing and organizing different concepts.[54]

As Andrews (2020, pp. 4–6) prudently suggested, completely abandoning this category of concepts could risk turning comparative psychology into a form of eliminative materialism or biological reductionism, where only sequences of physical processes are considered. Instead, by retaining this broader analytical framework, we can refine certain concepts, backing them with the most current experimental evidence, or discard them when they no longer serve their purpose.

The origins of anthropomorphism are ancient, rooted in mythology and early philosophy. In the context of comparative psychology, anthropomorphic tendencies are often linked to the idea of mental continuity between human and non-human animals, a view notably advocated by Charles Darwin and further developed by George Romanes (1848–1894). This tendency, when not taken to extremes, has helped overcome the more radical forms of anthropocentrism. The concept was later revived by Donald R. Griffin (1915–2003), who argued that animals possess their own forms of awareness and intelligence. Griffin is best known for his groundbreaking research, which experimentally confirmed that bats use echolocation to avoid obstacles (cf. Griffin, 1958).

Some authors suggest that the current wave of anthropomorphizing interpretations in biology may be driven by growing concerns about the anthropogenic impacts on ecosystems, which are accelerating global species extinction at alarming rates (cf. Taiz et al., 2019). This heightened awareness could have intensified the desire to reconnect with the "Nature" we have harmed, making it seem more similar to us, more human (in other words, anthropomorphizing it).

Anthropomorphism is not a monolithic phenomenon. There are various types, ranging from the most simple and naïve to the more justifiable forms. For instance, immature anthropomorphism, such as "anthropocentric" anthropomorphism, often involves projecting human experiences onto animals. This is commonly seen in literature and cinema, where animals are depicted as talking or behaving like humans, a tradition dating back to Aesop's fables. On a societal level, such representations often aim to entertain, educate, or moralize, sometimes reinforcing the notion that we, too, are animals (cf. De Waal, 1999).

[54] For instance, in the case of functionalism within the philosophy of mind, the concepts of common-sense psychology, understood as constructs, have causal power within the interpretative framework of the theory (cf. Lewis, 1972). Similarly, in the theory of intentional systems, such concepts are applied to models of observable behavior (cf. Dennett, 1993). These positions stand in contrast to those of certain authors, such as Churchland (1981), who advocated for completely replacing the concepts of naïve psychology, like belief and desire, with neuroscientific descriptions of processes.

Reference has also been made to more refined forms of anthropomorphism (cf. Timberlake, 2007), such as "animal-centric" (cf. De Waal, 1997, 2001), "heuristic" anthropomorphism (cf. De Waal, 1999), "biocentric" anthropomorphism (cf. Bekoff, 2000), and "constructive" anthropomorphism (cf. Arbilly & Lotem, 2017). These terms describe forms of anthropomorphism that are neither extreme nor naïve and can be grouped under the broader concept of "critical" anthropomorphism (cf. Burghardt, 1991). In all these cases, the accounts of animal behavior are still forms of anthropomorphism because our interpretations are inevitably mediated by concepts drawn from human cognition. However, these perspectives make a greater effort to consider the viewpoint of the organisms under investigation, aiming to interpret the world through the lens of the animal being analyzed, as far as possible. This more mitigating and corrective approach has helped clarify when anthropomorphism can be useful and represents a spontaneous and productive mode of reasoning, and when it is, by contrast, fundamentally flawed. It also opens the door to anti-anthropomorphism. The historical development of these debates has undoubtedly helped us reflect on the main advantages and drawbacks of anthropomorphism.

For scholars such as Daniel Dennett, Griffin, Burghardt, and De Waal, invoking mentalistic intentional states, often criticized as anthropomorphic, can, in the context of multifactorial explanations of animal behavior, sometimes offer a more parsimonious explanatory strategy. According to the principle of cognitive economy, this approach can be more efficient than providing a lengthy and potentially convoluted description of the complex physiological mechanisms underlying certain behavior. However, this does not preclude the possibility of clarifying the underlying processes; on the contrary, both approaches should be pursued. It is important to avoid the nominalist fallacy, which is the mistaken belief that merely naming something constitutes an explanation (cf. Wynne, 2004).

There are pros and cons to anthropomorphism. The main risk of extreme (anthropocentric) anthropomorphism is the unwarranted attribution of human-like qualities, particularly subjective, internal experiences, to other organisms without sufficient evidence.

From an eco-evolutionary perspective, as Colin Beer (1980) and John S. Kennedy (1992) observed, anthropomorphic thinking, which is learned from early childhood, may have served as a useful socio-culturally mediated tool. It likely aided in the attribution of beliefs, emotions, and intentions similar to those of humans, with the primary goal of predicting and managing (understanding) the behavior of other organisms, particularly animals. This tendency may have played a key role in processes such as domestication (cf. Timberlake, 2007; LeDoux, 2020, p. 384).

In line with these considerations, the tendency to anthropomorphize has been interpreted as an intuitive and frugal heuristic (cf. Gigerenzer et al., 2011), effectively useful in everyday life to quickly and simply categorize certain animal behavior (e.g., interpreting and responding promptly to a dog's threatening behavior) (cf. Arbilly & Lotem, 2017). On the other hand, this could also be explained as the ability to memorize and associate behavioral manifestations with their consequences for other organisms. Regarding this view, which sees (intuitive) anthropomorphism as a fast and frugal heuristic, some authors (cf. Dacey, 2017) have distinguished between two types of application: one legitimate and one more problematic. The legitimate use involves applying commonsense psychology to understand the minds of non-human animals, while the problematic use arises when anthropomorphism becomes a cognitive bias in experimental research, failing to meet the epistemic and methodological standards of scientific practices.

The choice of how to study the cognitive abilities of various organisms should not be reduced to mutually exclusive options. For example, Nick Humphrey (1976), perhaps to counter the excesses of procedural and definitional rigor, suggested using anthropomorphic explanations in the early stages of research, whether they seem correct or not. In practice, this approach would involve testing such hypotheses, with the option of confirming or discarding them as necessary.

One positive, or at least neutral, aspect of anthropomorphism is that it has led to concrete improvements in the living environments of the organisms studied, particularly in the field of ethology (cf. Boakes, 1984).[55]

Among its benefits is the evidence that has emerged in support of initially anthropomorphic hypotheses, such as Jaak Panksepp's (1943–2017) research on the behavior of so-called "laughing" rats (cf. Knutson et al., 1998). In this case, the verb "laugh" is used in a quasi-literal sense, traditionally reserved for human behavior. The ultrasonic vocalizations produced by these animals, around 50 kHz, have been correlated in recent years with positive emotional states, particularly in playful or socially rewarding contexts (cf. Panksepp & Burgdorf, 2003).

Despite these relatively favorable or benign aspects, anthropomorphism has been widely criticized. From this critique arises anti-anthropomorphism, which explicitly rejects the attribution of human qualities to other organisms without valid justification or evidential support (cf. Shettleworth, 2010).

In the field of comparative cognition, one of the most influential rejections of extreme anthropomorphism comes from Conwy L. Morgan (1852–1936), who is best known for having formulated the so-called Morgan's canon. This principle advises against attributing more mentally sophisticated processes to animals than is strictly necessary to explain their behavior.[56] As Morgan's canon states: "*In no case may we interpret an action as the outcome of the exercise of a higher psychical faculty, if it can be interpreted as the outcome of the exercise of one which stands lower in the psychological scale.*" (Morgan, 1894, p. 53).

---

[55] An example is the increase in comfort conditions for great apes in captivity, visible in the reduction of obsessive behaviors.

[56] In the late nineteenth century, there was a significant backlash against attributing complex or conscious mental capacities to nonhuman animals. The rise of behaviorism found support in the ideas commonly associated with Conwy L. Morgan, who played a key role in curbing the introspective approaches to animal behavior by advocating for the replacement of mentalistic explanations with mechanistic ones. For instance, in line with this objective, Edward L. Thorndike (1874–1949) in the United States emphasized the importance of replacing the anecdotal method in animal behavior studies with reproducible experiments conducted in controlled environments (cf. Wynne, 2004).

In analyzing the relationship between anthropomorphism and Morgan's canon, it is important to consider the various interpretations that have emerged over time. Indeed, the canon is one of the most frequently quoted and misrepresented statements in comparative psychology (cf. Dewsbury, 1984, p. 187; Thomas, 1998, p. 156). The "historical abuses" of the canon primarily revolve around issues of parsimony and anthropomorphism. Regarding the first issue, the canon is often mistakenly equated with Occam's razor, i.e. the principle that the simplest explanation, with the fewest assumptions, is usually the best. However, Morgan recognized that simplicity is not necessarily a criterion for truth. He acknowledged (1894) that sometimes attributing higher animal activities to the presence of a mind can be simpler than explaining them through a complex sequence of sensory experiences, though those experiences still exist at a different level of explanation. In some cases, anthropomorphism may offer the most parsimonious explanation (cf. Sober, 2005), or at least a simple way for us to refer to something to explain. Regarding the second issue, Morgan (1903, pp. 48–49) held a relatively open-minded stance toward anthropomorphism. He argued that psychologists could infer mental processes either through subjective inference (based on conscious introspection) or objective inference (based on observable evidence). This highlights that Morgan's stance was far less rigid and dogmatic than it has often been portrayed.

Today, in comparative psychology and cognition studies, a rigid observance of the canon would exclude too much and limit the scope of inquiry, making it an impractical approach to understanding complex behaviors. Particularly in the field of ethology, strict adherence to Morgan's canon and anti-anthropomorphism has been seen as a potential obstacle to understanding animal mental capacities, though the goal of mitigating excessive anthropomorphism remains valid. For instance, it would have been difficult to imagine Lorenz strictly following the canon while interacting with his geese or successfully communicating with Cioc, his pet jackdaw. A notable failure resulting from the rigid application of Morgan's canon was reported by Hebb (1946) during a two-year study at Yerkes Laboratories, where researchers attempted to describe animal behavior without anthropomorphizing. The results of this study led to an overwhelming number of isolated actions that lacked coherence or meaningful interpretation for the intended purpose. This highlights the practical value of cautiously applying concepts traditionally used to explain human cognition in understanding certain species' behavior (cf. Bruni et al., 2018).

On the other hand, one of the practical advantages of anti-anthropomorphism, as noted by Andrews (2020, p. 7), is that it helps young scholars avoid using concepts that fail to meet scientific standards, encouraging them to operationalize terminology and specify how certain concepts will be measured. Additionally, an anti-anthropomorphizing approach discourages the development of overly friendly relationships with study subjects, which could lead to the attribution of abilities or mental states not sufficiently supported by evidence.

An advance over anthropomorphizing attitudes, influenced by anti-anthropocentric reactions, can be seen in Timberlake's (1994, 2007) proposal to reason in "theromorphic" rather than "anthropomorphic" terms. Theromorphism refers to an animal-centered knowledge (cf. Timberlake & Lucas, 1989).[57] A theromorphic approach, therefore, aims to uncover and account for the sensory and motivational capacities of animals.

Recognizing our cognitive limitations, attempting to understand what other animals perceive and "know" (trying to see the world from the perspective of a given organism, presuming it has a distinct "point of view," even if anoetic) can help, and has already helped, in formulating more empirically sound questions about animal capacities. This approach further clarifies the traits different species share and the ways in which they are unique. For example, as Burghardt (1991) noted, a herpetologist studying snake behavior must try to "walk in the shoes of the serpent" (cf. Timberlake & Delamater, 1991, p. 39). It is both logical and beneficial to pursue this line of reasoning.

Building on the theromorphic approach, I propose that – by embracing even more counterintuitive and interpretively challenging perspectives – a further step would be to include other forms of life, not just animals, developing what could be called "Organismomorphism." This approach would focus on the functional and ecological interrelationships of different living beings, allowing for the explanation of plant behavior as well. The emphasis would be on the structural and agentive characteristics of different organisms – both neural and non-neural – in relation to their environment.

When taken to extremes, anti-anthropomorphism can degenerate into what has been termed "Anthropodenial." This term, coined by De Waal, refers to a priori rejection of any shared characteristics between humans and animals, particularly cognitive traits, even when such similarities may actually exist. Anthropodenial involves blindness not only to human-like traits in animals but also to animal-like traits in humans.

This stance has been both criticized and defended on evolutionary grounds. Historically, evolutionary theory has been used to support two opposing perspectives: one against and one in favor of anthropomorphizing inferences (cf. Sober, 1998). According to De Waal (1999, 2016), when it comes to species phylogenetically closest to us (and closely related species in general), the most parsimonious explanation is that they share various behavioral and cognitive processes. Conversely, this assumption is less likely for species that are more phylogenetically distant. For instance, in many behaviors exhibited by our closest relatives, chimpanzees, given the "recent" phylogenetic divergence that occurred between 5 and 7 million years ago, the greater risk is not anthropomorphizing but anthropodenial, despite the

[57] The root "thero," derived from the Greek "therio," traditionally referring to nonhuman animals, has been adapted by the author to signify that the approach he developed is intended for application to both nonhuman and human animals.

necessary caution demanded by scientific rigor to study their behavior (cf. De Waal, 1999).[58]

Today, with more refined observational and experimental methods, it is understood that the validity of comparisons depends not only on the species being compared but also on the specific traits being examined. Therefore, the best way to resolve this dilemma is to evaluate each case on its own merits, without veering too far in either direction.

Moreover, at a general level, difficulties can arise when scientists focus exclusively on either similarities or differences between species. In the first case, the risk lies in forcing the attribution of certain characteristics from one species to another. There is also the danger of perceiving similarities where none exist, simply because one expects to find them, thus overlooking the uniqueness and distinctive traits of other cognitive systems (cf. Povinelli & Bering, 2002). In the second case, structurally similar to the first, the risk is to overemphasize differences, particularly those that may not be as significant, if one anticipates their presence. This can lead to an exaggerated emphasis on evolutionary discontinuities while downplaying elements of continuity.

A fundamental problem with the concepts of similarity and difference is their inherent vagueness when considered in isolation. These concepts lack logical rigor and must always be evaluated within the relational context of a specific comparison. Similarities and differences make sense only considering the analytical context and the chosen terms of comparison. Any two portions of reality, if not identical, possess both similarities and differences (whether or not they share a characteristic for comparison). Therefore, the selection of traits to compare is a result of specific decisions, motivated by contextual factors, the objectives of the study, and the initial research questions (cf. Bianchi, 2021).

Consequently, in the use of the concepts of similarity and difference, since it would be highly complex or even impossible to eliminate them entirely, it is essential to clarify which aspects are being considered, under what approach, for what reasons, and in response to which questions (cf. Bianchi, 2024).

In the study of plants, the dual challenge – and indispensability – of using the concepts of similarity and difference in differentiating and comparing aspects of reality often leads to an oversimplification of relationships between phenomena. This has frequently resulted in the neglect of significant similarities and differences within the same plant class, as if no important distinctions existed between them (think of the enormous structural difference, of biological and ecological cycle,

[58] This position is not entirely new. A historical, pre-evolutionary precedent for such reasoning, specifically aimed at supporting interspecific uniformity in explaining behavior, can be found in the work of David Hume (1739). Hume argued that by observing the similarity between the external behavior of animals and our own, we can infer similarities in certain aspects of internal cognitive mechanisms and their causes (cf. Clatterbuck, 2016; Bruni et al., 2018). In contemporary terms, we might say that the cognitive life of another primate species can be inferred, either probabilistically or as actual evidence, since both humans and primates exhibit behavioral trait C, and humans produce behavior C due to mental trait M (cf. Sober, 2012).

between an herbaceous plant and a woody plant). On the other hand, once the shared and divergent traits are acknowledged, for practical reasons (e.g., time constraints in research and communication), it may be decided to refer broadly to the macrocategory "plants." From an eco-evolutionary perspective, members of this category, despite belonging to different species, are generally considered to share morphological and functional traits sufficiently distinct from those of animals or other living organisms (cf. Bianchi et al., 2025a).

In the case of plants, given their divergent evolutionary paths, which have led to the development of unique body systems and ecological relationships, it is crucial to avoid anthropomorphizing (or zoomorphizing) their activities. However, this necessary precaution should not lead us to overlook alternative modes of behavior, communication, or problem-solving abilities related to survival (cf. Bianchi, 2021, p. 164). This diversity should be understood as a source of richness, offering valuable insights that could contribute significantly to the advancement of human knowledge.

Notably, both the similarities and differences in behavioral and cognitive capacities between the animal (neural) and plant (non-neural) worlds, as well as between plants and other non-animal organisms, require further investigation. There are two primary reasons for this: first, to shed light on the common biological origins, however distant in evolutionary time, of certain molecular and electrical signaling processes underlying cognitive and behavioral capacities across various organisms; and second, in the case of differences, to explore the unique behavioral traits of diverse biological systems, particularly those that deviate significantly from the standard forms typically studied. This exploration can provide both theoretical insights (learning something new) by approaching the subject from a different perspective (according to a different starting point), and practical advancements by extending the boundaries of experimental research, with potential technological applications (ibid., pp. 64–65).

In summary, the main takeaway from this discussion of the simultaneous presence of tendencies to anthropomorphize and to anthroponegate is twofold. First, the critique of anthropomorphism is crucial to avoid overestimating the mental complexity of non-human organisms. Second, the critique of anthropodenial helps prevent the assumption of differences where none exist, or where they are negligible (cf. De Waal, 1999, 2016). To mitigate these biases, some have proposed that the "best prophylaxis" is empiricism (cf. Fitzpatrick, 2008, p. 242; Sober, 2005), emphasizing the importance of data collection, analysis, and evaluation. However, even inferences about what constitutes sufficient evidence may be subject to anthropomorphic intuitive biases (cf. Dacey, 2017), making it essential to interpret data within specific theoretical frameworks that are corroborated by multiple research groups. Ultimately, in the ongoing interplay between theory and experimental research – where theory guides experiments and experimental findings challenge theoretical assumptions – reducing rigid dichotomies is not about conceding to anthropomorphism, but about providing empirically grounded accounts of the biology and ecology of the organisms studied.

Given that many forms of anthropomorphism fail to meet the criteria of rigorous scientific analysis, the fact that we are animals gives us a significant advantage in studying other animals. However, the situation differs in the case of plants, where the risk of anthropomorphism is compounded by the risk of zoomorphism. For the purposes of this book, including plants in this discussion requires careful attention to the history of research on animal cognition, with all its challenges and successes.

In addition to the anthropocentric, anthropomorphizing, and anthropodenial tendencies discussed in this section, which can hinder scientific inquiry, any analysis of the behavior of living organisms, including plants, is bound to fail without sufficient knowledge of the organism's evolutionary history, biology, and ecology. Even if it will always remain a "human science," thanks to interdisciplinary and multilevel research, the lack of knowledge can be transformed into adequate knowledge to address the study of capacities differently expressed in "other" organisms.

2.7.4 Toward an Inclusive Approach to Evolutionary Cognition

Overall, the approach that informs this reflection is the "biogenic" approach to cognition (cf. Lyon, 2006; Lyon et al., 2021), which advocates for a continuity-based, dynamic-processual view of cognitive processes (cf. Bianchi, 2021, pp. 11; 17; 159).[59] It is important to recognize that this approach significantly differs in its theoretical and methodological underpinnings from the "anthropogenic" view,[60] which has historically dominated and still remains prevalent in cognitive science, aligning with certain assumptions of classical cognitivism. In the anthropogenic framework, human cognition serves as the reference model, typically explained by the presence of a relatively complex nervous system and related mental activity. Conversely, the biogenic approach, informed by empirical evidence, aligns with a post-cognitivist perspective, emphasizing a relational analysis of natural cognition that is not restricted to the study of nervous systems alone (although understanding neural dynamics remains crucial in cognitive studies). From this perspective, cognitive behavior emerges in relation to the real-world context of an organism's life, based on species-specific and individual-specific capacities. This conception does not preclude the possibility of differentiating between diverse cases. The extent to which an environment (whether a biotic or abiotic stimulus) influences an organism depends on the level of autonomy and behavioral flexibility exhibited by that organism (cf. Parise et al., 2020).

An empirically informed approach to cognition does not solely focus on the "peaks" of cognitive complexity (cf. De Waal, 2016, pp. 25–26), but rather intends to

---

[59] For a discussion on the potential connections with studies on the context-dependence of cognition and process ontologies, see Nicholson and Dupré (2018).

[60] In the anthropogenic approach it is typically assumed that human cognitive states, such as beliefs and desires, represent the defining characteristic of genuinely cognitive states (cf. Adams, 2018).

examine the diverse cognitive specializations found throughout nature.[61] These specializations result from a chain of processes shaped by physical, organizational, and contextual constraints, whose primary function is to promote survival by integrating various levels of information and coordinating behavioral responses based on the most pressing ecological needs. These constraints should be understood as possibilities within boundaries, representing a shift from the rigid idea of "determination" to a more nuanced concept of "regulation."

The concept of a constraint warrants special attention. There are various types of constraints, such as physical-biological and ecological. For instance, physical constraints limit cell growth to a certain extent in order to maintain a sustainable surface area capable of supporting the nutrient flow required for a larger interior (cf. LeDoux, 2020, p. 95). A similar logic applies to the size limits of organisms' bodies. Constraints operate at every level, beginning at the molecular and macromolecular scales. Gene regulatory networks, for example, establish boundaries within which cellular behavior patterns can be expressed through dynamic processes. Variations in gene expression trigger chains of processes that lead to structural and metabolic changes within cells.

Moreover, as we move beyond outdated notions of rigid genetic determinism, it becomes clear that many phenotypic traits are regulated by genes rather than strictly determined by them (cf. Buiatti & Longo, 2013). Genes are understood as constraints that shape the organism's development, offering a range of viable possibilities depending on environmental and developmental conditions (cf. Pievani, 2017, p. 50). At the selective level, vast regions of evolutionary potential have remained unexplored due to ontogenetic and biochemical constraints, which, for example, explain the absence of blue pigments in roses, green-eyed Drosophila, or herbivorous snakes. As Theodosius Dobzhansky (1900–1975) pointed out, the genetic mutation repertoire is finite, constrained by the composition of genes, meaning that not everything is evolutionarily possible. On the other hand, epigenetic processes demonstrate that developmental variation arises at multiple levels, highlighting the importance of the organism's ecological dimension (cf. Rispoli, 2012, p. 165; Pievani, 2017, p. 81).

An example of bioecological constraints that both limit and enable specific interactional possibilities can be found in the systems that have shaped intra- and extra-organismic plant communication (cf. Karban, 2017). Even at the behavioral and cognitive levels, the structural and functional organization of living organisms does not imply complete predictability; rather, it suggests that their constitution influences the course of their (inter)actions. Thus, like any other characteristic, each cognitive trait arises within the constraints that shape a given system. While constraints limit the range of possibilities, they do not directly cause the final

[61] For instance, spiders can weave webs because they possess specific skills and body structures suited to that task, just as humans are able to think in the way they do because of their brains, bodies, and social environments that make it possible. Therefore, every ability should be viewed as an (imperfect) bio-evolutionary adaptation.

procedural response (cf. Morabito, 2020, p. 8).[62] In the life sciences, the concept of "constraint" has therefore been closely tied to the idea of "possibility," moving beyond a simple framework of "causal" determination (cf. Ceruti, 1996). From this perspective, adaptation throughout an organism's life should not be seen as a result of environmental factors determining biological structures, but rather as the organism's active response to environmental constraints (cf. Nucara, 2014, pp. 86, 121). This shift represents a transition from a notion of determining causes to one of constraints that both limit and make possibilities achievable.

Returning to the main line of our argument, from an evolutionary standpoint, it is increasingly recognized that the mental realm emerges from the complex interaction of distinct subsystems of internal information organization in relation to external life contexts (cf. García Rodriguez & Calvo Garzon, 2010). This insight allows for a broader understanding of the origins and diversity of cognitive abilities, moving beyond cladistic classifications or the biological categories to which different species belong (cf. Heras-Escribano & Calvo, 2019). Current interdisciplinary research, which seeks to better connect and integrate cognitive sciences with the broader life sciences (cf. Keijzer, 2021), primarily aims to link the evolutionary stages of cognition, ranging from prokaryotes to metazoans with nervous systems, while also accounting for the unique characteristics of non-neural organisms such as plants and fungi.

This ongoing effort aims to establish a conceptually coherent research domain that remains pluralistic in its models, theories, and methodological approaches, as is typical in highly interdisciplinary research frameworks.

The concept of cognition presented here, which aims to explore the bio-ecological basis of cognitive phenomena across different species-specific adaptation pathways (and can be extended to discuss the cognitive capacities of plants), should be understood as "operational," flexibly open to future additions and modifications. For those who feel the need for more concrete definitions, to avoid concerns about excessive vagueness, it is possible to refer to the integrative work by Lyon (2020) and Lyon et al. (2021), which builds upon and expands Sara Shettleworth's (1998, p. 5) definition of cognition: "*Cognition refers to the mechanisms by which animals acquire, process, store, and act on information from the environment. These include perception, learning, memory, and decision-making*." Shettleworth focused on animal cognition, but the same could be true for other organisms, as these abilities have also been found in bacteria, single eukaryotic cells, fungi, and plants. Lyon et al. offer a refined working definiton, that goes beyond previous formulation: "*Cognition comprises the sensory and other information-processing mechanisms an organism has for becoming familiar with, valuing, and interacting productively with features of its environment in order to meet existential needs, the most basic of which are survival/persistence, growth/thriving, and reproduction*." (Lyon, 2020, p. 416).

[62] Gazzaniga (2019, p. 167) reported an example taken from everyday life to explain the constraint cone: when we have to choose how to dress for an event, the chosen clothing will be constrained by our wardrobe, but this does not mean that the constraint will establish exactly what clothing we will wear.

Based on decades of growing intersections between cognitive and biological research, this phylogenetically neutral definition emphasizes the adaptive value of cognition for an organism. It also clearly distinguishes cognitive abilities from fundamental metabolic functions (e.g., respiration, which involves integrating information but lacks the components of "familiarization" or the capacity to "evaluate" beneficial versus harmful elements, exploratory behaviors, exploitative actions, and obstacle avoidance). In this sense, cognition introduces novelty from both a species-wide perspective and an individual one (the specific relevance of cognition at the individual level).

In summary, an approach that draws on the diverse biological and ecological-relational characteristics of different species – including ongoing research into the cognitive behaviors of brainless organisms like plants – aims to progressively develop a comprehensive and inclusive theory of cognition. This theory should be based on empirically verifiable assumptions and allow detailed analysis of cognitive abilities, as is the case for other biological abilities (cf. Lyon, 2006).

More specifically, Lyon et al. (2021, p. 5) have developed and gathered some of the most inclusive and promising conceptual tools for examining the various skills that underpin behaviors generally considered cognitive across different forms of life, such as the ability to perceive and act with purpose, memory and learning capacities, decision-making, and communication (see Table 2.1). This "operational toolkit" provides a framework for describing the fundamental cognitive abilities and functions within the concept of "basal" cognition.

Table 2.1 illustrates the basal cognitive capacities and functions, which result from a series of biophysical mechanisms and signaling processes identified across various branches of the "tree of life." These capacities can be traced, in different forms, across organisms. The common thread among these capacities is that organisms must possess species-specific mechanisms to determine which behavior is effective, which promotes survival and reproduction, and which poses threats. By approaching the study of cognitive abilities from this perspective, or at least being inspired by it, it becomes possible to identify not only the basic characteristics likely shared by all living systems (non-random information-signaling networks) but also to specify the unique cognitive adaptations that have resulted from distinct evolutionary pathways. These adaptations range from those found in unicellular organisms without nervous systems to those in multicellular organisms with nervous systems. In this context, it will be particularly interesting to explore inter- and intra-specific differences in some of these capacities in plants.

In the field of cognitive biology, the rigid "either/or" distinction between cognitive and non-cognitive processes is gradually being replaced by an understanding of a differentiated *continuum* of specialization and complexity. This allows the recognition that certain behaviors may have emerged from physiological mechanisms and processes shared across species belonging to different kingdoms (cf. Linson & Calvo, 2020). A similar shift occurred in research on animal cognition, which led to the dissolution of the rigid boundary between what was once considered cognition and what was not (cf. Allen & Bekoff, 1997). Today, it would be difficult to argue that animals lack cognitive abilities (cf. Keijzer, 2001; Godfrey-Smith, 2002). Based

Table 2.1 Toolkit of basal cognitive capacities from Lyon et al. (2021), adapted from Lyon (2006, 2015)

Capacity	Function
Orienting response	Ability to selectively attend to a state of affairs to the exclusion of others. Biological basis of attention
Sensing/ Perception	The capacity to sense and recognize (re-cognize) existentially salient features of the external and internal milieux
Discrimination	The ability to determine that a state of affairs affords an existential opportunity or presents a challenge requiring a change in internal state or relation to the external environment
Memory	The capacity to retain information about the immediate (and possibly distant) past, and to calibrate the sensorium to take account of this information, at a minimum via habituation or sensitization
Valence	An organism's capacity to assign a value (advantage/good, harm/bad) to a particular stimulus or the summary of information about its surroundings relative to its own current state. The 'building block' of affect (cf. Barrett, 2006)
Decision making	The capacity to combine information from multiple sources and act, typically in furtherance of an implicit or explicit goal
Behaviour	An organism's capacity to adapt to changes in its internal or external milieux by changing its own structure and function (internal) and/or its spatial and interactive relations (external)
Problem solving	Behaviour selection in circumstances with multiple (possibly conflicting) parameters and high degrees of uncertainty
Error detection	The capacity to determine whether a behaviour has succeeded or failed. Together with homeostasis/allostasis, this is the basis of biological normativity
Motivation	Implicit goals arising from existential conditions. Teleonomic striving. The inclination of a thing to continue to exist and enhance itself
Learning	The capacity to adapt behaviour to salient stimuli according to past experience by altering the threshold for or the nature of a response
Anticipation	The capacity to predict what is likely to happen next based on an early stimulus
Communication	The capacity to interact productively with other organisms via forms of signalling, notably (but not limited to) conspecifics, including initiating collective action, which may or may not include an explicit means of differentiating 'like me' (us) from 'not-like me' (them)

on current evidence, cognition appears to be present, in varying degrees, across the behavior of organisms throughout the evolutionary tree, including beyond the animal kingdom. This recognition marks a significant achievement in the study of natural cognition, the result of a long and often tortuous path.

Cognitive scientists are increasingly extending their analyses and applying their models to systems beyond the standard multicellular animal model with a nervous system. This includes exploring behaviors emerging from groups of individuals, collective behaviors in colonies of organisms, and examining the cognitive abilities of microorganisms (organisms with fewer components for interaction but no less capable of engaging with their environments for survival purposes).

I can now provide some examples of cognitive abilities that, until recently, were not expected to be found in a wide range of organisms, such as invertebrates, bacteria, single-celled eukaryotes, fungi, and plants. These findings are not the result of anthropomorphizing speculation but are supported by models suited for the precise observation and experimental analysis of the species-specific multisensory capacities of these organisms at various levels (e.g., biochemical, cellular signaling, and whole-organism behavior) (cf. Bianchi et al., 2025a).

In this brief survey, I begin by recalling the less counterintuitive cases where cognitive abilities are justified by the presence of a nervous system, albeit with varying degrees of complexity and centralization. For example, forms of cognition in invertebrates, although not analyzed in detail here (cf. Keijzer, 2001; Carruthers, 2013; Schnell & Clayton, 2021), provide compelling examples. One striking example is the octopus, a mollusk with a highly modular and decentralized nervous system, in which roughly two-thirds of its neurons are distributed throughout its arms (cf. Gutnik et al., 2023; Godfrey-Smith, 2021, p. 152). Other examples pertain to the nervous systems of insects (cf. Giurfa, 2015), which, although distinct from one another and phylogenetically distant from us, still belong to the metazoans.

The study of these diverse biological systems in recent years has significantly broadened our understanding of non-human cognitive processes and behavior, which arise from different bodily structures and organizations within their respective ecological niches.

In identifying both common traits and significant differences, the benefits of extending cognitive research from well-known systems to less-studied ones are particularly evident in prokaryotic cells like bacteria. For instance, bacteria display a wide range of adaptations to changing environmental conditions, including the ability to sense their surroundings, move toward energy sources, selectively uptake carbon- and nitrogen-containing compounds, and engage in interactions with other cells (cf. Müller et al., 2001; Ben-Jacob et al., 2004; Gibbs et al., 2008; Lyon, 2015). These abilities have been studied not only at the level of individual cells but also within complex bacterial aggregations. An intriguing case studied by Süel's group (cf. Prindle et al., 2015) involves the "neural-like" behavior of *Bacillus subtilis* biofilms. These structures exhibit long-distance intercellular communication through electrical signaling, coordinate metabolism and growth under nutrient limitation and stress, and display forms of physiological "memory" at the cellular level, mediated by changes in membrane potentials. Given the possible evolutionary link between these phenomena, researchers have drawn parallels between neuronal functions related to the retention, integration, and transmission of signals and similar activities observed in bacterial groups (cf. Lyon et al., 2021).

An example of eukaryotic unicellular behavior can be observed in the fascinating *Physarum polycephalum*, a multinucleated protist lacking a neural architecture yet equipped with pseudopodia that can extend and form a reticulate structure. This remarkably flexible organism demonstrates capacities for sensory integration, memory, learning, and decision-making (cf. Dexter et al., 2019), as seen most notably in its ability to solve mazes (cf. Reid & Latty, 2016) and tackle relatively complex problems within short periods (cf. Boisseau et al., 2016; Ray et al., 2019; Smith-Ferguson & Beekman, 2020).[63]

Research on cognition is also expanding to fungi. These non-neural organisms exhibit remarkable mycelial flexibility, thanks to their rapid growth and the thinness of their hyphae. This allows them to grow and branch in multiple directions simultaneously, reaching areas of soil that are difficult for the thicker, more rigid plant roots to access due to distance and the compactness of the soil. Although fungi lack brains, they possess a high degree of perceptual sensitivity and the ability to exchange information. The directional growth of fungal filaments is highly responsive to microenvironmental conditions, which can either promote or inhibit further development. This is due to the extendable tips of the hyphae, which send signals to the nucleus that controls growth (cf. Etxebeste & Espeso, 2016). Sophisticated signaling occurs both within fungal bodies and in interactions with plants, such as when specialized fungal structures develop inside plant cells in arbuscular endomycorrhizae (cf. Parniske, 2008).[64]

Regarding plants, their inclusion in cognitive science research represents a renewed and theoretically enriched continuation of studies on plant behavior, which date back (at least) to Darwin (cf. Trewavas, 2003; Baluška et al., 2009; Mancuso & Viola, 2015; Van Loon, 2016). The following pages examine how various factors contribute to a cognitive interpretation of plant activities. This includes, for instance, a relational and context-dependent understanding of the growth processes of plant aerial and root systems, which are subject to numerous constraints and conflicting conditions. These conditions often force plants to prioritize survival factors and make decisions between competing needs (cf. Keijzer, 2021).

In the effort to extend the analysis of cognitive behaviors to organisms not traditionally considered, ranging from bacteria to plants, the multi-layered mechanisms by which organisms manage internal and external information channels should be recognized. This understanding also points to the difficult attempts to artificially replicate some of these behaviors, such as mimicking specific abilities in robots (cf. Pfeifer & Bongard, 2006). Even when focusing on perceptual-motor skills, which result from the organization of information in cognitive systems, these behaviors are embedded in interactional contexts of varying complexity (cf. Godfrey-Smith, 2002, p. 224).

---

[63] The behavioral and cognitive abilities of this protist (known as the "slime mould") have been extensively studied, particularly over the past two decades, following the famous experiment by Nakagaki et al. (2000). The experiment is especially noteworthy because it demonstrated the protist's ability to replicate the actual map of the Tokyo subway system – which was designed to optimally connect the city with surrounding areas – by selecting the shortest paths to reach food sources (cf. Bianchi, 2021, pp. 139–140). This remarkable example of path optimization has inspired engineers to draw on *Physarum*'s abilities to improve the planning of transport networks, such as railways and highways. Moreover, *Physarum* has been shown to find the shortest path to nutrient sources thanks to a form of "external chemical memory" based on its own secretions (cf. Reid et al., 2012).

[64] There are other forms of mycorrhizae, such as ectomycorrhizae, ericoid mycorrhizae and orchid mycorrhizae.

2.8 Ecology and Cognition: A Relational Approach to the Activities of Living Systems

As outlined in the introduction, one of the main objectives of these pages is to develop an ecological-relational approach to the study of living systems (in particular, an ecology of cognitive difference found in nature), which is also applicable to the study of plant behavior. In this way, the connection between current research on plant cognition and insights gained from ecological and eco-system studies can be deepened.

Ecological approaches to the phenomena of life, when considered in their fullest sense, offer a systemic-relational and multilevel perspective on natural processes. Over the decades, analyses aimed at specifying different sets of relationships have provided valuable frameworks for examining the epistemological question of how various modes of organizing heterogeneous processes are understood and addressed. These processes can be described on a *continuum* from less holistic to more holistic-integrated views (cf. Acot, 2009, pp. 452–453).

In this context, I review the development of different approaches to systemic, complexity, and organizational studies, ranging from physics to processual approaches in the study of living systems. Following this, I highlight the contribution of ecosystem studies to the understanding of various forms of behavioral interaction, an approach that can also be extended to research on plant behavior and cognition.

Before proceeding, it is worth asking: what exactly is meant by 'ecology'?

Ecology[65] is a discipline that is part of the broader field of biological studies that examines the relationships of organisms with their environment and with each other.

It is good to note from the outset that there is no strict boundary between organisms and their environments, or between "biotic" and "abiotic" factors. The distinction between these interacting aspects is often made for practical purposes. This is particularly evident in the etymology of the term 'ecosystem,' derived from the Greek *oikos* ("house" or "habitat") and *systema* ("organized whole" or "interconnected system"). The term refers to natural systems shaped by networks of functional relationships – such as the exchange of matter and the flow of energy, as seen in trophic dynamics – that connect living organisms with one another and with their physical surroundings.

Ecology's main focus is on ecosystems, the relationships within which exist in various environments. Arguably, the most significant contribution of ecological and ecosystem studies has been to deepen the understanding of the "contextual" aspect of life, from the internal environment of a single cell to all types of external

[65] The term "ecology" (from the Greek *οἶκος*, meaning "house," and *λόγος*, meaning "study") was introduced in 1866 by Ernst H. P. A. Haeckel (1834–1919) to refer to the science that examines the relationships between living organisms and their surrounding environment.

environments, up to the global ecosystem of the entire biosphere (cf. Umerez, 2001, p. 167).[66]

It is insightful to trace the evolution of the concept of "system," distinguishing its application to both artificial and living entities. A system is considered "complex" if it consists of numerous or diverse components with multiple interconnections. Among the resulting behaviors, some are more predictable than others (cf. Gazzaniga, 2019, p. 146). Over time, there has been a transformation in system theory, often linked to the increasing complexity of ideas first developed in cybernetics, followed by advancements in computer science and robotics related to artificial systems.

In models of cognition, systems range from "fully decomposable" (highly modular systems composed of independent subsystems) to "almost-decomposable" (systems with weak but non-negligible interactions between subsystems). Another type is "integrated" systems, where distinct parts perform different functions but work together to create a unified organization capable of nonlinear interactions. Finally, there are "holistic" systems, whose complexity is best explained at the level of overall organization, rather than at the level of subsystems or individual components (cf. Di Francesco et al., 2017, pp. 191–192).

In living systems, particularly in non-modular organisms, the levels of integration at the organ or whole-organism level are highly interconnected, with a greater degree of functional cohesion than in larger relational networks, where components are relatively more independent. From cellular systems to the entire biosphere, although energy flows and material recycling are always involved, the nature of the connections varies, with some bonds being more tightly close and others showing less interdependence (cf. Bianchi, 2021).

At a general level, interpreting organisms as "systems" highlights the constant modulation across various levels of organization, shaped by the observer's work in defining system "boundaries." For example, analyzing a single organ, such as the pancreas, under a microscope or in a test tube, differs significantly from studying its function in relation to a specific system (e.g., the digestive system) or the entire body. Both types of analysis are essential: the first provides insights into localized function, while the second reveals the organ's actual utility and its diverse connections to the rest of the body (cf. Gazzaniga, 2019, pp. 296–297).

Furthermore, an interesting metacognitive aspect is to remember that we ourselves are "thinking" systems, capable of renewing our constituent elements while maintaining our organizational processes, and of investigating other systems of information organization.

For the present purpose, to gain a more comprehensive understanding, particularly in the study of plant behavior guided by the basic cognitive abilities previously

[66] E.g., DNA must be organically integrated into a cell to express information. Isolated from its context, it remains an inert molecule. It is only within the cell-genome, or more accurately, the organism-genome ensemble, that genes can be activated and function properly (cf. Nucara, 2014, pp. 80–81).

mentioned,[67] focusing solely on the evolutionary origins of behavioral capacities, though essential, appears insufficient (Bianchi, 2022).

It is also necessary to consider the ecological relationships that shape the living environments of these organisms. These relationships influence the effective expression, modulation, and potential improvement of behavioral responses to environmental challenges. Such aspects are often overlooked, especially in the study of non-neural organisms such as plants, which have a highly distributed organization and in which the condition of coexistence, as rooted organisms, constitutes the basis for any form of exchange (cf. Bianchi, 2022). Plant activities cannot be fully analyzed and understood in isolation (cf. Bianchi, 2022). Therefore, an ecological approach to the study of behavior and cognition offers a more accurate and realistic explanation of these phenomena.

2.8.1 *From Systemic Theories to Different Levels of Organization in Nature*

I now briefly retrace the historical evolution of some passages on the modalities of systemic-relational reasoning. These issues can be examined today in a dialogue that considers both the advances and limitations of various approaches, often emerging from interdisciplinary hybridizations and theoretical expansions.

A text that helps to reconstruct the central aspects of this type of study is Giulia Rispoli's *Dall'empiriomonismo alla tectologia. Organizzazione, complessità e approccio sistemico nel pensiero di Aleksandr Bogdanov* (2012). I refer to this work not to analyze the contributions of its central figure, Aleksandr A. Bogdanov (1873–1928) – author of *Tectology: The Universal Organization Science*, where he developed a general theory of the principles of organization in natural, living, and social systems – but to highlight the historical development of the concepts of system, organization, and complexity, particularly as they apply to the biological sciences.

At a general level, the growing interest in systemic research – especially its progress toward the study of complex systems within the broader and more heterogeneous field of the epistemology of complexity – has not been solely the result of scientific and technological advances in the West (Europe and the United States), such as those linked to information theory, synergetics, and calculus of probability. It also reflects a shared approach, whose contributions came from early twentieth-century Russian thought, particularly in the natural sciences, emphasizing the need to consider multiple interacting factors for a comprehensive understanding of living systems.

[67] E.g. the ability to discriminate differences, make choices, forms of memory, learning and anticipation, orient growth to obtain light and nutrients, avoid obstacles and harmful substances, modulate signaling for defense against pathogens, and avoid sources of environmental stress, etc.

2.8.2 *Beyond the Concept of Interaction: Conservation and Transformation*

The work(s) of some authors has made it possible to move beyond the concept of interaction between different elements and to think in terms of co-constructive and co-determining relational processes. In contemporary terms, particularly in behavioral and cognitive studies, this has led to viewing the organism–environment relationship as a coupled system. This aspect is operative in the stratification of the multiple interrelations of historical and emergent processes involved in natural evolution, which are traceable in ecosystem analyses (cf. Gagliasso, 2012, p. 14).

From a historical point of view, it is now possible to identify some of the pioneers in the development of systemic theories. The advances in the broader field of analyzing "artificial" and "natural" organizations emerged independently in European and non-European contexts[68] and have generally addressed problems related to information transmission and, more importantly, the nature of organized systems. Starting from Bogdanov's early twentieth-century studies, which stand as precursors to later developments, the multidisciplinary field of systems theory evolved from the theories of the 1950s and 1960s to the complexity theories of the 1980s and 1990s. More specifically, the cybernetic approach is traditionally attributed to Norbert Wiener (1894–1964), whose work *Cybernetics: Or Control and Communication in the Animal and the Machine* (1948)[69] focuses on the internal processes of a system. Wiener placed particular emphasis on electrical communication signals, analyzed using statistical methods developed in probabilistic mathematics. The innovation was the quantification and calculation of information (cf. Rispoli, 2012, p. 120), which enabled systems to be controlled by the flow of information, incorporating feedback loops to dynamically modify the systems themselves (cf. Nucara, 2014, pp. 34–35). In general, cybernetics' greatest contribution lies in considering artificial machines as physical systems based on their organizational structure rather than their individual components. By contrast, basic engineering reductionism would have limited the possibility of a deeper exploration of concepts such as feedback, goal uncertainty, and information flow (cf. Morin, 2004, pp. 287–288).

Years later, Ludwig von Bertalanffy (1901–1972) advanced *General System Theory: Foundations, Development, Applications* (1968) with the primary aim of moving beyond a mechanistic model of linear causality. He proposed a multidisciplinary, anti-reductionist framework spanning physics, biology, and social sciences

[68] E.g. in Russia with Bogdanov, in Vienna with Bertalannffy, at Harvard with Wiener, at Princeton with von Neumann, in Cambridge with Craik and at the *Bell Telephone laboratories* with Shannon (cf. Bateson, 1976, p. 517).

[69] According to Bateson (1976, p. 324) the roots of the cybernetic revolution can already be traced in the physiology works of Cannon and Bernard, in the physics of Maxwell and in the mathematical philosophy of Russell and Whitehead.

to study the structures and networks of connections within systems. This approach contributed to the development of the concept of circular causality, offering a way to explain the complex interactions between the parts of a system (cf. Rispoli, 2012, pp. 106–107). Cybernetics and systems theory share several characteristics, including the treatment of elements (and their resulting processes) as active and material, and the systems' ability to maintain stability through self-regulation (ibid., p. 125). However, both approaches have limitations. In the case of cybernetics, especially in its early stages, the focus was primarily on signaling forms in codified languages, directly compared to the human mind. This led to the assumption that the ability to self-organize was the exclusive prerogative of highly sophisticated systems, thereby excluding many other living systems, particularly those without nervous systems, such as microorganisms and plants (ibid. p. 138). Nonetheless, as will be seen, this limitation is being gradually overcome through ecological studies of natural ecosystems. These studies aim to analyze complex feedback mechanisms within systems, as well as the behavior and cognition of various organisms, including plants, framed in relational terms. Today, however, there is growing awareness of these complexities. In the development of systems and complexity theories, greater attention has been given to the self-organization of living matter (cf. Maturana & Varela, 1980; Kauffman, 1995) and emergent processes. However, despite more sophisticated contributions and systematizations, the role of the environment was long regarded as merely an external perturbing factor for systems, rather than as a dynamic, interpenetrating factor capable of influencing evolutionary transformations.

An autopoietic system, for instance, has been understood as being capable of shielding itself from "external noise" by maintaining a stable functional organization. This conception of the environment as merely a selector or a source of inputs contrasts with an approach that views the environment as a context of coexistence where processes of trans-active evolution between systems occur (cf. Gagliasso, 2012, pp. 12–15; Gagliasso, 2013). In this perspective, the environment includes not only abiotic elements but also other interacting systems. Natural reality is conceivable as a densely intertwined set of systems capable of performing distinct functions. Richard C. Lewontin (1929–2021), for example, emphasized that the relationship between organisms and their environment is more than just an interaction of internal and external factors. Instead, it is a dialectical and processual development, involving conditioning responses and counterresponses. This conception has gained prominence through a shift towards understanding organization in a way that better reflects the reality of (living) systems.

At this point, it is useful to briefly explore the concepts of open/closed systems, equilibrium, and process irreversibility. Living beings have long been described as "open systems," that is, systems in dynamic equilibrium, able to sustain themselves through the continuous exchange of matter and energy with their surroundings. Bertalanffy and Wiener's insights on living systems as open systems, particularly their characterization of complex systems in non-reductionist terms, were further developed by Maturana and Varela (1980). They introduced the concept of autopoiesis to describe the self-organization of biological systems, as opposed to allopoietic systems (cf. Locker, 1981). Autopoietic systems are characterized by their ability to produce and maintain themselves as a unified whole through a circular organization of processes that are recursively dependent on each other within the system (cf. Zeleny, 1981).

Today, the concept of "operational closure" is viewed as limiting for realistically describing the functioning of living systems in their ecosystem interactions. The idea of independence or complete self-sufficiency does not accurately characterize the phenomenon of life. However, this does not negate the capacity of living systems for self-organization; rather, it suggests that their organization, formed by the set of processes that constitute them, is always relational and embedded within a specific context (cf. Nucara, 2014, pp. 62–63). Every system is continuously in relation to its environment, both influencing and being influenced by it (cf. Rispoli, 2012, p. 140). Alternatively, one could argue that instead of a direct relationship with the world, systems interact with a "model" of the physical world, which is always mediated in some way by the system itself (cf. Metzinger, 2017). In Maturana and Varela's work (1980, p. 79), living systems are described as autopoietic, meaning they are capable of self-production through continuous interaction with their environment. This process occurs via the regeneration of the network of processes that constitute their organization. This ongoing dynamic of adjustment constitutes the cognitive process. The concept of "closure" refers to the cyclical and self-sustaining organization of such systems and does not contradict their thermodynamic openness. Indeed, autopoietic systems are thermodynamically open – engaged in continuous exchanges of matter and energy with their environment – while remaining organizationally closed, meaning that the network of processes that constitute the system is self-producing. Thus, closure and openness are not mutually exclusive but complementary aspects of living systems (cf. Ceruti, 1994, pp. 89–90; Mingers, 1994; Razeto-Barry, 2012). To say that autopoietic systems are "autonomous" is not the same as saying they are "self-sufficient." Living systems require interaction with the world to preserve their organization, which allows them to develop an "individual" or specific perspective of their environment. This is one of the basis of what enactivists have referred to as sense-making, i.e., one of the characteristics that distinguishes cognitive systems from noncognitive ones, key to understanding biological agency (cf. Segundo-Ortin, 2020).

Furthermore, the theory of autopoietic systems is inherently relational, not only due to the nature of the systems themselves but also because their distinction as separate units from the environment (the domain in which they exist) is made by an observer. In any case, the core idea is that living systems, characterized as open systems, maintain themselves in a dynamic process of development, flexibly preserving their organization in changing environments and postponing entropy increase through metabolic processes (cf. Acot, 2009, p. 463). Metabolism, in its broadest sense, underpins the capacity for self-organization in living systems, allowing them to utilize internal energy resources and eliminate waste products externally (cf. Nucara, 2014, pp. 31–32). Every organism, through mechanisms of matter acquisition and expulsion, through which it gains energy and information, sustains its vital processes in a negentropic state.[70]

Living systems, including plant systems, are thus open and "historical," where behaviors are not strictly determined but are constrained by past experiences (cf. Maturana & Varela, 1980).

A closely related concept to the openness or closure of systems is equilibrium, which, if misunderstood, could lead to misconceptions. In living beings, equilibrium cannot be regarded as "static" or "final," as each system continuously transitions through different states. Over time, every living system undergoes a complete renewal and rearrangement of its constituent elements while maintaining its organization (cf. Rispoli, 2012, p. 130).[71] In addition, systems are conceivable in this dynamic of conservation/invariance and transformation.

Within this interrelated dynamic of continual loss and renewal under homeostatic conditions, complex systems evolve through irreversible processes that reflect their temporal unfolding (cf. Bianchi et al., 2025b). This perspective – now fundamental to the understanding of living systems – was applied by Ilya Prigogine (1917–2003) to microphysical systems in thermodynamics, enabling a deeper exploration of irreversibility as a core feature of natural processes. This can be tied to the idea of historical becoming, where the transformative interaction of each interdependent element influences the overall trajectory of the system (cf. Rispoli, 2012, pp. 35, 133). The irreversible timeline of biological evolution does not prevent us from tracing the common origin and shared elements of biological systems. As Monod (2017, p. 155) wrote, every living organism can be viewed as a "fossil," carrying within it the genetic traces of its ancestry. The historical dimension characterizes the dense web of connections that unites life forms (cf. Pievani, 2017, p. 7). This is the unity of difference in the path of multilevel variation. As with the historical evolution of species, the transformation of the biosphere is also the result of a set of irreversible processes that define a time direction (Monod, 2017, p. 99; 120).[72]

---

[70] Erwin R. J. A. Schrödinger (1887–1961) published what has since become a famous booklet (1944) in which he described living systems from a thermodynamic perspective. He observed that living systems, unlike inanimate matter, do not attain thermodynamic equilibrium (or maximum entropy) while alive, maintaining themselves in states of low entropy by exchanging energy and matter with their surroundings, thereby delaying, but not violating, the second law of thermodynamics (cf. Acot, 2009, p. 463).

[71] Consider that approximately every 8 weeks, the human body renews the matter it is composed of (cf. Gazzaniga, 2019, p. 144). Over the course of a year, the body replaces 98% of its atoms (cf. Trewavas, 2014, p. 256). These systems remain fundamentally the "same" despite being composed of entirely renewed material.

[72] From this perspective, in our epoch, marked by the anthropogenic impact on the environment, climate change, and the "sixth mass extinction," the historicity and irreversibility of life's developmental processes should draw even greater attention to our species' responsibility. We are causing harmful, irreversible changes that have the potential to disrupt the relational dynamics of communication, reproduction, and survival among different species, including plants and their interactions with other organisms within ecosystems.

In line with this argument, reference can be made to the epistemological approach of Bogdanov, a precursor to cybernetics and systems and complexity theory. Similar to Kuhn, Feyerabend, and Toulmin, this perspective emphasized the importance of the historical transformation of scientific knowledge, integrating various levels of interdisciplinary reasoning that bridge self-organization theory, Darwinian evolutionism, and eco-systemic thinking (cf. Gagliasso, 2012, p. 15). It holds that every process at a given level of reality possesses its own distinct form of organization, whether pertaining to inorganic or living matter. In the development of this science of dynamic organizations, the objective was to analyze systemic organization through a processual understanding of biochemical and physiological transformations. The focus is less on the organicistic aspects of systems and more on the historically channeled co-development of organized systems (cf. Rispoli, 2012, pp. 143–144). Instead of considering the networks of connections between system elements in a horizontal and interchangeable manner, this perspective emphasized the generative, constructive, and historical-temporal development of systems as dynamic processes (cf. Bianchi et al., 2025b). These processes emerge and evolve through exchanges with the environment (cf. Gagliasso, 2012, pp. 14–15).

Thus, even within the same overarching systemic research program, different stages of development have allowed for a distinction, as V. N. Sadovskij (1934–2012) noted, between two paradigmatic nuances. One focused on the equilibrium and stability of systems, exploring how systems maintain their organization in changing environments, while the other emphasized the role of imbalance, historical development, and emergent levels (cf. Rispoli, 2012, p. 136).[73]

2.8.3 *Relational Nature of Processes*

The systemic-relational approach proposed by the authors mentioned – and by many others not explicitly cited – has now permeated the concrete practices of scientific research and, as we shall see, the field of cognitive studies as well.[74]

The concept of relationship (whether physical or functional), though its meaning may vary, represents the "lifeblood" enabling the existence of any system. Due to the need to address the complexity of real-world organizations, relational reasoning has extended across various scientific disciplines, if not all. This influence is evident in many contemporary approaches.

[73] Proponents of the first approach include Ludwig von Bertalanffy, W. Ross Ashby, and Norbert Wiener, while the second approach includes figures like Ilya Prigogine, René Thom, Erich Jantsch, and, indeed, Alexander Bogdanov.

[74] Relational theoretical positions, in their effort to provide more comprehensive descriptions of reality, not only define various contemporary philosophical approaches but have also shaped philosophical systems throughout history. A notable example, despite their differences, can be found in the contributions of empiricist philosophers such as Locke, Berkeley, and Hume. These thinkers, each for different reasons, sought to understand perception, communication, behavior, and intelligence by examining the diverse forms of relationship between individuals and their environment.

While one of the primary missions of science, at least until the early twentieth century, was to reduce complexity and eliminate uncertainty to enhance control, in what has been called the "new science," there is a growing recognition of the inherent indeterminacy and complexity of phenomena (cf. Nucara, 2014, pp. 105–106; 28). For example, by recalling with Rovelli (2014, pp. 105; 118; 122) the "relational" aspect of quantum mechanics, it has become possible to meaningfully describe particle movement based on their positions at specific instants, namely, at the instants in which they interact with something else.[75] Perhaps running the risk of falling into an overgeneralization, one could say that all reality is related and that things or processes do not "are" but "happen" and "affect" one another in a more or less direct way. It is the relational aspect of processes, which are what they are according to possible interactions.[76]

Setting aside the complex transition from micro- to macro-level descriptions of behavior, systemic-relational approaches have inevitably taken root, with priority, in the branches of biological studies, given the intricate multifunctional organization of organisms and their multilevel interactions with their ecological niches. This broadening first followed a disciplinary integration of the framework of biological studies, which today is decidedly more ramified in its theoretical and methodological approaches. As highlighted in works on the philosophy of biology (cf. Gagliasso, 2001; Pievani, 2005; Borghini & Casetta, 2013; Gagliasso et al., 2016), since at least the mid-twentieth century, biological research, alongside the principles derived from systemic approaches, has deepened its focus on the organizational-functional findings of functional biology. These findings have been enriched by concepts of historical-population adaptation from evolutionary biology (cf. Mayr, 1982, pp. 73–74), as well as by insights from developmental systems theory, thermodynamics, and information theory (cf. Chandler & van de Vijver, 2000).

Clearly, the various scholars involved in the expansion and refinement of biological research have placed greater emphasis on the aspects most relevant to their respective specializations. A common element among these approaches to studying biological processes is the rejection of extreme or less verifiable views from the past, such as essentialism or vitalism, where the "metaphysical" component was strongly emphasized. Additionally, there is a more explicit critique and distancing from logical positivism and explanatory reductionism. The long-standing debate between reductionist, organicist, and holistic schools has also subsided, thanks to specialized advances that have enhanced the precision with which phenomena can be analyzed (cf. Monod, 2017, p. 79).

[75] From this perspective, electrons do not continuously "exist" in a specific location but only when they interact. In this sense, Heisenberg, by outlining the first quantum equations, did not specify the position and velocity of an electron. Instead, he had compiled tables of numbers representing its possible interactions (cf. Rovelli, 2014, p. 105).

[76] It would not seem an exaggeration to claim that the universe exists because difference exists (at least at a low energy level) and difference is possible on the basis of a different "behavior" of matter, which, in turn, is possible on the basis of different physical interactions.

However, a fundamental challenge remains, perhaps, unavoidable: the difficulty of achieving a unified theoretical-methodological framework across the increasingly specialized branches and subdisciplines of science. Nevertheless, in response to the need for more comprehensive explanations of how organisms function over time and in varying contexts, according to what could be termed "relational biology" (cf. Rashevsky, 1954; Gazzaniga, 2019, p. 144)[77] there has been a shift from "component thinking" to "systems thinking." This shift has required the development of new conceptual frameworks in biological and ecological studies, such as those seen in immunology and developmental biology (cf. Westerhoff & Palsson, 2004, pp. 1249, 1251). For instance, the biology of ecological development ("Eco-Evo-Devo") has focused on the intersection of environmental interactions, evolutionary processes, and developmental studies (cf. Gilbert, 2001; Sultan, 2021).

Recent advances in biology – particularly in molecular biology and bioinformatics – have led to the emergence of new disciplines characterized by the use of the suffix "-omics" and the corresponding root "-ome" to define their fields of study (e.g., genome, proteome, transcriptome, metabolome, etc.). These disciplines, though distinct in their focus, share a common core in systemic-relational reasoning, both functional and contextual, and rely on techniques capable of generating and analyzing vast amounts of data to investigate various biological systems. The "genomic revolution," for example, propelled molecular biology into the domain of systems biology, beginning with the first large-scale genomic analyses. A recent frontier in this evolution is phenomics (cf. Houle et al., 2010), which evaluates the broader effects of molecular changes on phenotypic variability, including environmental influences in its analysis.

It is increasingly clear that these disciplinary advances must be complemented by ecologically systemic studies that consider the relationship with the environment, encompassing behavioral and cognitive aspects as well. Ecology, rich in examples of relational dynamics, such as the trophic dynamics of cooperation and competition within and between species for resources, provides valuable insights into both individual and collective behavior. This allows for a deeper exploration of concepts such as interrelation, "relational action," mutualistic variation, and coadaptation between different systems, moving beyond simplistic behavioral input-output models.

Moreover, relational approaches offer the potential to articulate new conceptual metaphors that can guide and inspire research in specific directions, emphasizing novel connections (cf. Frezza & Gagliasso, 2014, p. 30). This approach is anti-reductionist and strongly interdisciplinary, drawing from different fields as epigenetics, bioinformatics, network theory, and artificial intelligence (cf. Alberghina, Hohmann & Westerhoff, 2005).

[77] A systemic-relational approach to complex organizations is enabling a more comprehensive investigation of eco-evolutionary processes, both from a genetic perspective (the structuring of genetic information across different levels) and from a behavioral perspective (the interrelationships and reciprocal coadaptations between various organisms).

2.8.4 Ecological Explanation of Relational Behaviors

Leaving aside the argumentative subtleties of these positions, and considering the nonlinearity of their development, the discussion so far serves to illustrate the transformative path and growing attention to the relational and functional complexity of phenomena, as well as to the different ways in which such complexity is expressed.

Systemic-relational studies of the various forms of organization found in nature are central to ecological and ecosystem studies, understood in their most profound sense as investigations into the conditions of origin and the concrete manifestation of exchanges within environments that include other organisms. This approach requires further development in the field of cognitive and behavioral studies. However, the ecological foundation of animal behavior research is not entirely unfamiliar. For instance, behavioral ecology is a discipline that examines how animals utilize environmental resources, among many other aspects (cf. Stuhrmann, 2022). The strategies they adopt are influenced by factors such as the type, quantity, and distribution of resources in the environment, which in turn affect reproductive behaviors, feeding strategies, territorial competition, and, for some species, social cooperation.

From a historical perspective, a form of "proto-ecological" thought, aside from its roots in ancient philosophy (cf. Glacken, 1970), gradually emerged through the growing recognition of interactions among organisms within specific environments. This perspective also appears in the travel diaries of explorers from the late 18th and early 19th centuries, such as Alexander von Humboldt (1769–1859) and Aimé J. A. Bonpland (1773–1858), whose focus shifted from merely collecting and cataloging species to understanding how organisms interact with their environment. Observations on the variability of vegetation with altitude are considered foundational to the emergence of plant geography, the science that studies the relationship between climate and vegetation, which has contributed to increasing the range of analysis from organisms to communities of organisms (cf. Acot, 2009, pp. 453–454).

Studies on different communities of organisms, such as those by Karl Möbius (1825–1908), who coined the term "biocenosis" in his work on oysters (1877), and by Victor Hensen (1835–1924), who focused on marine ecology, have been foundational in ecological research. In the field of plant ecology, Johannes Eugenius Bülow Warming (1841–1924) published the first formal study in 1895, *Plantesamfund* ("*Plant Communities*") (cf. Mayr, 1982, p. 120; Gagliano et al., 2017, p. VIII). Plant ecology, as a distinct and specialized discipline, gained greater autonomy by the late twentieth century (cf. Cittadino, 2009, p. 238).

Many of Darwin's observations also laid the groundwork for later ecological studies. Today, evolutionary theory necessarily incorporates ecological dynamics (cf. Pievani, 2017, pp. 57–58), given that the ecological context is where the survival of organisms is ultimately determined.

2.8.5 *Levels of Ecological Organization and Ways of Studying Ecosystems*

In ecology, different levels of organization are considered, ranging from individuals to populations, communities, and ecosystems. While the boundaries and interactions between these levels often overlap, this subdivision is mainly practical, helping to illustrate the growing complexity of living dynamics. The organism is the most basic unit that interacts with its surroundings to obtain shelter and acquire the resources necessary for survival and reproduction. The next level is the population, consisting of organisms of the same species living in a shared environment. Groups of populations from different species coexisting in the same area form a community. The biodiversity of a community is determined by numerical and functional aspects, that is, the number of species (genetic richness) and the roles or activities each species performs within the community. The ecosystem level derives from the consideration of the dynamics of the communities and their living environments. This includes the relationships between organisms and the environmental factors that influence physical, chemical, and biological transformations in terrestrial, aquatic, and marine ecosystems. Lastly, the global ecosystem, or the biosphere, encompasses all ecosystems on Earth (cf. Bianchi, 2021).

In general, ecological interactions are addressed from two main perspectives: bioenergetic, which focuses on the transfer and recycling of matter and energy transformation within ecosystems, and informational, which deals with the exchange of information within and between systems. Initially, these aspects were considered distinct and attributed to different units of analysis. For instance, physical boundaries were thought to be marked by cell membranes or the outermost layers of organisms (e.g., the epidermis), while channels of information flow extended beyond these physical limits, encompassing the modes of intra- and intersystem communication. Subsequently, the aim was to harmonize the various levels of analysis into coherent explanations (cf. Bateson, 1976, p. 501). Today, research has advanced to the point where it systematizes findings from different specialized fields to provide a more comprehensive understanding of ecological processes.

Historically, a major contribution to the development of a more integrated approach came from Eugene P. Odum (1913–2002) and, later, his brother Howard T. Odum (1924–2002). Eugene's foundational work *Fundamentals of Ecology* (1953) helped establish ecosystem ecology as a discipline, highlighting the links between the thermodynamic behavior of living systems and ecosystems. Together, their work promoted a scientifically grounded form of holism, far removed from metaphysical interpretations. From this perspective, the biosphere,[78] – understood as the interconnected network of ecosystems capable of sustaining life – maintains far-from-equilibrium states (cf. Acot, 2009, p. 464).

[78] The term "biosphere" was coined by Eduard Suess (1831–1914) in 1875, but it was Vladimir I. Vernadsky (1863–1945) who popularized the concept in a form more closely aligned with its modern understanding (cf. Acot, 2009, p. 465).

According to Vernadsky (1998), the continuous recycling of matter and the flow of energy through organism-environment systems make it difficult to draw clear boundaries between living and non-living entities. The biosphere, or global ecosystem, can be seen as a result of the relationship between solar energy and biological activity, with organisms, especially plants and other photosynthetic organisms, converting light energy into chemical energy (cf. Giovannetti, 2009, 2014, pp. 62–63; Rispoli, 2012, pp. 148–150; Bianchi, 2021, p. 19).

In this sense, the biosphere is the product of the coevolution of diverse life forms and Earth itself, highlighting the perspective that life's dynamics have the power to transform the planet that hosts them (cf. Pievani, 2017, pp. 24–25).

A fundamental aspect investigated in these interrelationships and exchange networks between species and environments is the pyramidal structure tied to the trophic dynamics of biocenoses,[79] which have helped to develop the concept of the ecological niche, shifting the attention from purely spatial or numerical aspects to functional dynamics, which govern population fluctuations within ecosystems (cf. Acot, 2009, p. 461).

In summary, when analyzing complex systems, nonlinear dynamics and emergent behaviors cannot be detected through isolated analysis of individual components. Instead, they must be critically examined through multifactorial reasoning about process dynamics (cf. Frezza & Gagliasso, 2014, pp. 30–31). This is one of the tasks of philosophical analysis.

It is therefore understandable that, at a general level, ecological-systemic reasoning has risen to the challenge of describing the increasing complexity of organizational levels in the living world. This approach goes beyond the simple idea of systemic connections to embrace reasoning in terms of "eco-evolutionary interpenetration" of processes, with the potential to span from the relationships between individual living systems to mutualistic associations, populations, multispecies communities, and ultimately to the organization of the entire biosphere (cf. Gagliasso, 2012, pp. 15–16; Rispoli, 2012, p. 152).

2.8.6 *Toward a Plant Cognitive Ecology: The Role of Context and Other Organisms*

More specifically, within the framework of systemic-relational analysis – where multiple functions hold reciprocal significance – it becomes clear that investigating the effectiveness of behavioral responses to environmental challenges across various organisms, including plants, is both feasible and meaningful. A thorough examination of plant bio-ecological reality requires the integration of eco-evolutionary

[79] The foundational contributions of Karl Semper (1832–1893) and Charles Elton (1900–1991) significantly shaped early ecological thought. Later, Raymond L. Lindeman (1915–1942) advanced the notion of the ecosystem as the fundamental ecological unit, emphasizing its analysis through trophic relationships and energy flow within food chains (cf. Acot, 2009, p. 463).

explanations with knowledge from plant biology and associated organism interactions. This will provide a more complete understanding of "bioecological cognition."

Moreover, in the case of plants – particularly given their strong connection to their life contexts – numerous physiological processes and activities result from interactions not only with conspecifics and other species within the plant community, but also with a wide array of organisms in microbial communities. The functional richness of microbiota, or the biodiversity of microorganisms in each environment, has reshaped the concept of "inside-outside" and debunked the once-presumed notions of organismal "independence" or "self-sufficiency." It is now well established that organisms are "co-habited" by myriad living microsystems, which sustain complex relationships within their microenvironments, both external and internal to the host. From an organizational perspective, endosymbionts, despite their genetic differences, cannot be considered "external" to the overall organization of the host.[80]

To grasp the importance of microorganism interactions, it is worth noting that, without its teeming microbiome – comprising billions of bacteria, archaea, fungi, viruses, and protozoa – a human being would be incapable of digesting food (cf. Pievani, 2017, p. 118). Although fetuses begin without microorganisms, their microbiota forms very early in life, with the host organism working to maintain a balance between the various microbial populations.

Plants, too, have their own microbiota, which is influenced by evolutionary factors, genotype, and dispersal mechanisms, as well as by climate, environmental conditions, and soil composition. Plants select and maintain certain microorganisms within them, such as those in the endosphere (e.g., in the root compartment) (cf. Bonfante, 2021, pp. 185–200). A prominent example of a symbiotic relationship essential to plant survival is that with nitrogen-fixing bacteria, typically of the genus *Rhizobium*, which convert atmospheric nitrogen into bioavailable forms that plants can assimilate (cf. Mauseth, 2017, p. 364).

In general, plants engage in various types of symbiotic relationships, which can be mutualistic, commensal, or parasitic. In mutualistic relationships, both partners benefit. In commensalism, one organism benefits while the other is neither helped nor harmed. In parasitism, one organism benefits at the expense of the other.

One of the most important symbiotic relationships between plants and other organisms is that of mycorrhizae, associations between plant roots and fungi.[81]

Mycorrhizas are typically classified into four main groups: Arbuscular, Ectomycorrhizal, Orchid, and Ericoid; and there are other types beyond these. The

---

[80] Sometimes, however, the organization of a system, such as that of cancer cells, which develops in contradiction to the vital connections of the organism, although spatially located within the host, is not conceivable as functionally cohesive with the rest of the regular functioning of the organism (cf. Rispoli, 2012, pp. 154–155).

[81] Fungi are organisms with a flexible metabolism. For example, many endophytes can switch from commensal to saprotrophic, pathogenic or symbiotic behavior, depending on the availability of nutrients in a given environment and the plant species (cf. Bonfante, 2021, p. 44). This metabolic flexibility is of great use to plant life processes.

two most extensively studied groups are arbuscular mycorrhizal fungi (AM), the most ancient type – dating back approximately 450 million years and present in about 72% of plant species – and ectomycorrhizal fungi (ECM), which emerged more recently, between 270 and 170 million years ago. The primary distinction lies in the mode of colonization: unlike AM fungi, ECM fungi do not penetrate root cortical cells, but instead grow intercellularly, forming a specialized network known as the Hartig net.[82] ECMs are also externally visible as a sheath or mantle surrounding the plant's fine root tips (cf. Bonfante, 2021, pp. 128–132).

In general, the presence of mycorrhizae is highly beneficial for plants. In exchange for sugars they cannot synthesize, fungi supply plants with water and essential nutrients, particularly phosphorus and nitrogen. The flexibility of fungal bodies, consisting of a dense network of rapidly growing and branching hyphae, enables them to extend over long distances and navigate around obstacles in the soil, infiltrating narrow spaces that rigid, cellulose-rich plant roots cannot easily access (cf. Bianchi, 2022).[83] In summary, mycorrhizae improve plant mineral nutrition, functioning as a natural "biofertilizer." They also act as "bioprotectors" and "biopesticides", enhancing plant resistance to biotic and abiotic stress, as well as pathogen attacks (cf. Bonfante, 2021, pp. 106; 123). On a broader ecological level, these associations facilitate the exchange of resources and signals across complex underground networks, often referred to as the "wood-wide-web" (cf. Giovannetti, 2014; Simard, 2018). Initial research on mycorrhizae primarily focused on their localized functions; however, it has since broadened within ecological studies to include the interdependence of aboveground and belowground processes (cf. Read, 1991).

The hypothesis of ecological interaction has been supported by early studies using carbon and phosphorus tracers, which demonstrated that carbon travels from plants to fungi and phosphorus from fungi to plants, respectively, suggesting the recirculation of substances (cf. Bonfante, 2021, p. 109). Mycorrhizae not only enhance the availability of resources but also facilitate communication and defense mechanisms, improving plants' ability to respond to environmental challenges. Indeed, mycelial networks may transport not only water and nutrients but also chemical and electrical signals, particularly those potentially involved in triggering defense responses against pathogens. These networks might even propagate such signals between different plants, possibly enhancing communal defenses against attacks (ibid., pp. 148–149; Simard et al., 1997, 2009). In any case, caution is needed when discussing the wood-wide web hypothesis. Its role is not sufficiently clear in the literature, and the specialist community remains highly divided on the issue (cf. Henriksson et al., 2023; Robinson et al., 2024).

In general, mycorrhizae are essential for the proper functioning of ecosystems, contributing significantly to soil stability and aggregation,[84] which helps prevent

[82] The Hartig net is a network of fungal hyphae that penetrates between the root cells of a plant, forming part of the symbiotic interface in ectomycorrhizal associations.

[83] Fungi have chitin in their cell walls.

[84] Mycorrhizal fungi specifically release glomalin, an aggregating glycoprotein, into the soil.

erosion and nutrient leaching. They also play an essential role in regulating climate factors, serving as natural carbon sinks. However, the anthropogenic reduction of mycorrhizal vegetation is leading to a decrease in terrestrial carbon reserves, causing the release of carbon into the atmosphere (cf. Mazzolai, 2021, p. 37).

From these considerations, echoing John Donne's (1624) famous lines, "*No man is an island, entire of itself; every man is a piece of the continent, a part of the main*" one could argue, as Bonfante (2021, p. 197) did, that "*a plant is not an island*," and this is not merely a metaphorical expression. Like any other organism – particularly rooted organisms, for which interaction with other life forms is an unavoidable condition for survival – a plant is inherently "relational." It can be described as a holobiont, a composite entity formed through the symbiotic coexistence of various species within it. The sum of information from these organisms, combined with the plant's own information and its interactions with the environment, constitutes an ecological system that extends from internal to external, interconnected through physical and metabolic interactions with its surroundings.

It is time to integrate this framework into studies of biological behavior and cognition, fully embedding them in biological and ecological research. There is growing demand for approaches capable of explaining the origins, expression, and modification of behaviors from an eco-systemic and relational perspective (cf. Bianchi et al., 2025b). In a network of relational interactions, each encounter in an organism's life opens the potential for new interactions (cf. Frazier, 2021). For example, in the realm of communication, where plant signaling is understood as a form of behavior, signals are not static but emerge in context during exchanges between organisms. The significance of a signal is shaped by the type of activity, context, and interaction, not merely by the presence of a specific chemical compound (cf. Witzany, 2016). Identifying the compound alone is not enough; it must be understood in relation to its context, circumstance, and dosage. This makes plant communication a truly "ecological process," where the signals shape behavior, which in turn foster new interactions and lead to new "meanings." Of course, this contextual flexibility is subject to certain limits, which are essential for the retention of environmental information and its transmission across generations (cf. Gagliano, 2017, p. 95).

In plant chemical communication, volatile organic compounds (VOCs) are involved, but they function as true signals only when recognized by specific receptors. It is now established that plants utilize signaling systems effective with other organisms as well, highlighting a fascinating ecological property: these signals can be recognized across species boundaries, and even across kingdoms, making them "chemically comprehensible" to a variety of organisms other than plants (cf. Bonfante, 2021, p. 26). This underscores the idea that the network of relationships within a system, at various levels of organization, is far more extensive and complex than that confined to a single organism. The system encompasses internal organization, the organism's interactions with other organisms, and its relationship with the wider environment. Without this broader understanding, it is impossible to fully comprehend individual behaviors, particularly plant behavior.

This approach, even at the behavioral-cognitive level, may appear familiar to those acquainted with methodologies employed in psychology and cognitive science. Its relational nature and focus on the joint analysis of the organism-environment system exhibit similarities to the approach of ecological psychology. For instance, the core assumption that perception and cognitive phenomena arise from the interaction between the animal and its environment (or organism/agent-environment) aligns with some principles of ecological psychology (cf. Gibson, 1950, 1966, 1979; Michaels & Carello, 1981, p. VII). However, this does not imply that an ecological approach to cognition, in the strict sense of ecological psychology, must be adopted in plant cognition studies, as if it were the only possible one. The approach discussed here is pluralist, as previously noted. Some aspects may benefit from further exploration within this framework, while others might be better explained using different positions.

From a perspective of relational codetermination, the goal is to better understand how plants manifest their abilities. Many of these abilities, if not directly caused by interspecific interactions, emerge and are enhanced through continuous ecological relationships (cf. Bianchi, 2022). In this view, the ability to acquire information from the "external" environment only makes sense in relation to the condition of the (cognitive) system in a specific context. For example, an organism will behave differently depending on whether it is in a state of hunger or satiety, or whether it is acting alone or in a group, pursuing different goals with varying intensities at different times (cf. Lyon et al., 2021).

Reasoning in relational terms is complex, which may explain why it is often overlooked in various fields. Despite its theoretical accuracy, stemming from its comprehensive scope, a systemic and relational approach that seeks to account for the multiplicity of contributing factors may initially be perceived as overly complex or less feasible in experimental contexts (cf. Gallagher, 2017, p. 21). However, once the practical challenges of managing interdisciplinary and multilevel complexity are addressed, it becomes possible to disentangle difficulties in accurately identifying the various behavioral activities of species within an ecological niche.

The recent emergence of interdisciplinary approaches such as cognitive ecology (cf. Hutchins, 2010; Mettke-Hofmann, 2014) – the diverse study of cognitive behaviors across different contexts, focusing on how various species acquire and use information about their environment to survive (cf. Dukas, 2004) – has highlighted the critical role that memory and learning processes play in eco-evolutionary dynamics of cognition (cf. Gagliano et al., 2014). Specifically, it pointed to the evolving network of interdependencies among the different elements that together form a cognitive ecosystem. An eco-evolutionary framework has the potential to provide a more complete and accurate understanding of natural cognition (cf. Morand-Ferron et al., 2016) and is also valuable for understanding certain plant capabilities.

Historically, various approaches – differing in their theoretical foundations, methodologies, and objectives – have adopted, or are increasingly exploring, ecological perspectives on cognition. Examples include Gibson's ecological psychology, Bateson's ecology of mind, Uexküll's concept of subjective environments, and more contemporary perspectives such as situated, embodied, distributed, and extended cognition (cf. Parise et al., 2020). These perspectives suggest that – albeit each in different ways – the study of cognitive ecology and cognitive ecosystems, due to their transversality and pervasiveness, will become an increasingly significant component of cognitive science, alongside the emerging field of ecosemiotics (cf. Farina, 2021). At the core of these perspectives is the idea that the environment itself can be seen as "cognitive," or more precisely, it acquires cognitive characteristics through the relationships that various organisms establish within it, as they perceive and solve different survival problems (cf. Bianchi, 2022).

The interaction between resources, functions, and behavioral responses, occurs within a space of differentiable relationships, which are recognizable and meaningful for a given organism. This approach is systemic-relational and considers the environment not as a static entity that remains unchanged over time, but as a dynamic reality that plays different roles depending on the species interacting with it. Contemporary discourse has expanded to include plants as well (cf. Heras-Escribano & Calvo, 2019).

One could argue that the study of ecosystem relationships goes hand in hand with the study of cognitive living environments. What constitutes an opportunity for interaction for one organism may not for another or may vary over the course of an individual's development. Information, therefore, serves to relate to other systems and emerges through the interaction between the organism and its environment; it acquires meaning only when it is recognizable. While this is a "coupled" conception, individual variability in the ability to perceive and interact is preserved. Information is not only species-specific but also individual-specific, shaped by each organism's unique history of ecological interactions (cf. Farina & Belgrano, 2006). Beyond the relationship with the environment, the relevance of studying the specific characteristics (possibilities and constraints) of different *organisms* remains essential for understanding forms of cognition.

In conclusion, it appears that eco-systemic and cognitive studies can form part of a broader framework of complementary explanations for the modes and relationships of living systems.

References

Acot, P. (2009). Ecosystems. In P. J. Bowler & J. V. Pickstone (Eds.), *The Cambridge history of science* (pp. 451–466). Cambridge University Press.

Adams, F. (2018). Cognition wars. *Studies in History and Philosophy of Science Part A, 68*, 20–30.

Akagi, M. (2018). Rethinking the problem of cognition. *Synthese, 195*(8), 3547–3570.

Alberghina, L., Hohmann, S., & Westerhoff, H. V. (2005). Systems Biology: necessary developments and trends. In *Systems Biology: Definitions and Perspectives* (pp. 389–402). Springer Berlin Heidelberg.

Allen, C., & Bekoff, M. (1997). *Species of mind: The philosophy and biology of cognitive ethology*. MIT Press.

Alpi, A., Amrhein, N., Bertl, A., Blatt, M. R., Blumwald, E., Cervone, F., et al. (2007). Plant neurobiology: No brain, no gain? *Trends in Plant Science, 12*(4), 135–136.

Andrews, K. (2020). *How to study animal minds*. Cambridge University Press.
Applewhite, P. B. (1975). Learning in bacteria, fungi, and plants. *Invertebrate Learning, 3*, 179–186.
Arbilly, M., & Lotem, A. (2017). Constructive anthropomorphism: A functional evolutionary approach to the study of human-like cognitive mechanisms in animals. *Proceedings of the Royal Society B: Biological Sciences, 284*(1865), 20171616.
Aristotele. (2008). *Le tre etiche* (A. Fermani, Ed.). Bompiani.
Ayres, P. (2008). *The aliveness of plants: The Darwins at the Dawn of plant science*. Pickering & Chatto.
Baker, E. N. (2017). The intelligence of plants and the problem of language: A Wittgensteinian approach. In M. Gagliano, J. C. Ryan, & P. Vieira (Eds.), *The language of plants: Science, philosophy, literature* (pp. 136–154). University of Minnesota Press.
Balding, M., & Williams, K. J. (2016). Plant blindness and the implications for plant conservation. *Conservation Biology, 30*(6), 1192–1199.
Baldwin, I. T., Halitschke, R., Paschold, A., Von Dahl, C. C., & Preston, C. A. (2006). Volatile signaling in plant-plant interactions «talking trees» in the genomics era. *Science, 311*(5762), 812–815.
Baluška, F. (2010). Recent surprising similarities between plant cells and neurons. *Plant Signaling & Behavior, 5*(2), 87–89.
Baluška, F., & Levin, M. (2016). On having no head: Cognition throughout biological systems. *Frontiers in Psychology, 902*.
Baluška, F., & Mancuso, S. (2009). Plant neurobiology: From sensory biology, via plant communication, to social plant behavior. *Cognitive Processing, 10*, 3–7.
Baluška, F., & Mancuso, S. (2021). Individuality, self and sociality of vascular plants. *Philosophical Transactions of the Royal Society B, 376*(1821), 20190760.
Baluška, F., Mancuso, S., Volkmann, D., & Barlow, P. (2009). The 'root-Brain'Hypothesis of Charles and Francis Darwin: Revival after more than 125 years. *Plant Signaling & Behavior, 4*(12), 1121–1127.
Baluška, F., & Reber, A. (2019). Sentience and consciousness in single cells: How the first minds emerged in unicellular species. *BioEssays, 41*(3), 1800229.
Baluška, F., Volkmann, D., & Menzel, D. (2005). Plant synapses: Actin-based domains for cell-to-cell communication. *Trends in Plant Science, 10*(3), 106–111.
Baluška, F., Volkmann, D., Hlavacka, A., Mancuso, S., & Barlow, P. W. (2006). Neurobiological view of plants and their body plan. In F. Baluška, S. Mancuso, & D. Volkmann (Eds.), *Communication in plants* (pp. 19–35). Springer.
Baluška, F., Yokawa, K., Mancuso, S., & Baverstock, K. (2016). Understanding of anesthesia–why consciousness is essential for life and not based on genes. *Communicative & integrative biology, 9*(6), e1238118.
Baluška, F., & Yokawa, K. (2021). Anaesthetics and plants: From sensory systems to cognition-based adaptive behaviour. *Protoplasma, 258*, 449–454.
Barlow, P. W. (2008). Reflections on 'plant neurobiology'. *Biosystems, 92*(2), 132–147.
Barrett, L. F. (2006). Valence is a basic building block of emotional life. *Journal of Research in Personality, 40*(1), 35–55.
Bateson, G. (1976). *Verso un'ecologia della mente*, Longo, G., & Trautteur, G. (Trans.). Adelphi (original work published 1972, *Steps to an ecology of mind: Collected essays in anthropology, psychiatry, evolution, and epistemology*. Chandler Publishing Company).
Bateson, G. (1979). *Mind and nature: A necessary Unity*. Dutton.
Bateson, M., Desire, S., Gartside, S. E., & Wright, G. A. (2011). Agitated honeybees exhibit pessimistic cognitive biases. *Current Biology, 21*(12), 1070–1073.
Bayne, T., Brainard, D., Byrne, R. W., Chittka, L., Clayton, N., Heyes, C., et al. (2019). What is cognition? *Current Biology, 29*(13), R608–R615.
Beer, C. G. (1980). Perspectives on animal behavior comparisons. In M. H. Bornstein (Ed.), *Comparative methodsin psychology* (pp. 17–64). Erlbaum.

Bekoff, M. (2000). Animal emotions: Exploring passionate natures: Current interdisciplinary research provides compelling evidence that many animals experience such emotions as joy, fear, love, despair, and grief—We are not alone. *Bioscience, 50*(10), 861–870.
Bellon, R. (2011). Inspiration in the harness of daily labor: Darwin, botany, and the triumph of evolution, 1859-1868. *Isis, 102*(3), 393–420.
Bennett, T. (2021). Plant–plant interactions. *Plant, Cell & Environment, 44*(4), 995–996.
Bianchi, M. (2021). *La vita ramificata.* Cognizione e comportamento nelle piante fra scienza e filosofia. Mimesis.
Bianchi, M. (2022). Networks and ramifications: Relational perspectives in plant cognition. *Rivista Internazionale di Filosofia e Psicologia, 13*(2), 157–168.
Bianchi, M. (2024). Plant cognition. Challenges and horizons in philosophical reflection. *Paradigmi*, (1), 21–46.
Bianchi, M. (2025). Oltre il Mondo Animale: i Darwin e la Botanica. In E. Gagliasso, S. Pollo & E. Severini (Eds.), *Origine e attualità dell'Origine dell'uomo. Scienza, Etica, Letteratura* (pp. 33–54). Edizioni ETS.
Bianchi, M., Avesani, S., Bonato, B., Dadda, M., Guerra, S., Ravazzolo, L., Simonetti, V., & Castiello, U. (2025a). Plant behavior: Theoretical and technological advances. *Current Opinion in Psychology, 64*, 102026.
Bianchi, M., Guerra, S., Bonato, B., Avesani, S., Ravazzolo, L., Simonetti, V., Dadda, M., & Castiello, U. (2025b). Temporal Dynamics With and Without a Nervous System: Plant Physiology, Communication, and Movement. *Cognitive Science, 49*(6), e70079.
Bianchi, M., & Castiello, U. (2023a). Il concetto di rappresentazione nelle scienze cognitive classiche e post-classiche: la sfida della cognizione vegetale. *Giornale Italiano di Psicologia, 50*(2), 349–385.
Bianchi, M., & Castiello, U. (2023b). Riflessioni corali nell'analisi interdisciplinare sul concetto di rappresentazione nella cognizione vegetale. *Giornale Italiano di Psicologia, 50*(2), 419–452.
Biedrzycki, M. L., Jilany, T. A., Dudley, S. A., & Bais, H. P. (2010). Root exudates mediate kin recognition in plants. *Communicative & Integrative Biology, 3*(1), 28–35.
Blumenbach, J. F. (1782). D. Joh. Friedr. Blumenbachs der Med. Prof. ord. zu Göttingen Handbuch der Naturgeschichte: Mit Kupfern. Dieterich.
Boakes, R. (1984). *From Darwin to Behaviourism: Psychology and the minds of Animals*. CUP Archive.
Boisseau, R. P., Vogel, D., & Dussutour, A. (2016). Habituation in non-neural organisms: Evidence from slime moulds. *Proceedings of the Royal Society B: Biological Sciences, 283*(1829), 20160446.
Bose, J. C. (1926). *The nervous mechanism of plants*. Longmans, Green & Co.
Bose, I., & Karmakar, R. (2003). Simple models of plant learning and memory. *Physica Scripta, 2003*(T106), 9.
Bonato, B., Simonetti, V., Bulgheroni, M., Wang, Q., Guerra, S., Quaggiotti, S., et al. (2023). Evidence of motor intentions in plants: A kinematical study. *Journal of Comparative Psychology, 137*(4), 228–237.
Boncinelli, E. (2020). *Il principio di Indeterminazione*. Il Mulino.
Bonfante, P., (2021). *Una pianta non è un'isola. Alla scoperta di un mondo visibile*. Il Mulino.
Borghini, A., & Casetta, E. (2013). Filosofia della biologia. *Carocci.*
Bouché, N., Lacombe, B., & Fromm, H. (2003). GABA signaling: A conserved and ubiquitous mechanism. *Trends in Cell Biology, 13*(12), 607–610.
Buiatti, M., & Longo, G. (2013). Randomness and multilevel interactions in biology. *Theory in Biosciences, 132*(3),139–158.
Boyko, A., & Kovalchuk, I. (2011). Genome instability and epigenetic modification—Heritable responses to environmental stress? *Current Opinion in Plant Biology, 14*(3), 260–266.
Brenner, E. D., Stahlberg, R., Mancuso, S., Vivanco, J., Baluška, F., & Van Volkenburgh, E. (2006). Plant neurobiology: An integrated view of plant signaling. *Trends in Plant Science, 11*(8), 413–419.

Bruni, D., Perconti, P., & Plebe, A. (2018). Anti-anthropomorphism and its limits. *Frontiers in Psychology, 9*, 2205.

Burghardt, G. M. (1991). Cognitive ethology and critical anthropomorphism: A snake with two heads and hognose snakes that play dead. In C. A. Ristau (Ed.), *Cognitive ethology: The minds of other animals: Essays in honor of Donald R. Griffin* (pp. 53–90). Lawrence Erlbaum Associates.

Cahill, J. F. (2015). Introduction to the Special Issue: Beyond traits: Integrating behaviour into plant ecology and biology. *AoB Plants, 7*, plv120.

Cahill, J. F., McNickle, G. G., Haag, J. J., Lamb, E. G., Nyanumba, S. M., & St. Clair, C. C. (2010). Plants integrate information about nutrients and neighbors. *Science, 328*(5986), 1657.

Calvo, P. (2007). The quest for cognition in plant neurobiology. *Plant Signaling & Behavior, 2*(4), 208–211.

Calvo, P., & Keijzer, F. (2009). Cognition in plants. In *Plant-environment interactions: From sensory plant biology to active plant behavior* (pp. 247–266). Springer.

Calvo, P., & Keijzer, F. (2011). Plants: Adaptive behavior, root brains and minimal cognition. *Adaptive Behavior, 19*(3), 155–171.

Calvo, P. (2016). The philosophy of plant neurobiology: A manifesto. *Synthese, 193*, 1323–1343.

Calvo, P., Baluška, F., & Sims, A. (2016). «feature detection» vs. «predictive coding» models of plant behavior. *Frontiers in Psychology, 7*, 1505.

Calvo, P. (2017). What is it like to be a plant? *Journal of Consciousness Studies, 24*(9–10), 205–227.

Calvo, P. (2018). Plantae. In *Encyclopedia of animal cognition and behavior* (pp. 1–8). Springer, Cham.

Calvo, P., & Friston, K. (2017). Predicting green: Really radical (plant) predictive processing. *Journal of the Royal Society Interface, 14*(131), 20170096.

Calvo, P., Sahi, V. P., & Trewavas, A. (2017). Are plants sentient? *Plant, Cell & Environment, 40*(11), 2858–2869.

Calvo, P., Gagliano, M., Souza, G. M., & Trewavas, A. (2020). Plants are intelligent, here's how. *Annals of Botany, 125*(1), 11–28.

Calvo, P., & Trewavas, A. (2020). Physiology and the (neuro) biology of plant behavior: A farewell to arms. *Trends in Plant Science, 25*(3), 214–216.

Calvo, P., & Segundo-Ortin, M. (2023). Plant sentience revisited: Sifting through the thicket of perspectives. Animal Sentience, 8(33), 32.

Carruthers, P. (2013). Animal minds are real,(distinctively) human minds are not. *American Philosophical Quarterly, 50*(3), 233–248.

Castiello, U. (2019). *La mente delle piante. Introduzione alla psicologia vegetale*. Il Mulino.

Castiello, U., & Guerra, S. (2020). La psicologia vegetale: alle radici della cognizione. *Giornale Italiano di Psicologia, 47*(3–4), 683–708.

Castiello, U. (2021). (Re) claiming plants in comparative psychology. *Journal of Comparative Psychology, 135*(1), 127.

Ceccarini, F., Guerra, S., Peressotti, A., Peressotti, F., Bulgheroni, M., Baccinelli, W., & Castiello, U. (2020). Speed-accuracy trade-off in plants. *Psychonomic Bulletin & Review, 27*, 966–973.

Ceccarini, F., Guerra, S., Peressotti, A., Peressotti, F., Bulgheroni, M., Baccinelli, W., et al. (2021). On-line control of movement in plants. *Biochemical and Biophysical Research Communications, 564*, 86–91.

Cellucci, C. (2008). *Perché ancora la filosofia*. Laterza.

Ceruti, M. (1996). *Constraints and possibilities*. Gordon and Breach.

Ceruti, M. (1994). *La danza che crea. Evoluzione e cognizione nell'epistemologia genetica (F. Varela preface)*. Feltrinelli.

Chandler, J. L., & Van de Vijver, G. (2000). *Emergent organizations and their dynamics*. New York Academy of Sciences.

Chemero, A. (2009). *Radical embodied cognitive science*. MIT Press.

Churchland, P. M. (1981). Eliminative materialism and the propositional attitudes. *The Journal of Philosophy, 78*(2), 67–90.

Cittadino, E. (2009). Botany. In P. J. Bowler & J. V. Pickstone (Eds.), *The Cambridge history of science: The modern biological and earth sciences* (Vol. VI, pp. 225–242). Cambridge University Press.
Clark, A. (1997). *Being there: Putting brain, body, and world together again*. MIT press.
Clark, A. (2008). *Supersizing the mind: Embodiment, action, and cognitive extension*. Oxford University Press.
Clark, A., & Chalmers, D. (1998). The extended mind. *Analysis, 58*(1), 7–19.
Clatterbuck, H. (2016). Darwin, Hume, Morgan, and the verae causae of psychology. *Studies in History and Philosophy of Science Part C: Studies in History and Philosophy of Biological and Biomedical Sciences, 60*, 1–14.
Colaço, D. (2022). Why studying plant cognition is valuable, even if plants aren't cognitive. *Synthese, 200*(6), 453.
Continenza, B. (2013). Confini Aperti. Il rapporto interno/esterno in biologia. In B. Continenza, E. Gagliasso, & F. Sterpetti (Eds.), *Sul confine e oltre* (pp. 126–139). Franco Angeli.
Cosentino, E., & Ferretti, F. (2011). Il Linguaggio. In Marraffa, M., & Paternoster, A. (Eds.), *Scienze cognitive*. Un'introduzione. Carocci.
Crisp, P. A., Ganguly, D., Eichten, S. R., Borevitz, J. O., & Pogson, B. J. (2016). Reconsidering plant memory: Intersections between stress recovery, RNA turnover, and epigenetics. *Science Advances, 2*(2), e1501340.
Crump, A., Browning, H., Schnell, A., Burn, C., & Birch, J. (2022). Sentience in decapod crustaceans: A general framework and review of the evidence. *Animal Sentience, 7*(32), 1.
Dacey, M. (2017). Anthropomorphism as cognitive bias. *Philosophy of Science, 84*(5), 1152–1164.
Darwin, C. R. (1859). *Origin of species by means of natural selection, or the preservation of favoured races in the struggle for life*. John Murray.
Darwin, C. R. (1862). *On the various contrivances by which British and foreign orchids are fertilised by insects, and on the good effects of intercrossing*. John Murray.
Darwin, C. R. (1871). *The descent of man, and selection in relation to sex*. John Murray.
Darwin, C. R. (1875a). *Insectivorous plants*. John Murray.
Darwin, C. R. (1875b). *The Movements and Habits of Climbing Plants* (2nd ed.). John Murray. (Originally published as an article in the *Journal of the Linnean Society* in 1865).
Darwin, C. R. (1877). *The different forms of flowers on plants of the same species*. John Murray.
Darwin, C. R., & Darwin, F. (1880). *The power of movement in plants*. John Murray.
De Jong, H. L. (2002). Levels of explanation in biological psychology. *Philosophical Psychology, 15*(4), 441–462.
De la Fuente, I. M. (2015). Elements of the cellular metabolic structure. *Frontiers in Molecular Biosciences, 2*, 16.
De Waal, F. B. M. (1997). The chimpanzee's service economy: Food for grooming. *Evolution and Human Behavior, 18*(6), 375–386.
De Waal, F. B. M. (1999). Anthropomorphism and anthropodenial: Consistency in our thinking about humans and other animals. *Philosophical Topics, 27*(1), 255–280.
De Waal, F. (2001). *The Ape and the Sushi Master: Cultural Reflections by a Primatologist*. Basic Books.
De Waal, F. B. M. (2016). *Are we smart enough to know how smart animals are?* Granta Books.
Debono, M. W., & Souza, G. M. (2019). Plants as electromic plastic interfaces: A mesological approach. *Progress in Biophysics and Molecular Biology, 146*, 123–133.
Dener, E., Kacelnik, A., & Shemesh, H. (2016). Pea plants show risk sensitivity. *Current Biology, 26*(13), 1763–1767.
Dennett, D. C. (1993). *Consciousness explained*. Penguin Books.
Demey, M. L., Mishra, R. C., & Van Der Straeten, D. (2023). Sound perception in plants: from ecologicalsignificance to molecular understanding. *Trends in Plant Science, 28*(7), 825–840.
Dewsbury, D. A. (1984). *Comparative psychology in the twentieth century*. Hutchinson Ross Publishing Company.
Dexter, J. P., Prabakaran, S., & Gunawardena, J. (2019). A complex hierarchy of avoidance behaviors in a single-cell eukaryote. *Current Biology, 29*(24), 4323–4329.

Di Francesco, M., Marraffa, M., & Tomasetta, A. (2017). *Filosofia della Mente: Corpo, Coscienza, Pensiero*. Carocci.

Di Francesco, M., & Piredda, G. (2012). *La mente estesa. Dove finisce la mente e comincia il resto del mondo?* Mondadori Università.

Di Paolo, E., Buhrmann, T., & Barandiaran, X. (2017). *Sensorimotor life: An enactive proposal*. Oxford University Press.

Dicke, M., & Hilker, M. (2003). Induced plant defences: From molecular biology to evolutionary ecology. *Basic and Applied Ecology, 4*(1), 3–14.

Dominey, P. F., Prescott, T. J., Bohg, J., Engel, A. K., Gallagher, S., Heed, T., et al. (2016). Implications of action-oriented paradigm shifts in cognitive science. In A. K. Engel, K. J. Friston, & D. Kragic (Eds.), *The pragmatic turn—Toward action-oriented views in cognitive science, 18* (pp. 333–356). MIT Press.

Dröscher, A. (2008) *Biologia, Storia e Concetti*. Carocci.

Dukas, R. (2004). Evolutionary biology of animal cognition. *Annual Review of Ecology, Evolution, and Systematics, 35*, 347–374.

Etxebeste, O., & Espeso, E. A. (2016). Neurons show the path: Tip-to-nucleus communication in filamentous fungal development and pathogenesis. *FEMS Microbiology Reviews, 40*(5), 610–624.

Farina, A. (2021). *Ecosemiotic landscape: A novel perspective for the toolbox of environmental humanities*. Cambridge University Press.

Farina, A., & Belgrano, A. (2006). The eco-field hypothesis: Toward a cognitive landscape. *Landscape Ecology, 21*, 5–17.

Favre, P., & Agosti, R. D. (2007). Voltage-dependent action potentials in Arabidopsis thaliana. *Physiologia Plantarum, 131*(2), 263–272.

Fay, M. F., & Chase, M. W. (2009). Orchid biology: From Linnaeus via Darwin to the 21st century. *Annals of Botany, 104*(3), 359–364.

Fechner, G. T. (1848). *Nanna, oder, Über das Seelenleben der Pflanzen*. Leipzig, L. Voss.

Felle, H. H., & Zimmermann, M. R. (2007). Systemic signalling in barley through action potentials. *Planta, 226*, 203–214.

Feyerabend, P. (1975). 'Science'. The myth and its role in society. *Inquiry, 18*(2), 167–181.

Fitzpatrick, S. (2008). Doing away with Morgan's canon. *Mind & Language, 23*(2), 224–246.

Frazier, P. A. (2021). On the possibility of plant consciousness: A view from ecointeractivism. *Mind and Matter, 19*(2), 229–259.

Frezza, G., & Gagliasso, E. (2014). Fare metafore e fare scienza. *Aisthesis. Pratiche, Linguaggi e Saperi dell'Estetico, 7*(2), 25–42.

Friston, K. J. (2005). A theory of cortical responses. *Philosophical Transactions of the Royal Society B: Biological Sciences, 360*(1456), 815–836.

Fukano, Y., & Yamawo, A. (2015). Self-discrimination in the tendrils of the vine Cayratia Japonica is mediated by physiological connection. *Proceedings of the Royal Society B: Biological Sciences, 282*, 20151379.

Fuller, H. J. (1934). Plant behavior. *The Journal of General Psychology, 11*(2), 379–394.

Gagliano, M. (2017). The mind of plants: Thinking the unthinkable. *Communicative and Integrative Biology, 10*(2), e1288333.

Gagliano, M., Renton, M., Depczynski, M., & Mancuso, S. (2014). Experience teaches plants to learn faster and forget slower in environments where it matters. *Oecologia, 175*(1), 63–72.

Gagliano, M., Ryan, J. C., & Vieira, P. (Eds.). (2017). *The language of plants. Science, philosophy, literature*. University of Minnesota Press.

Gagliasso, E. (2001). *Verso un'epistemologia del mondo vivente, evoluzione e biodiversità tra legge e narrazione*. Edizioni Guerini.

Gagliasso, E. (2012). *Ragionare sui sistemi tra bolscevismo e cibernetica*. In G. Rispoli, *Dall'Empiriomonismo alla Tectologia. Organizzazione, Complessità e Approccio Sistemico nel Pensiero di Aleksandr Bogdanov*. Aracne.

Gagliasso, E. (2013). Ambiente. In F. Michelini & J. Davies (Eds.), *Frontiere della biologia*. Prospettive filosofiche sulle scienze della vita.

Gagliasso, E., Morganti, F., & Passariello, A. (2016). *Percorsi evolutivi.* Lezioni di filosofia della biologia. Franco Angeli.
Gallagher, S. (2017). *Enactivist interventions: Rethinking the mind.* Oxford University Press.
Galviz, Y. C., Ribeiro, R. V., & Souza, G. M. (2020). Yes, plants do have memory. *Theoretical and Experimental Plant Physiology, 32*(3), 195–202.
Garcia Rodriguez, A., & Calvo Garzon, P. (2010). Is cognition a matter of representations? Emulation, teleology, and time-keeping in biological systems. *Adaptive Behavior, 18*(5), 400–415.
Gazzaniga, M. S. (2019). *La coscienza è un istinto. Il legame misterioso tra il cervello e la mente* (Trans. It.). Cortina Editore (Original work published 2018, The consciousness instinct: Unraveling the mystery of how the brain makes the mind. Farrar, Straus and Giroux).
Giaculli, F. (2019). Interpretare e raffigurare la natura: Charles Darwin e la vita delle piante. *Filosofia, 64*, 79–92.
Gibbs, K. A., Urbanowski, M. L., & Greenberg, E. P. (2008). Genetic determinants of self identity and social recognition in bacteria. *Science, 321*(5886), 256–259.
Gibson, J. J. (1950). *The perception of the visual world.* Houghton Mifflin.
Gibson, J. J. (1966). *The senses considered as perceptual systems.* Houghton Mifflin.
Gibson, J. J. (1979). *The ecological approach to visual perception.* Lawrence Erlbaum Associates.
Gigerenzer, G. E., Hertwig, R. E., & Pachur, T. E. (2011). *Heuristics: The foundations of adaptive behavior.* Oxford University Press.
Gilbert, S. F. (2001). Ecological developmental biology: Developmental biology meets the real world. *Developmental Biology, 233*, 1–12.
Gilroy, S., Białasek, M., Suzuki, N., Górecka, M., Devireddy, A. R., Karpiński, S., & Mittler, R. (2016). ROS, calcium, and electric signals: Key mediators of rapid systemic signaling in plants. *Plant Physiology, 171*(3), 1606–1615.
Gilroy, S., & Trewavas, T. (2023). Agency, teleonomy and signal transduction in plant systems. *Biological Journal of the Linnean Society, 139*(4), 514–529.
Ginsburg, S., & Jablonka, E. (2009). Epigenetic learning in non-neural organisms. *Journal of Biosciences, 34*(4), 633–646.
Ginsburg, S., & Jablonka, E. (2021). Evolutionary transitions in learning and cognition. *Philosophical Transactions of the Royal Society B, 376*(1821), 20190766.
Giovannetti, M. (2009). *Reti di interazioni nel funzionamento degli ecosistemi.* In *Il futuro di Darwin, L'ecosistema*, L. Calabi (Ed.). UTET.
Giovannetti, M. (2014). *Microbi domestici e addomesticati. Come l'uso consapevole e inconsapevole dei microrganismi ha cambiato la nostra vita.* Edizioni ETS.
Giurfa, M. (2015). Learning and cognition in insects. *Wiley Interdisciplinary Reviews: Cognitive Science, 6*(4), 383–395.
Glacken, C. J. (1970). Man against nature: An outmoded concept. *The Environmental Crisis*, 127–142.
Glock, H. J. (2019). Agency, intelligence and reasons in animals. *Philosophy, 94*(4), 645–671.
Godfrey-Smith, P. (1994). Spencer and Dewey on life and mind. In R. Brooks & P. Maes (Eds.), *Artificial life IV* (pp. 80–89). MIT Press.
Godfrey-Smith, P. (2002). Environmental complexity and the evolution of cognition. *The Evolution of Intelligence*, 233–249.
Godfrey-Smith, P. (2016). *Other minds: The octopus, the sea, and the deep origins of consciousness.* Farrar, Straus and Giroux.
Godfrey-Smith, P. (2021). *Metazoa* (I. C. Blum, Trans.). Adelphi. (Original work published 2020, Metazoa: Animal life and the birth of the mind. Farrar, Straus and Giroux).
Gomila, A., & Calvo, P. (Eds.). (2008). *Handbook of cognitive science: An embodied approach.* Elsevier.
Grémiaux, A., Yokawa, K., Mancuso, S., & Baluška, F. (2014). Plant anesthesia supports similarities between animals and plants. *Plant Signaling & Behavior, 9*(1), e27886.
Griffin, D. R. (1958). *Listening in the dark: The acoustic orientation of bats and men.* Yale University Press.

Groszmann, M., Greaves, I. K., Albert, N., Fujimoto, R., Helliwell, C. A., Dennis, E. S., & Peacock, W. J. (2011). Epigenetics in plants—vernalisation and hybrid vigour. *Biochimica et Biophysica Acta (BBA)-Gene Regulatory Mechanisms, 1809*(8), 427–437.

Griffin, D. R. (1984). Animal thinking. *American Scientist, 72*(5), 456–464.

Guerra, S., Peressotti, A., Peressotti, F., Bulgheroni, M., Baccinelli, W., et al. (2019). Flexible control of movement in plants. *Scientific Reports, 9*(1), 16570.

Gutnick, T., Neef, A., Cherninskyi, A., Ziadi-Künzli, F., Di Cosmo, A., Lipp, H. P., & Kuba, M. J. (2023). Recording electrical activity from the brain of behaving octopus. *Current Biology, 33*(6), 1171–1178.

Hebb, D. O. (1946). Emotion in man and animal: An analysis of the intuitive processes of recognition. *Psychological Review, 53*(2), 88–106.

Heil, M., & Karban, R. (2010). Explaining evolution of plant communication by airborne signals. *Trends in Ecology & Evolution, 25*(3), 137–144.

Henriksson, N., Marshall, J., Högberg, M. N., Högberg, P., Polle, A., Franklin, O., & Näsholm, T. (2023). Re-examining the evidence for the mother tree hypothesis—Resource sharing among trees via ectomycorrhizal networks. *New Phytologist, 239*(1), 19–28.

Heras-Escribano, M., & Calvo, P. (2019). The philosophy of plant neurobiology. In S. Robins, J. Symons, & P. Calvo (Eds.), *The Routledge companion of philosophy of psychology* (pp. 529–548). Routledge.

Heisenberg, W. (1958). *Physics and philosophy: The revolution in modern science*. Harper & Brothers.

Hiernaux, Q. (2021). Differentiating behaviour, cognition, and consciousness in plants. *Journal of Consciousness Studies, 28*(1–2), 106–135.

Hodge, A., Berta, G., Doussan, C., Merchan, F., & Crespi, M. (2009). *Plant root growth, architecture and function*, 153–187.

Houle, D., Govindaraju, D. R., & Omholt, S. (2010). Phenomics: The next challenge. *Nature Reviews Genetics, 11*(12), 855–866.

Hoyle, F., & Wickramasinghe, N. C. (1976). Primitive grain clumps and organic compounds in carbonaceous chondrites. *Nature, 264*(5581), 45–46.

Hoyle, F., & Wickramasinghe, C. (1981). *Comets–A vehicle for panspermia*. Comets and the Origin of Life, 227–239.

Huber, A. E., & Bauerle, T. L. (2016). Long-distance plant signaling pathways in response to multiple stressors: The gap in knowledge. *Journal of Experimental Botany, 67*(7), 2063–2079.

Humphrey, N. (1976). The social function of intellect. In P. P. G. Bateson & R. A. Hinde (Eds.), *Growing points in ethology* (pp. 303–317). Cambridge University Press.

Hutchins, E. (2010). Cognitive ecology. *Topics in Cognitive Science, 2*(4), 705–715.

Hutto, D. D., & Myin, E. (2013). *Radicalizing Enactivism*. The MIT Press.

Ippoliti, E., & Cellucci, C. (2016). *Logica*. Egea.

Isnard, S., & Silk, W. K. (2009). Moving with climbing plants from Charles Darwi's time into the 21st century. *American Journal of Botany, 96*(7), 1205–1221.

Jacob, E. B., Becker, I., Shapira, Y., & Levine, H. (2004). Bacterial linguistic communication and social intelligence. *Trends in Microbiology, 12*(8), 366–372.

Jonas, H. (1966). *The phenomenon of life: Toward a philosophical biology*. Harper and Row.

Kahneman, D. (2012). *Pensieri lenti e veloci*. Edizioni Mondadori.

Kaplan, D. M. (2012). How to demarcate the boundaries of cognition. *Biology and Philosophy, 27*, 545–570.

Karban, R. (2008). Plant behaviour and communication. *Ecology Letters, 11*(7), 727–739.

Karban, R. (2015). *Plant sensing and communication*. University of Chicago Press.

Karban, R. (2017). Plant communication increases heterogeneity in plant phenotypes and herbivore movement. *Functional Ecology, 31*(5), 990–991.

Karban, R., & Shiojiri, K. (2009). Self-recognition affects plant communication and defense. *Ecology Letters, 12*(6), 502–506.

Kauffman, S. A. (1995). *At home in the universe: The search for laws of self-organization and complexity*. Oxford University Press.

Keijzer, F. (2001). *Representation and behavior*. The MIT Press.
Keijzer, F. (2021). Demarcating cognition: The cognitive life sciences. *Synthese, 1*, 137–157.
Kennedy, J. S. (1992). *The new anthropomorphism*. Cambridge University Press.
Khait, I., Lewin-Epstein, O., Sharon, R., Saban, K., Goldstein, R., Anikster, Y., et al. (2023). Sounds emitted by plants under stress are airborne and informative. *Cell, 186*(7), 1328–1336.
Klein, C., & Barron, A. B. (2016). Insects have the capacity for subjective experience. *Animal Sentience, 1*(9), 1.
Knutson, B., Burgdorf, J., & Panksepp, J. (1998). Anticipation of play elicits high-frequency ultrasonic vocalizations in young rats. *Journal of Comparative Psychology, 112*(1), 65–73.
Knuuttila, T. (2011). Modelling and representing: An artefactual approach to model-based representation. *Studies in History and Philosophy of Science Part A, 42*(2), 262–271.
Kuhn, T. S. (1962). *The structure of scientific revolutions*. The University of Chicago Press.
Kumar, A., Memo, M., & Mastinu, A. (2020). Plant behaviour: An evolutionary response to the environment? *Plant Biology, 22*(6), 961–970.
Kutschera, U., & Briggs, W. R. (2009). From Charles Darwin's botanical country-house studies to modern plant biology. *Plant Biology, 11*(6), 785–795.
La Vergata, A. (1988). Charles Darwin. In P. Rossi (Ed.), *Storia della Scienza. La Scienza Classica: da Faraday a Darwin* (Vol. IV, pp. 205–238). UTET.
LeDoux, J. (2020). *Lunga storia di noi stessi: Come il cervello è diventato cosciente* (G. Guerriero, Trans.). Raffaello Cortina Editore (Original work published 2019, *The deep history of ourselves: The four-billion-year story of how we got conscious brains*. Viking).
Lewis, D. (1972). Psychophysical and theoretical identifications. In N. Block (Ed.), *Readings in the philosophy of psychology* (Vol. 1). Harvard University Press.
Linson, A., & Calvo, P. (2020). Zoocentrism in the weeds? Cultivating plant models for cognitive yield. *Biology & Philosophy, 35*(5), 1–27.
Locker, A. (1981). Meta-theoretical presuppositions for autopoiesis. Self-reference and 'autopoiesis'. *Autopoiesis: A Theory of Living Organization, 3*, 209–233.
Lorenz, K. Z. (1952). *King Solomon's ring* (M. K. Wilson, Trans.). Methuen.
Lucas, W. J., Groover, A., Lichtenberger, R., Furuta, K., Yadav, S. R., Helariutta, Y., et al. (2013). The plant vascular system: Evolution, development and functions f. *Journal of Integrative Plant Biology, 55*(4), 294–388.
Lyon, P. (2006). The biogenic approach to cognition. *Cognitive Processing, 7*(1), 11–29.
Lyon, P. (2015). The cognitive cell: Bacterial behavior reconsidered. *Frontiers in Microbiology, 6*, 264.
Lyon, P. (2020). Of what is "minimal cognition" the half-baked version? *Adaptive Behavior, 28*(6), 407–424.
Lyon, P., Keijzer, F., Arendt, D., & Levin, M. (2021). Reframing cognition: Getting down to biological basics. *Philosophical Transactions of the Royal Society B, 376*(1820), 20190750.
Maher, C. (2017). *Plant minds: A philosophical defense*. Routledge.
Maher, C. (2020). Experiment rather than define. *Trends in Plant Science, 25*(3), 213–214.
Mallatt, J., Blatt, M. R., Draguhn, A., Robinson, D. G., & Taiz, L. (2021). Debunking a myth: Plant consciousness. *Protoplasma, 258*(3), 459–476.
Mancuso, S. (2017a). *Plant revolution*. Le piante hanno già inventato il nostro futuro. Giunti.
Mancuso, S. (2017b). *Botanica*. Viaggio nell'universo vegetale. Aboca.
Mancuso, S. (2018). *L'incredibile viaggio delle piante*. Gius. Laterza & Figli Spa.
Mancuso, S., & Viola, A. (2015). *Verde brillante*. Sensibilità e intelligenza del mondo vegetale. Giunti.
Marder, M. (2017). Voiceless speech, or vegetal logos without logos. *Qui Parle: Critical Humanities and Social Sciences, 26*(2), 367–369.
Margolis, J. (1996). *Life without principles: Reconciling theory and practice*. Blackwell.
Margulis, L. (2001). The conscious cell. *Annals of the New York Academy of Sciences, 929*(1), 55–70.
Martin, L., Leblanc-Fournier, N., Julien, J. L., Moulia, B., & Coutand, C. (2010). Acclimation kinetics of physiological and molecular responses of plants to multiple mechanical loadings. *Journal of Experimental Botany, 61*, 2403–2412.

Maturana, H. R., & Varela, F. J. (1980). *Autopoiesis and cognition: The realization of the living*. D. Reidel Publishing Company.
Mauseth, J. D. (2017). *Botanica, fondamenti di biologia delle piante*. Idelson-Gnocchi.
Mayr, E. (1982). *The growth of biological thought: Diversity, evolution, and inheritance*. Harvard University Press.
Mazzolai, B. (2021). *Il futuro raccontato dalle piante. Cosa possiamo imparare dal regno vegetale e dal suo percorso sul pianeta*. Longanesi.
McCauley, R. N., & Bechtel, W. (2001). Explanatory pluralism and heuristic identity theory. *Theory & Psychology, 11*(6), 736–760.
Mettke-Hofmann, C. (2014). Cognitive ecology: Ecological factors, life-styles, and cognition. *Wiley Interdisciplinary Reviews: Cognitive Science, 5*(3), 345–360.
Metzinger, T. (2017). The problem of mental action: Predictive control without sensory sheets. In T. Metzinger & W. Wiese (Eds.), *Philosophy and predictive processing* (pp. 1–26). MIND Group.
Michaels, C. F., & Carello, C. (1981). *Direct perception*. Prentice-Hall.
Michmizos, D., & Hilioti, Z. (2019). A roadmap towards a functional paradigm for learning & memory in plants. *Journal of Plant Physiology, 232*, 209–215.
Miguel-Tomé, S., & Llinás, R. R. (2021). Broadening the definition of a nervous system to better understand the evolution of plants and animals. *Plant Signaling & Behavior, 16*(10), 1927562.
Mikhalevich, I., & Powell, R. (2020). Minds without spines: Evolutionarily inclusive animal ethics. *Animal Sentience, 5*(29), 1.
Monod, J. (2017). *Il caso e la necessità: saggio sulla filosofia naturale della biologia contemporanea*, A. Busi (Trans.). Mondadori (original work published 1970, *Le hasard et la nécessité: Essai sur la philosophie naturelle de la biologie moderne*. Éditions du Seuil).
Morabito, C. (2020). *Il motore della mente: il movimento nella storia delle scienze cognitive*. Laterza.
Morand-Ferron, J., Cole, E. F., & Quinn, J. L. (2016). Studying the evolutionary ecology of cognition in the wild: A review of practical and conceptual challenges. *Biological Reviews, 91*(2), 367–389.
Morgan, C. L. (1894). *An introduction to comparative psychology*. Walter Scott.
Morgan, C. L. (1903). *An introduction to comparative psychology* (New Ed., Revised). Walter Scott Publishing.
Morin, E., (2004), *Il metodo 2. La vita della vita*, G. Bocchi & A. Serra (Trans.). Cortina (original work published 1980, *La vie de la vie*).
Müller, B. S., di Primio, F., & Lengeler, J. W. (2001). Contributions of minimal cognition to flexibility. In *Proceedings of the 5th World Multi-Conference on Systemics, Cybernetics and Informatics* (Vol. 15, pp. 93–98).
Murphy, G. P., & Dudley, S. A. (2009). Kin recognition: Competition and cooperation in impatiens (Balsaminaceae). *American Journal of Botany, 96*(11), 1990–1996.
Nagel, T. (1974). What is it like to be a bat. *The Philosophical Review, 83*(4), 435–450.
Nakagaki, T., Yamada, H., & Tóth, Á. (2000). Maze-solving by an amoeboid organism. *Nature, 407*(6803), 470–470.
Nasrallah, J. B. (2002). Recognition and rejection of self in plant reproduction. *Science, 296*(5566), 305–308.
Neisser, U. (1967). *Cognitive psychology*. Prentice Hall.
Nicholson, D. J., & Dupré, J. (Eds.). (2018). *Everything flows: Towards a processual philosophy of biology*. Oxford University Press.
Nick, P. (2021). Intelligence without neurons: A Turing test for plants? *Protoplasma, 258*(3), 455–458.
Novoplansky, A. (2016). Future perception in plants. In R. Dillmann, Y. Nakamura, & D. Vernon (Eds.), *Anticipation across disciplines* (pp. 57–70). Springer.
Nucara, L. (2014). *La filosofia di Humberto Maturana*. Le Lettere.
Oborny, B. (2019). The plant body as a network of semi-autonomous agents: A review. *Philosophical Transactions of the Royal Society B, 374*(1774), 20180371.

Panksepp, J., & Burgdorf, J. (2003). "Laughing" rats and the evolutionary antecedents of human joy? *Physiology & Behavior, 79*(3), 533–547.
Parise, A. G., Gagliano, M., & Souza, G. M. (2020). Extended cognition in plants: Is it possible? *Plant Signaling & Behavior, 15*(2), 1710661.
Parniske, M. (2008). Arbuscular mycorrhiza: The mother of plant root endosymbioses. *Nature Reviews Microbiology, 6*(10), 763–775.
Parsley, K. M. (2020). Plant awareness disparity: A case for renaming plant blindness. *Plants, People, Planet, 2*(6), 598–601.
Pellegrino, G., & Di Paola, M. (2019). *Etica e Politica delle piante*. DeriveApprodi.
Perconti, P. (2017). Filosofia della mente. *Il Mulino.*
Pfeifer, R., & Bongard, J. (2006). *How the body shapes the way we think: A new view of intelligence*. MIT press.
Piazza, T. (2017). Che cos'è la conoscenza. *Carocci.*
Pievani, T. (2005). *Introduzione alla filosofia della biologia*. Laterza.
Pievani, T. (2017). *La teoria dell'evoluzione. Attualità di una rivoluzione scientifica*. Il Mulino.
Poincaré, H. (1914). *Science and Method* (F. Maitland, Trans., B. Russell, Preface). T. Nelson (Original work 1908, *Science et méthode*. Ernest Flammarion).
Pollo, S. (2019). Il sentimentalismo etico e il mondo vegetale. In G. Pellegrino & M. Di Paola (Eds.), *Etica e politica delle piante*. DeriveApprodi.
Povinelli, D. J., & Bering, J. M. (2002). The mentality of apes revisited. *Current Directions in Psychological Science, 11*(4), 115–119.
Preuss, T. M., & Wise, S. P. (2022). Evolution of prefrontal cortex. *Neuropsychopharmacology, 47*(1), 3–19.
Prindle, A., Liu, J., Asally, M., Ly, S., Garcia-Ojalvo, J., & Süel, G. M. (2015). Ion channels enable electrical communication in bacterial communities. *Nature, 527*(7576), 59–63.
Raja, V., Silva, P. L., Holghoomi, R., & Calvo, P. (2020). The dynamics of plant nutation. *Scientific Reports, 10*(1), 1–13.
Ramsey, W. M. (2007). *Representation reconsidered.* Cambridge University Press.
Rashevsky, N. (1954). Topology and life: In search of general mathematical principles in biology and sociology. *The Bulletin of Mathematical Biophysics, 16*, 317–348.
Ray, S. K., Valentini, G., Shah, P., Haque, A., Reid, C. R., Weber, G. F., & Garnier, S. (2019). Information transfer during food choice in the slime mold Physarum polycephalum. *Frontiers in Ecology and Evolution, 7*, 430494.
Razeto-Barry, P. (2012). Autopoiesis 40 years later. A review and a reformulation. *Origins of Life and Evolution of Biospheres, 42*(6), 543–567.
Read, D. J. (1991). Mycorrhizas in ecosystems. *Experientia, 47*, 376–391.
Reid, C. R., & Latty, T. (2016). Collective behaviour and swarm intelligence in slime moulds. *FEMS Microbiology Reviews, 40*(6), 798–806.
Reid, C. R., Latty, T., Dussutour, A., & Beekman, M. (2012). Slime mold uses an externalized spatial "memory" to navigate in complex environments. *Proceedings of the National Academy of Sciences, 109*(43), 17490–17494.
Reid, C. R., Garnier, S., Beekman, M., & Latty, T. (2015). Information integration and multiattribute decision makingin non-neuronal organisms. *Animal Behaviour, 100*, 44–50.
Repici, L. (2015). *Nature Silenziose. Le piante nel pensiero ellenistico e romano*. Il Mulino.
Rispoli, G. (2012). *Dall'empiriomonismo alla tectologia: organizzazione, complessità e approccio sistemico nel pensiero di Aleksandr Bogdanov*. Aracne.
Robinson, D. G., Ammer, C., Polle, A., Bauhus, J., Aloni, R., Annighöfer, P., et al. (2024). Mother trees, altruistic fungi, and the perils of plant personification. *Trends in Plant Science, 29*(1), 20–31.
Robinson, D. G., Draguhn, A., & Taiz, L. (2020). Plant "intelligence" changes nothing. *EMBO Reports, 21*(5), e50395.
Rovelli, C. (2014). *La realtà non è come ci appare. La struttura elementare delle cose*. Raffaello Cortina. (English translation published in 2016 by S. Carnell & E. Segre, *Reality is Not What It Seems: The Journey to Quantum Gravity*, Riverhead Books).

Rowlands, M. J. (2010). *The new science of the mind: From extended mind to embodied phenomenology*. Mit Press.
Schlicht, T. (2018). Does separating intentionality from mental representation imply radical enactivism? *Frontiers in Psychology, 9*, 1497.
Schnell, A. K., & Clayton, N. S. (2021). Cephalopods: Ambassadors for rethinking cognition. *Biochemical and Biophysical Research Communications, 564*, 27–36.
Schrödinger, E. (1944). *What is life? The physical aspect of the living cell.* Cambridge University Press.
Segundo-Ortin, M. (2020). Agency from a radical embodied standpoint: An ecological-enactive proposal. *Frontiers in Psychology, 11*, 1319.
Segundo-Ortin, M., & Calvo, P. (2019). Are plants cognitive? A reply to Adams. *Studies in History and Philosophy of Science A, 73*, 64–71.
Segundo-Ortin, M., & Calvo, P. (2022). Consciousness and cognition in plants. *Wiley Interdisciplinary Reviews: Cognitive Science, 13*(2), e1578.
Segundo-Ortin, M., & Calvo, P. (2023). Plant sentience? Between romanticism and denial: Science. *Animal Sentience, 8*(33), 1.
Semchenko, M. S., Lepik, S., & A. (2014). Plant root exudates mediate neighbour recognition and trigger complex behavioural changes. *New Phytologist, 204*(3), 631–637.
Shemesh, H., Ovadia, O., & Novoplansky, A. (2010). Anticipating future conditions via trajectory sensitivity. *Plant Signaling & Behavior, 5*(11), 1501–1503.
Shettleworth, S. J. (1998). *Cognition, evolution, and behavior*. Oxford University Press.
Shettleworth, S. J. (2010). Clever animals and killjoy explanations in comparative psychology. *Trends in Cognitive Sciences, 14*(11), 477–481.
Simard, S. W. (2009). The foundational role of mycorrhizal networks in self-organization of interior Douglas-fir forests. *Forest Ecology and Management, 258*, S95–S107.
Simard, S. W. (2018). Mycorrhizal networks facilitate tree communication, learning, and memory. In F. Baluška, M. Gagliano, & G. Witzany (Eds.), *Memory and learning in plants* (pp. 91–213). Springer.
Simard, S. W., Perry, D. A., Jones, M. D., Myrold, D. D., Durall, D. M., & Molina, R. (1997). Net transfer of carbon between ectomycorrhizal tree species in the field. *Nature, 388*(6642), 579–582.
Smith-Ferguson, J., & Beekman, M. (2020). Who needs a brain? Slime moulds, behavioural ecology and minimal cognition. *Adaptive Behavior, 28*(6), 465–478.
Sober, E. (1998). Morgan's canon. In D. D. Cummins & C. Allen (Eds.), *The evolution of mind* (pp. 224–242). Oxford University Press.
Sober, E. (2005). *Comparative psychology meets evolutionary biology* (p. 85). New perspectives on anthropomorphism.
Sober, E. (2012). Anthropomorphism, parsimony, and common ancestry. *Mind & Language, 27*(3), 229–238.
Souza, G. M., Ferreira, A. S., Saraiva, G. F., & Toledo, G. R. (2017). Plant «electrome» can be pushed toward a self-organized critical state by external cues: Evidences from a study with soybean seedlings subject to different environmental conditions. *Plant Signaling & Behavior, 12*(3), e1290040.
Souza, G. M., Toledo, G. R., & Saraiva, G. F. (2018). Towards systemic view for plant learning: Ecophysiological perspective. In F. Baluska, M. Gagliano, & G. Witzany (Eds.), *Memory and learning in plants* (pp. 163–189). Springer.
Stenhouse, D. (1974). *The evolution of intelligence—A general theory and some of its implications*. Barnes & Noble Books.
Stolarz, M., Żuk, M., Król, E., & Dziubińska, H. (2014). Circumnutation tracker: Novel software for investigation of circumnutation. *Plant Methods, 10*(1), 1–9.
Struik, P. C., Yin, X., & Meinke, H. (2008). Plant neurobiology and green plant intelligence: Science, metaphors and nonsense. *Journal of the Science of Food and Agriculture, 88*(3), 363–370.

Stuhrmann, C. (2022). "It Felt More like a Revolution." How Behavioral Ecology Succeeded Ethology, 1970–1990. *Berichte zur Wissenschaftsgeschichte, 45*(1–2), 135–163.
Sultan, S. E. (2015). *Organism and environment: Ecological development, niche construction, and adaptation*. Oxford University Press.
Sultan, S. E. (2021). Eco-evo-devo. *Evolutionary developmental biology: A reference guide*, 1165–1177.
Taiz, L., Alkon, D., Draguhn, A., Murphy, A., Blatt, M., Hawes, C., et al. (2019). Plants neither possess nor require consciousness. *Trends in Plant Science, 24*(8), 677–687.
Thelen, E., & Smith, L. B. (1994). *A dynamic systems approach to the development of cognition and action*. The MIT Press.
Thellier, M. (2017). *Plant responses to environmental stimuli: The role of specific form of plant memory*. Springer.
Thellier, M., & Lüttge, U. (2013). Plant memory: A tentative model. *Plant Biology, 15*(1), 1–12.
Thomas, R. K. (1998). Lloyd Morgan's canon. In G. Greenberg & M. M. Haraway (Eds.), *Comparative psychology: A*. Garland Publishing Co.
Thompson, E. (2007). *Mind in life: Biology, phenomenology, and the sciences of mind*. Harvard University Press.
Thompson, K. (2018). *Darwin's Most wonderful plants. A tour of his botanical legacy*. University of Chicago Press.
Timberlake, W. (1994). Behavior systems, associationism, and Pavlovian conditioning. *Psychonomic Bulletin & Review, 1*, 405–420.
Timberlake, W. (2007). Anthropomorphism revisited. *Comparative Cognition & Behavior Reviews, 2*.
Timberlake, W., & Delamater, A. R. (1991). Humility, science, and etiological behaviorism. *The Behavior Analyst, 14*, 37–41.
Timberlake, W., & Lucas, G. A. (1989). Behavior systems and learning: From misbehavior to general principles. In S. B. Klein & R. R. Mower (Eds.), *Contemporary learning theories: Instrumental conditioning theory and the impact of biological constraints on learning* (pp. 237–275). Erlbaum.
Tompkins, P., & Bird, C. (1973). *The secret life of plants. A fascinating account of physical, emotional, and spiritual relations between plants and man*. Penguin Books.
Topham, A. T., Taylor, R. E., Yan, D., Nambara, E., Johnston, I. G., & Bassel, G. W. (2017). Temperature variability is integrated by a spatially embedded decision-making center to break dormancy in Arabidopsis seeds. *Proceedings of the National Academy of Sciences, 114*(25), 6629–6634.
Toyota, M., Spencer, D., Sawai-Toyota, S., Jiaqi, W., Zhang, T., Koo, A. J., et al. (2018). Glutamate triggers long-distance, calcium-based plant defense signaling. *Science, 361*(6407), 1112–1115.
Trewavas, A. (2002). Plant intelligence: Mindless mastery. *Nature, 415*(6874), 841–841.
Trewavas, A. (2003). Aspects of plant intelligence. *Annals of Botany, 92*(1), 1–20.
Trewavas, A. (2005). Green plants as intelligent organisms. *Trends in Plant Science, 10*(9), 413–419.
Trewavas, A. (2014). *Plant behaviour and intelligence*. Oxford University Press.
Trewavas, A. (2017). The foundations of plant intelligence. *Interface Focus, 7*(3), 20160098.
Umerez, J. (2001). Howard Pattee's theoretical biology—A radical epistemological stance to approach life, evolution and complexity. *Biosystems, 60*(1–3), 159–177.
Van Gelder, T., & Port, R. F. (1995). *It's about time: An overview of the dynamical approach to cognition*. Mind as motion: Explorations in the dynamics of cognition, 1, 43.
Van Loon, L. C. (2016). The intelligent behavior of plants. *Trends in Plant Science, 21*(4), 286–294.
Van Volkenburgh, E., & Brenner, E. D. (2024). Novel perspectives on plant behavior. *Plant Signaling & Behavior, 19*(1), 2419673.
Varela, F. J., Thompson, E., & Rosch, E. (1991). *The embodied mind: Cognitive science and human experience*. The MIT Press.

Vernadsky, V. I. (1998). *The biosphere*. (D. B. Langmuir, Trans.; M. A. S. McMenamin, Ed.). Springer.

Viola, A. (2019). *Diritti dei vegetali: nuovi diritti per una giustizia ecologica*. In G. Pellegrino, M. Di Paola, *Etica e Politica delle Piante*. DeriveApprodi.

Vodeneev, V. A., Katicheva, L. A., & Sukhov, V. S. (2016). Electrical signals in higher plants: Mechanisms of generation and propagation. *Biophysics, 61*, 505–512.

Volkov, A. G. (Ed.). (2012). *Plant electrophysiology: Signaling and responses*. Springer Science & Business Media.

Volkov, A. G. (2017). Biosensors, memristors and actuators in electrical networks of plants. *International Journal of Parallel, Emergent and Distributed Systems, 32*(1), 44–55.

Wandersee, J. H., & Schussler, E. E. (1999). Preventing plant blindness. *The American Biology Teacher, 61*(2), 82–86.

Wang, Q., Guerra, S., Bonato, B., Simonetti, V., Bulgheroni, M., & Castiello, U. (2023). Decision-making underlying support-searching in pea plants. *Plants, 12*(8), 1597.

Warden, C. J. (1928). The development of modern comparative psychology. *The Quarterly Review of Biology*, *3*(4), 486–522.

Warden, C. J., Jenkins, T. N., & Warner, L. H. (1935). Comparative psychology: A comprehensive treatise. *Principles and Methods*, Vol. 1.

Westerhoff, H. V., & Palsson, B. O. (2004). The evolution of molecular biology into systems biology. *Nature Biotechnology*, *22*(10), 1249–1252.

Wheeler, M. (1997). *Cognition's coming home: The reunion of life and mind*, in *Proceedings of the fourth European conference on artificial life* (pp. 10–19). MIT Press.

Wheeler, M. (2005). *Reconstructing the cognitive world: The next step*. MIT Press.

Wickramasinghe, C. (2003). Panspermia According to Hoyle. *Astrophysics and Space Science*, *285*(2), 535–538.

Witzany, G. (2016). The Biosemiotics of Plant Communication. *The American Journal of Semiotics*, *24*(1–3), 39–56.

Wynne, C. D. (2004). The perils of anthropomorphism. *Nature*, *428*(6983), 606–606.

Yakura, H. (2020). Cognitive and memory functions in plant immunity. *Vaccines*, *8*(3), 541.

Yokawa, K., Kagenishi, T., Pavlovič, A., Gall, S., Weiland, M., Mancuso, S., & Baluška, F. (2018). Anaesthetics stop diverse plant organ movements, affect endocytic vesicle recycling and ROS homeostasis, and block action potentials in Venus flytraps. *Annals of Botany*, *122*(5), 747–756.

Yoneyama, K., & Bennett, T. (2023). Whispers in the dark: Signals regulating underground plant–plant interactions. *Current Opinion in Plant Biology*, 75, 102456.

Žárský, V. (2015). Signal transduction: GABA receptor found in plants. *Nature Plants*, *1*(8), 1–2.

Zeleny, M. (1981). *Autopoiesis: A theory of living organization*. North Holland.

References

Vernadsky, V. I. (1998). *The biosphere* (D. B. Langmuir, Trans.; M. A. S. McMenamin, Ed.). Springer.

V[illegible], A. (20[illegible]). [illegible] In [illegible] M. D. [illegible] [illegible].

Vodeneev, V. A., Katicheva, L. A., & Sukhov, V. S. (20[illegible]). Electrical signals in higher plants: Mechanisms of generation and propagation. *Biophysics, 61*, [illegible].

[illegible] (20[illegible]). [illegible]

W[illegible] (20[illegible]). [illegible] *Frontiers* [illegible]

W[illegible] (1[illegible]). [illegible] 3–26.

Wang, C., [illegible] (20[illegible]). [illegible] in pea plants. *Plants*, [illegible].

W[illegible] (19[illegible]). The development [illegible] 420–522.

W[illegible] (20[illegible]). [illegible]

W[illegible] (20[illegible]). The evolution [illegible] 12–[illegible].

Wheeler, M. (19[illegible]). [illegible] MIT Press.

W[illegible], M. (1[illegible]). [illegible] MIT Press.

W[illegible] (20[illegible]). [illegible]

Witzany, G. (20[illegible]). The [illegible] of Plant Communication. [illegible]

W[illegible] (20[illegible]). [illegible] *Nature*, [illegible].

Y[illegible], H. (20[illegible]). [illegible] in plant immunity. [illegible]

Y[illegible], [illegible], M., [illegible] S., & [illegible] (20[illegible]). [illegible]

[illegible] and [illegible] potential in [illegible]. [illegible]

Z[illegible] (20[illegible]). [illegible] signals [illegible] *Biochemistry*, [illegible].

Z[illegible], V. (20[illegible]). Signal [illegible] ABA [illegible] in plants. [illegible]

Z[illegible], S. (1981). [illegible] North Holland.

Chapter 3
Plant Characteristics and Activities

Contents

3.1 Specificities in Plant Organization 116
3.1.1 Body Structure and Function: The Difference Between Sessile and Vagile Modes of Life 117
3.1.2 Phenotypic Plasticity 119
3.1.3 Levels of Complexity 120
3.2 Senses, Perception, Action 121
3.2.1 Signaling in Plants 122
3.2.2 Plant Perceptual Sensitivity 124
3.2.3 Plant Movements 128
3.2.4 Movements of Climbing Plants 129
3.2.5 Motor Intentions? 133
3.3 On Behavior 134
3.3.1 Some Peculiarities of Plant Behavior 136
3.3.2 Signaling and Communication in Plants 141
3.4 Processes of Memory, Learning, Decision-Making, Anticipation, and Attention 145
3.4.1 Types of Memory 145
3.4.2 Memory Processes in Plants 148
3.4.3 Learning Processes 150
3.4.4 Ability to Learn in Plants 152
3.4.5 Anticipatory Behavior 156
3.4.6 Decision-Making 159
3.4.7 Response Selection Processes in Plants 160
3.4.8 Attentional Processes 162
3.4.9 Attention in Plants? 163
3.5 Individuality and Consciousness 165
3.5.1 Individuality in Divisible Organisms: The Problem of Integration and Coordinated Responses 166
3.5.2 Conceptions of Consciousness 169
3.5.3 Consciousness and Self-Consciousness 174
3.5.4 Awareness in Plants? 175
3.6 Intelligence and Cognition in Plants 179
3.6.1 On Intelligence 180
3.6.2 Plant Intelligence? 184
3.6.3 Cognitive Behaviors in Plants 187
References 191

M. Bianchi, *Philosophical Implications of Plant Cognition*, Philosophical Studies Series 158, https://doi.org/10.1007/978-3-032-01328-6_3

Abstract This chapter is devoted to the current scientific literature on the main features and capabilities of plants. It offers a detailed and example-rich exploration of plant biology and ecology, with particular attention to behavioral dynamics and communication processes.

Keywords Anticipation · Attention · Behavioral flexibility · Biological and ecological complexity · Decision-making · Memory and learning · Multimodality · Multisensory perception · Phenotypic plasticity · Plant awareness · Plant behavior · Plant cognition · Plant intelligence · Plant movements · Plant senses · Plant signaling and communication · Plant structure and functional organization

In this chapter, I revisit the main characteristics and abilities of plants, as analyzed extensively in scientific literature, particularly over the past few decades. By deepening the analysis, it becomes possible to gain a clearer understanding of both the nature of plants and their interactions, as well as clarify, in light of the most recent evidence (observational studies and experimental data), the theoretical framework and objectives outlined in the previous chapter. This sets the stage for the potential theoretical developments in the next chapter, particularly regarding the impact of this research on various areas of philosophical analysis. As such, this chapter serves as a crucial bridge between Chaps. 2 and 4, offering concrete examples to illustrate plant capacities.

I begin by explaining the distinctive features of plants' structural and functional organization, focusing on their sensitivity, perception, action, and movement, particularly the movements of climbing plants. I then introduce the broader topic of behavior, detailing the behavior in plants, with a special emphasis on plant communication. Subsequently, I focus on memory processes, learning, decision-making, anticipatory behavior, and the possible presence of attention-like processes. I also explore the controversial topics of plant individuality and "consciousness" (awareness). Finally, I synthesize these investigations through the concepts of intelligence and cognition in plants.

3.1 Specificities in Plant Organization

In this section, I introduce the peculiarities of the structural and functional organization of plants, which underlie the behavioral possibilities of these organisms. I begin by summarizing the evolutionary history of plants, before proceeding to an examination of their cellular and physiological organization, as well as their developmental and ecological characteristics.

Plants evolved from unicellular green algae, with evidence of terrestrial colonization by the first multicellular pioneer plants dating back to approximately 470 million years ago, as suggested by the discovery of fossilized spores (cf. Trewavas, 2014, p. 43). The Devonian period marked the emergence of the first vascular plants and symbiotic fungi. During the Carboniferous, the earliest seed-bearing plants appeared, with conifers becoming prominent by the Triassic. The most recent major group, the angiosperms (flowering plants) arose in the Early Cretaceous, with around 260,000–270,000 species identified to date (cf. Pievani, 2017, pp. 21–24; Godfrey-Smith, 2021, p. 237).

Although plants are still often seen as "archaic" organisms, the reality is more complex. Green algae—the precursors of the first plants—did evolve before the earliest animals. However, terrestrial plants emerged after the first jawless fish, although they predate most vertebrate lineages. In contrast, flowering spermatophytes appeared well after the first mammals and birds (cf. Maher, 2017, p. 36).

At the cellular level, plants originated through primary endosymbiosis, whereby an ancestral cyanobacterium was incorporated into a eukaryotic cell, eventually evolving into the chloroplast (cf. Margulis, 1992, 1993; Smalley & Coats, 2002; Mauseth, 2017, pp. 510–511). Compared to animal cells, plant cells can be up to five times larger; they lack lysosomes and centrioles but possess unique structures such as a large central vacuole, an outer cellulose cell wall, and chloroplasts (cf. Trewavas, 2014, p. 50). From a physiological perspective, plant and animal systems share several fundamental traits, including chemical–electrical and hormonal signaling (cf. Chamovitz, 2018; Castiello, 2021), as well as immune resistance mechanisms involving recognition and memory, enabling plants to respond to subsequent pathogenic attacks with increased speed and intensity (cf. Conrath, 2011; Trewavas, 2014, p. 264).

From a development standpoint, plants possess an important characteristic: an open developmental program. Thanks to the continued activity of meristems,[1] even in mature individuals, plants display indeterminate organogenesis, allowing them to grow and form new organs throughout their lifespan, despite operating within a defined genetic framework. For example, while a chestnut tree is clearly identifiable as a distinct species, it does not possess a fixed number of branches or a standardized branch shape. In contrast, animals such as cats typically develop a fixed number of limbs, each with a consistent form (cf. Godfrey-Smith, 2021, pp. 112–114).

3.1.1 *Body Structure and Function: The Difference Between Sessile and Vagile Modes of Life*

In terms of body structure and organization, there is a fundamental difference between a vagile mode of life, typical of animals capable of locomotion (the reference here is not to the mere ability to move, but to the capacity for locomotion from

[1] Meristems are tissues located in the buds, at the root apices, and in the vascular cambium of stems and roots. They are characterized by their capacity to remain in an undifferentiated, embryonic state. The cells within these regions can undergo continuous mitotic division and subsequently differentiate to perform various functions within the plant's structure.

one point to another), and a sessile mode of life, characteristic of plants. This distinction stems from the different conditions of heterotrophy and autotrophy, although the boundary is not as clear-cut as often assumed, as intermediate forms like parasitic and semi-parasitic plants exist. In any case, while heterotrophs, particularly animals, must move to obtain food and cope with various survival challenges, autotrophs, through chlorophyll-based photosynthesis (which includes C_3, C_4, and CAM metabolic pathways [cf. Mazzolai, 2021, pp. 113–114]), do not need to actively seek nutritional sources. Instead, they synthesize the organic compounds required for their survival, primarily carbohydrates, by using water and atmospheric carbon dioxide in the presence of solar energy. This distinct mode of transforming substances and acquiring energy underpins the unique structure and modular organization of plant bodies.[2]

The challenges of nutrient acquisition and predation appear to have been the major evolutionary pressures driving the development of modular growth and the branched structures characteristic of sessile organisms (cf. Trewavas, 2003, p. 10). In biology, this refers to a modular body plan, where an organism increases in size by repeatedly replicating a specific structural unit, thereby avoiding the need for more complex requirements (cf. Godfrey-Smith, 2021, pp. 112–114). Compared to "unitary" or more integrated organisms, such as many animals (though some animals, like segmented worms or millipedes, also exhibit modularity), modular organisms, such as plants, consist of numerous repeating units that are functionally semi-autonomous (cf. Trewavas, 2014, pp. 78–79).[3]

This characteristic has long been observed and commented on by various authors, such as Theophrastus in antiquity—although he did not use the term 'modularity'—and, many centuries later, Erasmus Darwin, who argued that a tree could be conceived as a 'swarm' of individual plants. In contemporary terms, numerous researchers have explicitly referred to plant modularity to explain the unique developmental features and interactions of these organisms. For example, Beata Oborny (2019) linked modularity to network theory analysis, highlighting it as a key to understanding certain aspects of plant behavior in close interaction with the local environment. As she noted: "*(A) Each module can be considered as an individual point of interaction. The locations of these points can be actively influenced by the plant through its environment-dependent (i.e. plastic) rules of growth* [...]. *(B) According to the environment at these locations, the same genotype can be expressed in various phenotypes simultaneously. For example, a genet can contain vegetative and reproductive ramets. (C) The modules can share resources and information through their connections.*" (Oborny, 2019, p. 6). Thus, plant modularity enables growth with variable shapes and adaptability to local conditions.

---

[2] Thanks to their modular structure and the presence of meristems, many plants can be propagated vegetatively through stem cuttings, while others can be propagated using root or leaf cuttings.

[3] There are also structures with modular characteristics, such as capillaries, which are composed of repeated units. However, the concept of a repeated unit differs from that of the semi-autonomy of modules, as seen in plants, where each module has the potential to reproduce an entire organism with differentiated organs.

In their interactions with the environment, plant modules, which are produced and die during different stages of plant development, are capable of transmitting resources and information, and they can specialize in different functions (e.g., leaves for photosynthesis, flowers for reproduction). At the structural level, the various branching forms of plants reflect the history of stimuli encountered, obstacles navigated, and the "decisions" made, such as where the plant found sunlight, nutrients, or where it avoided harmful substances (cf. Bianchi, 2022).

However, it would be an overstatement to emphasize this adaptive flexibility in development and behavior. Plants, especially woody trees, possess relatively rigid body structures due to their thick, cellulose- and lignin-rich cell walls. As a result, while their chemical-level behaviors are highly sophisticated, their physical capabilities related to movement are limited (cf. Oborny, 2019).

3.1.2 Phenotypic Plasticity

Another term commonly associated with the modular development of plant bodies is phenotypic plasticity. In general, even in non-plant organisms, phenotypic plasticity refers to the feature of an organism, even with the same genotype (e.g., in monozygotic twins), to develop different phenotypic traits in response to environmental conditions, which themselves may change over the course of the organism's life. Phenotypic plasticity results from the fact that living systems evolve in dynamic environments where an optimal and stable state of "fitness" cannot be permanently achieved. Organismic life unfolds through the continuous repetition and variation of processes in response to changing conditions (cf. Trewavas, 2003). In sessile organisms like plants, which are well anchored to their ecological niche, phenotypic plasticity provides adaptive opportunities akin to those afforded by locomotion in vagile organisms for solving survival challenges (cf. Pievani, 2017, pp. 50–51).

Phenotypic plasticity manifests at structural, metabolic, and behavioral levels (cf. Trewavas, 2003), enabling organisms to modify aspects of their morphology and function during development in response to both internal (e.g., physiological) and external (e.g., environmental) stimuli. This capacity is particularly pronounced in plants. For instance, in *Critica Botanica* (1737), Linnaeus described how certain aquatic species, such as *Ranunculus aquatilis* L., produce finely divided, capillary leaves below the water surface and broader leaves above it, each optimized for photosynthetic efficiency in their respective environments (cf. Bradshaw, 1965).

In recent years, phenotypic plasticity has been increasingly linked to discussions surrounding "plant intelligence," as the relatively flexible and context-dependent development of plants suggests a form of environmental discernment, enabling the selection of adaptive responses (cf. Pellegrino & Di Paola, 2019, pp. 90–91). This trait has also been explored in light of potential future scenarios, in which plants display a "forward-looking" capacity to modify their growth in anticipation of environmental stressors or pathogenic threats, or to grow directionally before actually securing access to light or nutrients (cf. Trewavas, 2002; Novoplansky, 2016).

However, it is important to avoid overgeneralizing this concept. Phenotypic plasticity has limits; it is not an "all-or-nothing" trait. While some characteristics exhibit plasticity, others remain more stable, particularly those vital for reproduction, which must remain as consistent as possible. Additionally, the degree of plasticity varies across plant divisions (for instance, it is less pronounced in gymnosperms compared to angiosperms) and depends on the species (cf. Trewavas, 2003, pp. 12–13).[4]

3.1.3 Levels of Complexity

When comparing the characteristics of animals and plants, there is a common tendency to frame the discussion not in terms of bioecological diversity, but rather in terms of relative complexity (which is not problematic) and, often, in terms of a greater or lesser degree of "perfection" (which is problematic), linking the argument to a value-based perspective. According to this second trend, evolution is seen as a progression of transformative stages, implying that some currently existing species are more "evolved" than others. Instead, from a systemic standpoint, more complex organisms exhibit multiple levels of emergence.

The question of how to define biological complexity is itself a fascinating, and inherently complex, topic. The answer varies depending on the criteria used.

For instance, at the structural and behavioral levels, a vertebrate may be considered more complex than a single-celled organism, but this does not necessarily correlate with superior adaptive success, quite the contrary. In fact, "simpler" organisms, like bacteria, often exhibit remarkable functional effectiveness (cf. Pievani, 2017, p. 120).[5]

Considerations of this kind are not new in theoretical reflections on biology. For example, Daniel W. McShea (1996) proposed several criteria for addressing the issue of biological complexity, including the number of genes, the diversity of cell types and organs, and, at the ecosystem level, the number and variety of species. Today, these parameters are often evaluated quantitatively (e.g., based on the number of elements and processes involved), by the intensity and nature of the interactions among elements, and by the degree of hierarchical (and sometimes heterarchical) organization within the system, ranging from molecules to tissues to populations (cf. Trewavas, 2014, p. 67). Increased complexity has been associated with a greater capacity to acquire and process environmental information, as well as

[4] Compare with the rule of reaction (Schlichting & Pigliucci, 1998).

[5] Some have wondered why such well-adapted organisms eventually evolved into multicellular eukaryotes. The main reasons include the increase in synergistic functional interactions between cells and the corresponding reduction in the autonomy of individual cells. Additionally, predation mechanisms tend to inflict less damage on multicellular organisms compared to unicellular ones, and competitive pressures may have driven the emergence of more complex behaviors, both among members of the same species and between different species, for securing resources (cf. Trewavas, 2014, pp. 43–66).

with an increased independence from environmental conditions (cf. McShea, 1996). These considerations do not conflict with recognizing the complex web of historical contingencies that have shaped the evolution of different organisms, leading to their specific morphological and functional traits (cf. Trewavas, 2014, p. 48).

As far as plants are concerned, a frequent question has been whether they, particularly the more recent division of angiosperms, are more or less complex than certain animal classes, especially vertebrates. There is a common tendency to regard the plant kingdom as a whole—often overlooking its internal diversity—as simpler and, in a certain sense, "less evolved" than the animal kingdom. From an evolutionary timeline perspective, this view is incorrect and overly reductive, as it often relies on arbitrary selection of characteristics without considering the diverse ecological roles and needs of different species (cf. Bianchi, 2022).

The primary distinction traditionally made between plants and animals stems from the absence of a nervous system and locomotor abilities. Organisms that have developed this specialized (and undeniably sophisticated) tissue for motor control and behavioral coordination are generally considered more complex than self-organizing control systems like plants. However, rooted organisms, such as plants, do not require locomotion.

The intent here is not to diminish the unique complexity of animal nervous systems, but rather to acknowledge that each group of organisms, with their distinct body structures and corresponding "mode of life," is well adapted to its ecological niche (cf. Trewavas, 2014, pp. 71–72). The complexity of communication and behavioral strategies in each species, which can be seen as highly evolved within its own evolutionary branch, manifests at different organizational levels. For plants, for instance, this complexity likely lies in the synthesis, recognition, and modulation of chemical substances.

Therefore, unless we wish to perpetuate human-centric biases, a bio-ecologically informed approach suggests that all organisms are equally evolved, at least in the sense that they are adapted to their environments (cf. Pellegrino & Di Paola, 2019, pp. 12–13). The Darwinian Tree of Life (or more aptly, the Darwinian Coral of Life) does not place *Homo sapiens* at the top (cf. Pievani, 2012). Differences arise from evolution in distinct ecological niches, and complexity should be understood as existing at different levels, depending on the specific traits being examined. This, naturally, does not preclude the possibility of identifying greater complexity from a certain point of analysis, once the criteria for comparison have been established (cf. Bianchi, 2021).

3.2 Senses, Perception, Action

Given the variability and unpredictability of environmental conditions—often associated with scarcity or fluctuating availability of resources, as well as with diverse interspecific cooperative and competitive strategies—an organism's capacity to perceive and respond to its surroundings is essential for survival. Every living being must detect and recognize the key features of its environment (cf. Lyon et al., 2021). Sensitivity to external stimuli, even in its simplest forms, has accompanied the evolution of cellular life since its origins, from bacteria to multicellular eukaryotes, and is evident in the various ways organisms respond to environmental challenges (cf. Godfrey-Smith, 2021, p. 49).

Organisms actively gather information about their environment through sensory mechanisms that detect physical stimuli, including light (electromagnetic radiation of different wavelengths for vision), chemical stimuli (volatile and soluble molecules related to smell and taste), and mechanical stimuli (pressure or vibration for touch, and sound waves for hearing) (cf. Castiello, 2019, p. 37).

3.2.1 Signaling in Plants

Before discussing the perceptual abilities of plants, it is compelling to consider that plants use various forms of signaling, particularly chemical, hormonal, and electrical signals (cf. Brenner et al., 2006; Huber & Bauerle, 2016; Souza et al., 2017). Within their bodies, plants have two primary pathways—the apoplastic and symplastic—through which water, nutrients, and informational signals are transported (cf. Reece et al., 2015).

Chemical signaling will be examined in greater detail in the section dedicated to plant communication (see *Signaling and Communication in Plants*). However, it is important to emphasize here that specialized signaling molecules, hormones, play a crucial role in plant physiology. These should be understood in terms of their overall effects, as hormonal balance, rather than the concentration of a single hormone, is critical. Hormones coordinate cellular activities to achieve specific biological functions; at the level of different organs and in response to varying contexts, they modulate a wide range of processes. Plant phytohormones include auxins—the first-discovered and among the most abundant phytohormones—which regulate various aspects of growth and development; cytokinins, which promote cell division and delay senescence; gibberellins, produced primarily in young tissues of shoots and seeds, that stimulate stem elongation, seed germination, and flowering; abscisic acid (ABA), which plays a key role in maintaining seed dormancy and mediating stress responses, particularly to drought; brassinosteroids, which promote cell elongation and division and are essential for vascular development; ethylene, a gaseous hormone involved in fruit ripening, leaf abscission, and response to mechanical stress; jasmonates, which are central to defense responses against herbivores and pathogens; and strigolactones, which influence shoot branching (apical dominance), stimulate seed germination, and mediate symbiotic interactions with mycorrhizal fungi (cf. Reece et al., 2015, p. 148).

It is also pertinent to recall the role of reactive oxygen species (ROS) in plant chemical signaling. These are highly reactive molecules derived from oxygen metabolism, including hydrogen peroxide (H_2O_2), superoxide anion (O_2^-), and hydroxyl radical (·OH). While high concentrations of ROS can cause oxidative stress and cellular damage, at low concentrations they function as crucial signaling molecules involved in numerous cellular processes. In this context, ROS are not merely toxic metabolic by-products, but key mediators of both local and systemic signaling in plants. Among their primary roles is the activation of defense pathways in response to biotic stressors (e.g., pathogen attack) and abiotic stressors (e.g., drought, high temperatures). ROS also contribute to long-distance signaling by acting as mobile signals that travel through tissues, enabling communication between different organs (e.g., from roots to leaves). They interact with various hormonal pathways—including abscisic acid, ethylene, and auxins—modulating physiological responses such as growth, stomatal movement, and senescence. Furthermore, ROS play a role in plant development, participating in cell differentiation, root growth, and other physiological processes (cf. Waszczak et al., 2018; Mittler et al., 2022; Li et al., 2022).

In recent years, electrical signaling has gained considerable attention (cf. Parise et al., 2021, 2022). It is important to note that electrical signaling cannot be considered independently from chemical signaling, as it involves the movement of ions, electrically charged molecular entities, across cell membranes, either as a trigger or result of a process. Electrical signaling occurs during an organism's normal physiological activities and varies in response to its ecological interactions. These signals, driven by action potentials and mediated by calcium ion fluxes and cytosolic Ca^{2+} signals, propagate across plant cell membranes and can be triggered by stimuli such as herbivory, physical damage, mechanical stress, or rapid temperature changes (cf. Zhu et al., 2009). This type of electrical signaling is transmitted through the phloem's vascular tissues (cf. Bose & Guha, 1922; Fromm & Lautner, 2007) and is capable of long-distance conduction, enabling systemic responses throughout the organism. Such signaling contributes to more integrated physiological coordination, even in modular organisms like plants (cf. Baluška, 2010).

The study of electrical activity in plants traces back to the pioneering work of John S. Burdon-Sanderson (1828–1905) and Sir Jagadish Chandra Bose (1858–1937). Burdon-Sanderson (1873, 1899) was among the first to investigate electrical impulses in *Dionaea muscipula* (the Venus flytrap), initially assuming that only plants with rapid, visible movements relied on electrical signaling. He observed that mechanical stimulation or wounding of the leaves induced measurable electrical changes, and that touching two trigger hairs within a brief interval generated an electrical impulse that propagated along the leaf and preceded the trap's rapid closure.

Sir Jagadish Chandra Bose—later recognized as the father of plant electrophysiology, despite having been unjustly criticized by Burdon-Sanderson (cf. Shepherd, 2012)—is best remembered for his pioneering measurements of what are now known as action potentials, in *Mimosa pudica*, and for extending the study of electrical signaling to plants that do not exhibit rapid or visible movements (cf. Bose, 1907, 1913, 1926). Bose sought to investigate the "nerve-like" properties of remotely transmitted impulses in plants, and demonstrated that they possess a network of electrically excitable, non-neural tissues (cf. Linson & Calvo, 2020). In his seminal work, *The Nervous Mechanism of Plants* (1926), Bose illustrated the vascular system of *Papaya*, revealing a complex reticulated network of connections. In adult stems, particularly trunks, the vascular architecture becomes highly dense and interconnected, featuring numerous anastomoses between bundles. Initially, it was believed that the vascular system solely mediated the transport of water and nutrients, but it was later found to also facilitate electrical signaling (cf. Calvo et al., 2017), as evidenced by signal disruption following vascular severing.

Over the years, several types of electrical potentials have been identified in plant cells. Among these are action potentials (APs), which bear similarities to those found in animals in terms of their all-or-nothing nature and ion-mediated transmission (cf. Baluška & Mancuso, 2009), and variation potentials (VPs), which are much slower and specific to plant cells, often triggered by stimuli such as wounding or environmental stress (cf. van Bel et al., 2014; Calvo et al., 2017). Additional types of electrical signals include local electrical potentials (LEPs), which are confined to specific tissues, and system potentials (SPs), which are involved in long-distance signaling throughout the plant.

Ongoing research on the interplay between electrical and chemical signaling holds promise for advancing our understanding of plant behavioral activities.

3.2.2 Plant Perceptual Sensitivity

Concerning sensory and perceptual capacities, both animals and plants (as well as other forms of life) face temporally and spatially heterogeneous environments, leading to the evolution of relatively flexible behavioral response systems.

In nature, a plurality of experiential modalities exists, shaped by sensory systems that are similar across species yet not identical (cf. Bianchi, 2024). Each species exhibits sensory specialization or "dominance," relying more heavily on certain senses than on others (cf. Bianchi et al., 2025a).

Plant organisms have faced many of the same selective pressures that animals have encountered throughout their evolutionary history (cf. Karban, 2008, p. 736). Some proponents of plant intelligence, such as Stefano Mancuso and Alessandra Viola (2015), have collected information about plant sensory abilities, highlighting plant analogous of the animal senses of "sight," "smell," "taste," "touch," and "hearing," along with a variety of additional sensory capabilities, such as detecting changes in atmospheric humidity, electromagnetic fields, and gravity.

The difference between plant and animal sensory systems, both of which are efficient, is that plant senses are organized and distributed differently (cf. Bianchi et al., 2025a). Instead of being centralized in specific organs, plant senses are diffused throughout their entire body. On a cellular level, plants are highly sensitive: the speed of signaling reactions within plant cells is comparable to that of animal cells. However, the slower response movements seen at the organismal level in plants are due to the involvement of growth processes in organ movement (cf. Trewavas, 2014).

At a more integrated level of organization, plants possess proprioception, the sense (nonreflective awareness) of the position of their body parts in space (cf. Okamoto et al., 2015; Moulia et al., 2019). At the cellular level, this ability appears to be mediated by long actin filaments within elongated fiber cells (cf. Hamant & Moulia, 2016). On the organismal level, proprioception is facilitated by the capacity of root apices to sense gravity via statoliths, which are starch granules located in specific cells of the root cap. These specialized plant graviceptors form the basis of plants' geotropic sensitivity: as they move within plant cells, they transmit information regarding positional changes (cf. Chamovitz, 2012; Bianchi, 2021, pp. 73–75).[6]

In terms of visual perception, plants are extremely sensitive to light, which they rely on to synthesize the organic compounds that form the very structure of their bodies. By employing various protein photoreceptors—such as phytochromes, cryptochromes, and phototropins—plants are capable of detecting a broad spectrum of electromagnetic radiation, including red and far-red light (approximately 600–750 nm), as well as blue (400–500 nm), UV-A (320–400 nm), and UV-B (280–320 nm) wavelengths (cf. Mawphlang & Kharshiing, 2017). Red and blue light play critical roles in regulating physiological processes like germination, growth, and flowering, and are the most efficient wavelengths for driving photosynthesis. As early as 1864, von Sachs observed that blue light induces phototropism, triggering directional movement in plants. In contrast, wavelengths corresponding to green light are largely reflected rather than absorbed, making them less useful for photosynthesis: hence the green appearance of chlorophyll-containing plant parts (cf. Castiello, 2019, p. 39).

One of the most highly developed sensory capacities in plants is their ability to detect a wide range of chemical compounds, both through aerial tissues and via their roots in the soil. Depending on the organ involved and whether the substances are volatile or soluble, this detection ability has been likened to plant forms of "smell" and "taste," due to the abundance of receptors that initiate signaling cascades (ibid., p. 57). More broadly, the ability to detect chemicals in the environment enables plants to perceive their surroundings, identifying the availability of nutrients, the presence of harmful substances, or other species, as in the case of parasitic plants of the genus *Striga* (known for identifying host plants by sensing the strigolactone hormone released by plant roots to attract beneficial mycorrhizal fungi [cf. Trewavas, 2014, p. 86]).

In terms of tactile perception, plants rely on mechanosensitive ion channels, with receptors primarily located between the cell wall and the plasma membrane, particularly in cells exposed to the external environment (cf. Chamovitz, 2012). Remarkably, in some species, plant sensitivity to touch can be up to ten times greater than that measured in human skin (cf. Stahlberg, 2006; Linson & Calvo, 2020). Unlike humans, who possess specialized mechanoreceptors such as Pacinian, Meissner, and Krause corpuscles for sensing pressure and temperature, plants lack

[6] The role played by statoliths was confirmed by the observation of lack of geotropic sensitivity in cells lacking statoliths (cf. Mancuso, 2017).

this kind of receptor diversity. Instead, for example, mechanical stimulation in plant cells activates signaling pathways involving the phytohormone jasmonic acid, which can trigger the generation of electrical signals that propagate to neighboring cells via plasmodesmata, specialized cytoplasmic channels that connect plant cells (cf. Volkov et al., 2010).

The study of tactile sensitivity in plants gained renewed attention with the observations of Frank B. Salisbury in 1963. While investigating the effects of floral-inducing chemicals in *Xanthium strumarium* L., Salisbury measured the daily growth of tassel leaves using a ruler. Unexpectedly, this seemingly harmless contact altered growth patterns: touched leaves failed to reach their normal length and, in some cases, showed yellowing or necrosis. This mechanically induced growth inhibition was later termed "thigmomorphogenesis" by Mark Jaffe in 1973, from the Greek *thígma* (touch) and *morphogenesis* (formation of form) (cf. Castiello, 2019, pp. 48–52). Today, thigmomorphogenesis refers specifically to structural changes in response to mechanical stimuli—such as the thickening of pea stems due to physical contact or persistent wind—and is distinct from reversible movement responses, such as the thigmonastic movements observed in *Mimosa pudica* or *Dionaea muscipula* following mechanical stimulation (cf. Maher, 2017, pp. 55–56).

Research on touch sensitivity advanced significantly in 1990, when Janet Braam and Robert W. Davis identified a group of touch-inducible genes—known as TCH genes—in the model plant *Arabidopsis thaliana*. Their work also highlighted the critical role of calcium ions (Ca^{2+}) in the perception and transduction of mechanical stimuli. Upon tactile stimulation, plants exhibit a rapid increase in cytosolic calcium levels, which leads to the upregulation of TCH gene expression. This response involves calmodulin, a calcium-binding messenger protein that plays a central role in translating calcium signals into downstream physiological responses (cf. Braam, 2005). The exploration of plant tactile abilities dates back at least to Darwin's 1880 observations of root movements that avoid obstacles in the soil. Subsequent studies have focused on the responses of the so-called "sensitive plants" and "carnivorous plants", the adaptation of tendrils in climbing plants, and the increased secretion of nectar by flowers in the presence of pollinators (cf. Karban, 2017).

Concerning the "sense of hearing", it is now known that plants are capable of perceiving and distinguishing between different types of mechanical vibrations (and sound waves of varying frequencies). These vibrations are detected through numerous mechanosensitive channels distributed throughout the plant body, much like in tactile perception (cf. Monshausen & Gilroy, 2009; Chamovitz, 2012).

Plant bioacoustics is an emerging field of research that explores how certain sounds can trigger or enhance defensive strategies in plants (cf. De Melo, 2023; Calvo & Segundo-Ortin, 2023). Studies have shown that exposing seeds and young seedlings to ultrasounds can affect their metabolism, root growth, and germination rates. Additionally, plants are not only capable of detecting various vibrations, but under stress, it appears that they can also emit ultrasounds that are detectable in the air from more than a meter away (cf. Khait et al., 2023; Waqas et al., 2023). Initially, these sound emissions were attributed to cavitation, a process involving a sudden drop in pressure in the water transport system of the xylem, leading to the expansion of air bubbles (a form of embolism). However, recent studies suggest that not all plant sound emissions are solely due to cavitation. Other possible sources of acoustic emissions include physiological processes such as cell growth, stress-related responses, stomatal closure, and tissue movements (cf. Zweifel & Zeugin, 2008; Mishra et al., 2016; Robert, 2023).

Regardless of the exact processes involved, the ability to perceive and emit sound vibrations may provide adaptive advantages to plants, manifesting in various behavioral changes depending on the frequency of the sound (cf. Gagliano et al., 2012, 2017; Gagliano & Renton, 2013). For instance, certain acoustic stimuli can activate genes associated with stress responses. More generally, some sound frequencies can influence plant growth, either stimulating, inhibiting, or altering it (cf. Yiyao et al., 2002; Creath & Schwartz, 2004). This phenomenon is evident in the different growth patterns of trees located along busy roads compared to those in quieter areas (cf. Martens & Michelsen, 1981).

An example of the ability to distinguish between different types of vibrations is provided by Gagliano and Renton (2013), who observed enhanced growth in chili pepper plants (*Capsicum annuum* L.) when placed near basil (*Ocimum basilicum* L.), a plant whose proximity typically inhibits the growth of other species and offers protection against pathogenic attacks. After excluding other possible signaling mechanisms, such as chemical communication, the researchers hypothesized that the interaction between the two plants occurred through acoustic signals. In other studies on *Arabidopsis thaliana*, researchers observed that the plant's trichomes (small epidermal hairs that act like antennas) can detect vibrations caused by caterpillars chewing on the plant's leaves. In response, the plant synthesizes defense toxins, even distinguishing between the specific vibrations caused by caterpillars and irrelevant stimuli like wind (cf. Appel & Cocroft, 2014).

Research into plant-pollinator interactions has also provided fascinating insights. For example, Veits et al. (2019) reported that the flowers of *Oenothera drummondii* increase nectar sugar concentration when exposed to sound frequencies within the range produced by the buzzing of pollinating bees. This knowledge has been applied in agricultural treatments that use sound waves to enhance stomatal opening, increasing the plant's ability to absorb fertilizers and herbicides, thus reducing the need for chemical products. Sound perception in plants is particularly prominent at the root level. For instance, the roots of young *Zea mays L.* plants produce clicking sounds (around 220 Hz) related to the breaking of cell walls during growth. It has been hypothesized that these sounds may play a role in helping root tips perceive the presence of nearby roots from neighboring plants (cf. Gagliano et al., 2012; Mancuso & Viola, 2015). This clicking may represent a form of communication that requires no significant energy expenditure (cf. Castiello, 2019, p. 111). Ongoing research is exploring how plants might emit sounds to inhibit the growth of unrelated plants (cf. Demey et al., 2023). Additionally, other studies have focused on directional phonotropic responses, such as plant reactions to ecologically significant sounds like the sound of water, which can guide root growth (cf. Rodrigo-Moreno et al., 2017).

These examples illustrate how plants, through their ability to detect and distinguish various signals, can respond selectively and "meaningfully" to different types of environmental information, exhibiting responses that allow them to adapt to their surroundings, thus being able to "behave."

3.2.3 Plant Movements

Closely related to sensory analysis is the issue of movement. As discussed earlier, plants are not capable of "locomotion," and one of the main reasons they have historically been excluded from the cognitive domain is the tendency to associate locomotion, that is, the ability to move from one place to another, with cognitive abilities (cf. Calvo & Keijzer, 2011). However, this does not mean that plants are incapable of displacement and generally moving. As Castiello (2019, p. 59) noted, plant movement is intricately tied to their growth processes, and within the spatial limits of their environment, movement is crucial for exploring their surroundings, and ultimately, for survival. For instance, plants need to establish the direction in which to grow, whether moving toward nutrients, away from harmful substances or stressors, or orienting their leaves to capture sunlight. It is expected that a plant would perish if it ceased movements related to its physiological functions and behavioral activities (cf. Castiello, 2021).

The mechanisms underlying plant movement are now understood. Unlike animals, plants lack contractile muscle fibers but are capable of movement through the bending, elongation, or folding of their organs. The processes at the basis of spatial orientation in plants can be broadly classified into four main categories: growth, primarily driven by phytohormones like auxins, which promote organ elongation; turgidity, which involves temporary changes in cell size due to osmotic pressure differences between the vacuoles and cytosol; hygroscopy, which refers to the ability of certain tissues to absorb moisture from the environment; and desiccation, the opposite process in which tissues release water and contract as they dry out. The first two processes are considered "active" movements, as they involve metabolic changes, while the latter two are "passive" movements, as they occur without direct metabolic involvement or energy expenditure (cf. Castiello, 2019, p. 61).

More specifically, various types of plant movements have been identified, each serving distinct roles in the plant's overall physiology. These include both spontaneous movements, such as nutations, and directional movements in response to biologically relevant stimuli, known as tropisms. In technical terms, the category of nastic movements encompasses non-directional responses to stimuli, mediated either by differential growth or by changes in turgor pressure within motor cells, often located in structures such as the pulvinus. These movements can involve different organs—including stems, leaves, and branches—and are frequently reversible. Nastic responses are governed by internal physiological mechanisms and are not oriented by the direction of the stimulus. Stimuli triggering such responses include variations in light (photonasty), mechanical contact (thigmonasty), sudden shock or trauma (seismonasty), temperature changes (thermonasty), and circadian rhythms, particularly in leaf movements (nyctinasty).

Some types of movement are widespread across the plant kingdom. One well-known example is the circumnutation movement, identified by Darwin (cf. Darwin & Darwin, 1880), which refers to the circular movement of the apical sections of young, growing plants, particularly evident in the shoots of climbing plants. This motion is caused by alternating growth patterns along the stem, which cause the plant to curve where elongation is less pronounced. Another category, tropisms, involves movements directed toward external stimuli, such as light (phototropism) or gravity (gravitropism). Tropic movements can be classified as positive (toward the stimulus), negative (away from the stimulus), or transverse (perpendicular to the direction of the stimulus). An example of transverse tropism is transverse gravitropism, as observed in the horizontal growth of branches or lateral roots (ibid., pp. 62–69).

Plant movements can also be classified based on the specific organs involved, such as leaves (e.g., solar tracking or reorientation), young stems (e.g., the circumnutatory movements of climbing plants), or roots (e.g., the oscillatory movements of root tips during growth and obstacle avoidance) (cf. Roy & Bassham, 2014; Castiello, 2019, p. 77).[7]

3.2.4 *Movements of Climbing Plants*

Since Darwin's pioneering investigations into the movements of insectivorous and climbing plants (1875, 1880), researchers from diverse disciplines—including botany, psychology, philosophy, bioengineering, chemistry, and physics—have been increasingly engaged in the study of these phenomena (cf. Bianchi, 2025b). This interest continues today, largely because climbing plants, such as *Pisum sativum* L. or *Phaseolus vulgaris* L., provide valuable insights into behavioral variability in non-neural organisms (cf. Thompson, 2018). The study of plant climbing movements not only explores the biomechanical causes of these motions but also investigates the "reasons" behind movement planning, approaching, and reaching different types of supports. These studies reveal differences in how plants modulate their behavior, depending on the stimulus type and the context of interaction, including social, collaborative, or competitive conditions (cf. Castiello, 2019).

Research on climbing movements is currently conducted in controlled laboratory settings, where environmental variables such as light, temperature, and humidity are precisely regulated. One such facility is the phytotron recently established at the Mind(the)Plant Laboratory, University of Padua. Here, researchers from diverse disciplinary backgrounds—including biology, chemistry, experimental psychology,

[7] Root movements are thought to result from the combination of circumnutation with processes related to contact and the perception of gravity (cf. Migliaccio & Piconese, 2001).

motor neuroscience, engineering, and philosophy—investigate plant behavior and forms of communication. Within this specialized environment, experimental conditions can be systematically controlled to better understand the mechanisms underlying climbing movements and their variability. Studying plant behavior under such conditions offers valuable insights into the factors that influence movement, both at the level of individual plants and in social contexts involving interactions among multiple plants (cf. Guerra et al., 2019, 2024; Bonato et al., 2023, 2024; Bianchi et al., 2025a).

In general, one might ask why climbing plants need to climb. In natural environments, the movement exhibited by these plants—regardless of differences in their climbing organs or the specific strategies employed—serves the function of locating climbable structures in order to reach light as efficiently as possible, thereby avoiding the shade cast by taller neighboring vegetation (cf. Gianoli, 2015). As Darwin (1875, pp. 12–13) observed in his study of the twining movements of *Ceropegia* around a stick, these processes of growth, selection, and attachment (an elaboration of circumnutation) are not merely automatic reflexes or fixed responses. Rather, they appear to be ecologically motivated and flexibly regulated behaviors.

Furthermore, research has confirmed that climbing plants possess the selective capacity to reject unsuitable supports, whether due to their dimensions or surface characteristics, for example, supports with excessively large diameters or those made of smooth materials such as glass (cf. Wang et al., 2023). Darwin himself noted that the tendrils of *Wisteria sinensis* were unable to coil around supports with a diameter of 15 cm. Subsequent field studies conducted in tropical, subtropical, and temperate rainforests have corroborated this observation, showing that the abundance of certain climbing species declines as the diameter of available tree supports increases (cf. Guerra et al., 2022).

Darwin (1875, 1880) also observed that the tendrils of climbing plants appear to conform to the shape of the surfaces they contact, suggesting an ability to recognize the physical characteristics of a potential support before making full contact with it (cf. Trewavas, 2017). This behavior implies that plant movement may involve a form of anticipation, possibly even a rudimentary form of motor planning aimed at achieving specific goals (cf. Calvo & Friston, 2017). Such observations suggest that climbing plants are capable of perceiving the properties of a stimulus they are about to encounter and adjusting their behavior accordingly (cf. Castiello, 2019, p. 73). From this perspective—that is, the dynamic interaction between the plant and its environment—the process may involve detecting ecologically relevant features and responding with a coordinated sequence of motor actions (cf. Bianchi, 2025).

Today, there is increasing recognition that plant behavior, evident in both their chemical interactions and movements, is not rigidly programmed or purely automatic. Instead, it is relatively flexible and variable depending on the context of interaction (Bianchi, 2021, p. 79).

A growing body of research suggests that climbing plants, such as *Pisum sativum* L., are capable of detecting supports and modulating their movements accordingly, based on contextual cues. These studies (cf. Guerra et al., 2019) employed three-dimensional kinematic analyses—quantitative descriptions of motion based on mathematical modelling—alongside time-lapse imaging to examine changes in circumnutation patterns (cf. Simonetti et al., 2021). The findings indicate that once a support is detected, climbing plants adjust both the aperture of their tendrils and the velocity of their approach in relation to the characteristics of the target structure. Specifically, tendrils opened earlier in the trajectory when the support was thicker, and the approach velocity decreased with increasing stimulus diameter (e.g., a wooden pole). A deceleration, functionally analogous to "braking," was observed as the plant neared the support, seemingly to enable more precise identification of contact points (cf. Castiello, 2019, pp. 73–76). When the potential support was presented in two-dimensional (2D) form—i.e., as a flat, unclimbable surface—no significant modulation of movement was observed.

In subsequent studies on the same plant species, Ceccarini et al. (2020) investigated the speed–accuracy trade-off (SAT), a principle widely documented in animals, which describes the inverse relationship between movement speed and accuracy—namely, that faster movements tend to result in reduced accuracy. This concept, first formalized by Fitts (1954), is based on the idea that movement time increases with task difficulty, such as when targeting smaller or more distant objects, reflecting a trade-off between speed and accuracy in motor performance. Ceccarini et al. (2020) found that the tendrils of *Pisum sativum* L. modulate their movement speed in response to the diameter of the available support. However, in contrast to what is typically observed in animals (cf. Heitz & Schall, 2012), the highest movement speeds were recorded when grasping thinner supports, rather than thicker ones. This suggests that pea plants may encounter greater biomechanical challenges when attempting to grasp larger support, likely due to the increased energy demands required to stretch the tendrils and generate sufficient force to overcome gravitational constraints (cf. Getter & Rowe, 2006).

In line with these findings, Ceccarini et al. (2020) demonstrated that *Pisum sativum L.* perform a series of sub-movements during the approach phase, adjusting their tendril positions more frequently when the support is thicker. Generally, plant movements can be divided into two phases: an initial movement phase and a subsequent phase of control or correction (cf. Dipietro et al., 2011; Ceccarini & Castiello, 2018). As observed in animals, this characteristic reflects the need to fine-tune movements through secondary corrective sub-movements to enhance accuracy (cf. Fradet et al., 2008). In summary, tasks that require greater accuracy result in more secondary adjustments, an observation that applies to plant movements as well.

The emerging evidence suggests that climbing plants (at least based on results from pea plants) possess the ability to flexibly and contextually modulate both the speed and accuracy of their movements, depending on the type of support and the difficulty of the task. This behavior appears to be goal-directed, achieved through sensory abilities and perception-action processes within a given environment (cf. Castiello, 2020, pp. 9–10).

However, much remains to be clarified regarding how climbing plants perceive, locate, and differentiate among potential supports (cf. Guerra et al., 2022). Several hypotheses, some of which remain speculative, have been proposed concerning the mechanisms and sensory-perceptual capacities involved in stimulus identification.

One suggestion is proprioception, whereby the plant perceives its own spatial configuration, enabling it to modulate the aperture of its tendrils in accordance with the thickness of the support. Another hypothesis explores the possibility that plants can detect electromagnetic fields, while others suggest a form of echo-based sensing, whereby plants emit clicking sounds and interpret the returning echoes to gather spatial information (cf. Gagliano et al., 2012).

To identify the position and features of potential supports, plants may also rely on a highly sophisticated light-sensing system, possibly analogous to a rudimentary form of vision, through which they may generate a form of environmental "representation." Plants are highly sensitive to light and can detect a wider portion of the electromagnetic spectrum than humans. However, the extent and nature of their capacity to "see" remains an open question. Some researchers (Haberlandt, 1905; Baluška & Mancuso, 2016) have speculated that plants, despite lacking a nervous system, might also be capable of image perception. This could be facilitated by the anatomical structure of their leaves, particularly the upper lamina and the underlying epidermis, which contain cells that may function like "ocelli," eye-like structures that act as lenses (cf. Gianoli, 2017). A notable example often discussed in this context is *Boquila trifoliata*, a climbing plant capable of mimicking the size, shape, and color of the leaves of neighboring host plants.[8] Such mimicry might enable *Boquila* to resemble species that are typically avoided by herbivores, thereby reducing predation risk (cf. Gianoli & Carrasco-Urra, 2014; Mancuso, 2017, pp. 63–71). Ongoing research suggests that this imitation may even occur in the absence of direct physical contact with the host, although the underlying mechanisms remain unknown.

In addition to the vision hypothesis, several other explanations have been proposed to account for the perceptual abilities of (climbing) plants. One line of inquiry focuses on chemical recognition of the environment, including aerial communication via volatile organic compounds (VOCs). A related hypothesis concerns chemoreception, that is, the ability to detect and respond to ecologically informative chemical cues (cf. Runyon et al., 2006; Karban, 2015; Ceccarini et al., 2020; Guerra et al., 2022). Another hypothesis involves the possibility of horizontal gene transfer as a mechanism that could facilitate certain adaptive responses in plants (cf. Chen et al., 2021).

The extent to which a plant's aerial structures (such as stems and tendrils) interact with the root system in these perceptual processes remains an open question. For example, Guerra et al. (2022) investigated how *Pisum sativum* integrates above- and below-ground information to guide its movement toward and twining behavior around the support. Their findings suggest that the root system contributes to

[8] This instance of plant mimicry, although less frequently studied and observed than animal mimicry (cf. Lev-Yadun, 2016; Niu et al., 2018), refers to the adaptive resemblance between an organism and its model. It is particularly intriguing from a cognitive perspective, as it involves a dynamic process that occurs during the individual development of plants, allowing them to adapt to current environmental conditions.

sensing the thickness of nearby supports, thereby informing the plant's motor responses and goal-directed movement.

In this broad, multilevel research on the perceptual-cognitive abilities of climbing plants (and plants in general), it is important to note that these organisms possess multiple sensory modalities (cf. Karban, 2015). It may be more realistic to hypothesize that their behavior results from multisensory perception and signaling (cf. Bianchi et al., 2025a), involving the integration of several processes and abilities rather than a single sensory capacity (cf. Betti et al., 2021).

This fascinating field of research, with much still to be discovered, is expected to advance both through laboratory investigations in controlled environments and through studies conducted in more natural ecological contexts.

3.2.5 Motor Intentions?

A fascinating aspect of research on climbing plant movement—one that may point to the presence of cognition in plants—is the possibility of motor intentions. Indeed, both the type and velocity of plants' movements appear to vary in response to the characteristics of the support and the specific context of interaction (cf. Wang et al., 2021).

The question of whether plants exhibit intentional behavior has a much earlier origin. For instance, von Sachs (1887) proposed that all adaptations of an organism are "intentional," in the sense that they contribute to its survival (Trewavas, 2014, p. 256). However, in these contexts, the reference is not to mentalistic intentionality, reflective or self-conscious, which would imply the presence of mental content associated with subjective will (cf. Maher, 2017, p. 108). Furthermore, there is a distinction between the term "intention" and "intentionality." Generally, the term *intention* refers to goal-directed actions performed by an agent, whereas *intentionality* has traditionally been associated with mental states, specifically, the property of being directed toward or about something. For instance, Franz C.H.H. Brentano (1838–1917), in his *Psychologie vom empirischen Standpunkt* (1874), defined intentionality as the directedness of the mind toward an object, which he considered the defining characteristic of all psychic phenomena. According to Brentano, it is impossible to believe, desire, or perceive without believing in, desiring, or perceiving *something*, that is, all mental acts are necessarily about or oriented toward an object (cf. Di Francesco et al., 2017, pp. 64–65).

When applying the concept of "intention" or "motor intention" to plant behavior, it refers to bodily activities directed toward a goal, such as preparing to interact with an object, that are significant for the plant's condition and survival. In the case of climbing plants, a significant issue is whether they always move toward a support structure in the same way or if their movements are adjusted and articulated based on the final goal. As noted by Castiello (2019, p. 79) and Wang et al. (2021), who focused on these very topics, the biomechanical constraints related to the structural and physical limitations of plants do not seem sufficient on their own to explain the variation in movements and the diversity of plant behavior. From this perspective, plants appear to share similarities with many animal species in their ability to interact with objects in their environment through a form of cognitive control over action, which implies the ability to distinguish, modulate and flexibly vary their behavior (cf. Castiello, 2021). This suggests that plants may possess a capacity for selective motor control and anticipatory planning (anticipating action-effects) in their movements. Wang et al. (2021) suggested that the motor behaviors of climbing plants might meet the criteria for acts as investigated in the experimental paradigm of Libet (1985), later discussed in the literature on intentional actions: (I) the action originates endogenously in response to environmental context, rather than being a stereotyped mechanical reaction to external stimuli (e.g., plants moving toward a support to access light); (II) the movement is not triggered or governed by rigid external constraints or impulses (climbing plants appear to regulate their interaction with supports); and (III) the organism may initiate, sustain, or interrupt a behavior (plants may begin or stop movement when no suitable supports are present or existing ones prove inadequate). Together, these findings open new paths for research into whether the motor behavior of climbing plants is modulated by the intentions, collaborative or competitive, of neighboring plants. For instance, are these behaviors affected by the presence of other plants (cf. Bonato et al., 2023)? Several complex questions remain to be addressed: to what extent, and how, can plants plan their movements? Can these non-neural organisms recognize and "interpret" the motor intentions of other organisms? If so, how do they accomplish this? Does the intention to engage socially involve movement patterns, chemical signals, or epigenetic mechanisms shared by plants and other life forms, or are some of these processes unique to the plant kingdom? (cf. Wang et al., 2021).

3.3 On Behavior

Discussing plant activities necessarily requires a more detailed examination of the concept of behavior to determine which conception of "behavior" can be applied to the study of plants. This type of analysis is particularly significant as current approaches are more inclined to interpret plant interactions through the lens of behavior rather than notions of intelligence or cognition. Plant cognition is often conceptualized as a form of behavior—arising from a constellation of processes including communication, memory, learning, anticipation, and problem-solving—which manifest through observable variations in behavioral expression (cf. Bianchi, 2021).

Furthermore, from a practical standpoint, the study of plant behavior, approached from various perspectives, can provide deeper insights into a wide range of ecological consequences. Due to the dynamics of interdependence, these consequences impact not only the plant kingdom but also the entire biosphere (Karban, 2008).

From a broad and inclusive perspective, applicable to all species, behavior in the bio-ecological field has been understood as the ability to adapt to environmental changes through various modifications, selecting and adjusting responses to different challenges encountered during development (cf. Lyon et al., 2021). Behavior, therefore, represents the active part of a living organism's functioning, engaging in actions and exchanges with the world. Like any biological trait, behavior is linked to genomic specificities, which are relatively stable and resistant to change. However, as a phenotypic characteristic shaped by a specific ecological niche, a key feature of behavior is its capacity to vary, within its range of possibilities, under the influence of environmental factors and developmental processes (cf. Trewavas, 2014, p. 62). The fact that behaviors do not have a single, unchangeable relationship with a given goal is central to understanding behavioral flexibility. The variety of behavioral options reflects the different responses organisms have to environmental challenges, and contrary to common belief, these responses may or may not involve muscular movement.

Many survival challenges, such as obtaining food, finding mates, evading or defending against predators, and responding to pathogens, are shared across species. The main differences in behavioral strategies are not determined by the type of goal or survival problem but rather by the means used to address it. These means, which differ depending on the strategies employed, are shaped by the anatomical and functional organization of each living system. As Maturana and Varela (1980) observed, anatomical structure places limits on the range of possible behaviors. Both structure and function must be considered in terms of the conditions that define a system as a unit of interaction, while at the same time allowing variability and preserving its organization.

Behavior can be analyzed at several levels. For instance, in evolutionary biology, a distinction is made between ultimate causes (why a behavior evolved) and proximate causes (how and under what circumstances a behavior is expressed). Recent years have seen a growing emphasis on ecological and relational aspects of behavior (LeDoux, 2020, p. 73). An additional aspect to consider, of a meta-reflexive nature (referring to the ability to observe one's own modes of observation), is the relational dynamic between the observer and a given behavior. This means that an organism's set of responses across various situations is identified and interpreted by an observer (cf. Nucara, 2014, pp. 95–96). From the perspective of the organism itself, this is simply a matter of different possibilities, whether self-conscious or not, through which it achieves a goal or solves a problem.

Some scholars argue that certain elements are common to all forms of behavior. One such feature is the "teleonomic aspect." On this point, Trewavas (2014, p. 27) noted that it is hard to envision life without teleonomy. According to him, only a teleonomic organism can survive in environments characterized by resource scarcity. However, it is important not to confuse the modern use of this term in biology and cognitive-behavioral studies with teleology or finalism, which suggests that nature is organized with a specific ultimate purpose, rather than through processes involving randomness and contingency. Teleonomy, on the other hand, refers to an organism's capacity to act in non-stereotyped ways in order to maintain, and, in some cases, enhance, its viability or living condition (cf. Lyon et al., 2021). Thus, teleonomy should be understood as the characteristic of living beings to be directed, through performance and self-regulation processes, toward goals essential to their survival, sometimes in seemingly intentional ways (cf. Monod, 1971; Trewavas, 2014, p. 256). From these general considerations, behaviors across different organisms can be seen, broadly speaking, as context-dependent, variable, and reversible responses (cf. Silvertown & Gordon, 1989), though they operate within a limited set of options throughout the organism's lifespan. This understanding of behavior also allows for the inclusion of plant behavior in such analyses.

Despite the existence of broader and more inclusive ways of understanding behavior across the diversity of life, common sense still tends to associate behavior, especially that learned during individual development, primarily with the activity of a nervous system and the use of muscles for action. This perspective, however, would lead to the exclusion of organisms such as unicellular life forms, which recent studies suggest exhibit behavioral traits, including memory and learning abilities (cf. LeDoux, 2020, p. 25), as well as non-neural organisms like plants. Excluding unicellular organisms, plants, fungi, and certain sessile animals from behavioral considerations, due to their differences from animal organisms (which are often used as the reference model), seems unjustified today.

By focusing on these differences—not as evidence against the possibility of behavioral activity in plants, but as distinctive traits that offer valuable insights—we can first ask how animal and plant behaviors differ and then examine the unique characteristics of plant behavior.

As previously noted, plants lack the sensorimotor organization found in animals for regulating and coordinating behavior (cf. Calvo & Keijzer, 2009). Their behaviors, as will be explored in more detail, can involve movement and are largely driven by changes in metabolism and growth rates that enable the organism to adapt its phenotype to varying environmental conditions.

Just as plants are capable of perceiving and carrying out vital functions without organs organized in the manner typical of animals (e.g., breathing without lungs or obtaining nutrients without a stomach), they are also capable of exhibiting behavior in their unique way.

3.3.1 Some Peculiarities of Plant Behavior

According to Trewavas' historical reconstruction (2014), one of the earliest discussions of plant behavior was presented in the mid-twentieth century by Agnes Arber (1879–1960). Reinterpreting certain Goethean ideas, Arber argued that the variability of plant morphology could be likened to forms of behavior typically explored in zoological research.

It is now increasingly recognized that, despite lacking locomotion, plants exhibit flexibility in responding to various stimuli, exploring both aerial and subterranean environments (cf. Silvertown & Gordon, 1989). The growth-related changes they undergo are not entirely random; instead, they reflect a kind of "sense" (e.g., consider the directed growth and increased root mass in areas with more abundant resources [Trewavas, 2014, pp. 73–76]).

In terms of efficiency, plant responses to environmental stimuli must be relatively quick to meet specific demands in a timely manner. Although growth processes are irreversible, plants responses should be quite flexible or variable at least on a metabolic-molecular level (Karban, 2008, p. 727). For instance, the production of various compounds by plants, often in response to herbivore or pathogen attacks, can be initiated, modulated, or halted as needed (cf. Metlen et al., 2009). Consequently, plant behavior is not always irreversible, as was once believed (Trewavas, 2014, pp. 83; 91). Additionally, it appears that plant behavior includes some form of discrimination and recognition of relevant differences, enabling plants to select certain factors under changing conditions (cf. Baker, 2017, p. 145). A closely related aspect is the plant's ability to respond to the costs and benefits associated with a particular behavioral trajectory, responses that involve energy expenditure and require trade-offs in the redistribution of internal resources among tissues. What is allocated to one tissue will not be available, or will be available in reduced amounts, to another (Trewavas, 2014, p. 90).

A relevant question, one that has prompted considerable reflection, particularly since the concept of plant cognition or intelligence has gained attention in recent years, is whether behavior occurs at the level of individual modules or at the level of the whole organism. It is likely that behavior operates at both levels, depending on the task at hand. Indeed, meristematic activity enables plants to change during development and perform various functions. Due to their modular structure, many plant responses are localized, though there is some signal integration, especially through electrical signaling (cf. Karban, 2008, p. 727). In addition, when considering information exchange via biological signaling networks, it is plausible that problem-solving behavior may emerge either from localized processes or as a systemic property of the plant as a whole, depending on the context (cf. Parise et al., 2022).

Lastly, from a broader perspective, it is important to consider that plant behavior is not always perfectly regular, even though general patterns often remain predictable. As Mazzolai (2021, p. 142) observed, certain behaviors may deviate from typical plant predispositions. For instance, some roots have been shown to grow against gravitational cues in sterile soils in order to access nutrients applied to the surface by agricultural practices, such as fertilization.

Although less frequently criticized than cognitive interpretations or the concept of plant intelligence, the notion of plant behavior has also faced some skepticism. In some cases, when plant activity has been framed as a form of behavior, it has been interpreted as purely mechanical and reactive to environmental stimuli, often likened to the functioning of automatic doors equipped with motion sensors. However, as Segundo-Ortin and Calvo (2019) argued in response to Adams (2018), current empirical evidence does not support such a reductionist analogy. Unlike automatic systems, which operate based on simple input-output mechanisms with no internal modulation or flexibility, many plant behaviors appear to be goal-directed and context-sensitive. Plants have been shown to prioritize among competing stimuli and to integrate multiple biotic and abiotic cues simultaneously. Rather than displaying fixed or stereotyped responses, they often exhibit dynamic, selective, and temporally extended patterns of activity that reflect ongoing environmental monitoring and resource recognition.

To explore plant behavior in greater detail, a few illustrative examples will be presented here, although not all cases can be addressed at this stage. Additional examples will be discussed in subsequent sections.

One example involves processes related to movements in response to light and dark, which is essential for the synthesis of compounds necessary for plant growth and development (cf. Seluzicki et al., 2017). Plants not only follow and anticipate the sun's movement, bending their stems accordingly, but also adjust the position of their leaves to maximize exposure to light, thereby promoting growth. This ability is supported by the ability to store information about the location of dawn for several days, even in the absence of light signals.

The importance of light has also been studied in relation to how plants space themselves when growing close together. For instance, Gruntman et al. (2017) reported that the phenotype of *Potentilla reptans* can be altered based on the height and density of competing plants. Plants can detect variations in the ratio between red light and far-red light. Direct sunlight contains more red light, whereas far-red light becomes dominant in shaded conditions. Because plants absorb red light and reflect far-red light, an increase in far-red light signals to plants that they are surrounded by competitors (cf. Frazier, 2021). In general, plant behavior related to light can be categorized as either light-seeking or light-avoiding, and shade-tolerant or shade-avoiding. For instance, shade-avoidance responses in plants growing close together often involve structural changes, such as increased vertical growth to enhance light exposure. Shade tolerance, on the other hand, involves morphological adaptations like increasing leaf area and metabolic adjustments to maintain various functions even in low-light conditions. In cases of competition-driven shade avoidance or light acquisition, an increase in horizontal branch growth has also been observed (cf. Segundo-Ortin & Calvo, 2019). These examples underline the varying degrees of involvement of different plant systems. In light-seeking or shade-avoiding behavior (cf. Nozue et al., 2015), the aerial parts of the plant, especially the branches and leaves, are directly involved. In addition to expanding their surface area, leaves can adapt by increasing the rate of new leaf production, regulating internal temperature, and adjusting stomatal openings to optimize photosynthesis. Stems, particularly those containing chlorophyll, can also respond to shading, much like branches and leaves.

At the root level, typical processes include the ability to absorb minerals, sensitivity to nutrient deficiencies, avoidance of obstacles or harmful substances in the soil, and responses to gravity. Additionally, roots can send signals to attract mycorrhizal fungi or initiate symbiotic relationships with certain species of bacteria (cf. Trewavas, 2014, p. 272).

To some, the previously mentioned examples might appear less like actual behaviors and more like simple modifications of physiological activity in response to certain environmental stimuli, with visible structural effects. One way to approach this issue could be to examine behaviors resulting from metabolic and structural changes not just at the localized level of a single organ or system, but more broadly at the level of the entire organism. Examples that could address this concern include defense behaviors against herbivores and pathogens, as well as foraging behavior. In the case of defense, plants employ various strategies, often activated simultaneously, such as structural defenses, chemical production, programmed cell death, and even the recruitment of predators that attack the "plant's enemies" (cf. Appel & Cocroft, 2014; Campbell et al., 2015). The most common active, modular defense strategy against herbivores and pathogens is the production and release of substances that function as natural pesticides. It is estimated that plants produce thousands of these compounds. For example, when an insect begins feeding on a plant's leaf, the plant can respond by producing phytohormones that signal the production of toxins to defend itself. These signals are transmitted within the plant's cells and tissues, and, in some cases, can even be detected by neighboring plants. These adjacent plants may then activate their own defense mechanisms to protect themselves from an impending attack (Toyota et al., 2018).[9]

In the case of root foraging behavior, depending on local needs and conditions, it can be either fast and imprecise or slower and more precise. Foraging can occur over a relatively small or large area, and under conditions of resource scarcity, it may trigger defense mechanisms to protect the occupied territory or lead to the exploration of new areas. It may also result in the sharing of space and resources, particularly among genetically related plants. Understanding these mechanisms is crucial for comprehending ecosystem diversity and stability (cf. Oborny, 2019).

Plants' ability to coordinate information from various sources plays a key role in their capacity to adapt their nutrient acquisition strategies (cf. de Kroon et al., 2009). Foraging behaviors are influenced by different conditions. For instance, Cahill et al. (2010) found that the root growth patterns of *Abutilon theophrasti* differ depending on whether the plants are grown in isolation or with other species. When grown in isolation, the plants exhibited maximum root distribution and width, regardless of resource availability in the soil. However, when in the presence of other species, they adjusted their foraging strategy based on resource distribution. Thus, the presence of neighboring plants and the availability of resources significantly influence foraging behavior. Moreover, root activity directly affects the behavior and

[9] It is also possible to distinguish between different effects produced by various plant compounds. Some substances directly kill the pathogen, while others hinder or discourage predators from feeding on the plant. For example, mucilage, gum, or sticky substances can make the plant difficult to digest (cf. Trewavas, 2014, pp. 50–51). A specific example is provided by the geranium (*Pelargonium* spp.), which produces a compound in its petals that paralyzes certain beetle species within about thirty minutes. Another example comes from some acacia populations in the savannah. When attacked by giraffes, these acacias increase the production of unpalatable compounds, causing the giraffes to reduce their consumption of the leaves and move on to the next tree that has not yet begun this defense process (cf. Zinn et al., 2007). For plants, this allows the leaves to recover quickly, while from the giraffes' standpoint, it ensures the plant population is not entirely depleted, thus preserving a future food source (cf. Castiello, 2019, p. 87).

functioning of the aerial parts of the plant. Plants employ various sensory mechanisms to identify other organisms and initiate foraging strategies. For example, a study by Latzel and Münzbergova (2018) showed that *Fragaria vesca* L. clones can associate nutrient locations in unoccupied soil with specific light intensities, emphasizing the relationship between light perception and nutrient foraging in photosynthetic plants. In contrast, the foraging strategy of parasitic plants like *Cuscuta* (commonly known as dodder) follows the Charnov model, typically applied to animals, which describes the investment and gain of energy in changing environments. This model is relevant to dodder's behavior, and it can be divided into distinct phases: detection of volatile organic compounds (VOCs) emitted by host plants, an encounter phase, and a decision phase related to nutrient acquisition (cf. Runyon et al., 2006). During the search phase, dodder appears capable of individuating suitable hosts based on potential resource availability, anticipating nutrient acquisition in advance (Kelly, 1992; Trewavas, 2014, pp. 267, 275).

Thus, it appears that plants exhibit active modification behaviors, which can also include mimetic abilities, such as the "manipulation" or "deception" of non-plant species (cf. Mancuso, 2017; Karban, 2017). One example of "manipulation" comes from the mutualistic relationship between ants of the genus *Pseudomyrmex* and a species of acacia. In this relationship, the ants receive food and shelter from the plant, and in return, they defend it from pests and other insects, as well as clear out shoots of competing plants in the surrounding area (cf. Campbell & Reece, 2009, p. 1235). Some studies suggest that this interaction involves a form of modulation by the plant, as it can increase the sugar content of its nectar, an energy source for ants, when under attack by herbivores, thereby inciting the ants to repel enemies with greater intensity.

Similarly, some plants have evolved to modify the digestive capabilities of insects, forcing them to rely solely on the plant's nectar for nourishment. Alternatively, certain plants add small amounts of nicotine to their nectar to deter unwanted visitors, such as "nectar thieves" (cf. Bonato et al., 2021).

An example of a "exploitative" or deceptive strategy comes from the pollination mechanisms of certain orchids, such as *Ophrys speculum* and *Ophrys apifera*. These orchids attract male wasps by mimicking the appearance, scent (pheromones), and color of female wasps. As the male wasps attempt to "mate" with the flower, the orchid's pollen attaches to their bodies and is transferred to other flowers. Another form of deception involves brood-site mimicry, where flowers imitate the smells of decaying organic matter (e.g., feces, urine, carrion) that female insects use as oviposition sites (cf. Raguso & Kessler, 2017). Such dynamics and others like them allow plants to regulate the emission of certain substances or control the opening and closing of flower components as needed. For example, *Arum palestinum* traps *Drosophila* flies inside its flower, coating them with pollen while attracting them through the scent of fermented fruit (cf. Reece et al., 2015, p. 95; Mancuso, 2017, pp. 96–98).

Across all of these strategies, the behavioral and communicative aspect of plants may lie in their ability to flexibly modulate their signaling responses in relation to the presence of different elements in their ecological environment.

3.3.2 *Signaling and Communication in Plants*

Arguably, the most sophisticated form of plant behavior involves different levels of signaling and, according to some scholars, genuine "communication," primarily—though not exclusively—mediated by chemical compounds. Including plants in discussions on communication requires an ecological-relational perspective, interpreting communication in a broad sense as the ability to engage in productive interactions with both conspecifics and members of other species. This, at a general level, seems to include the capacity to differentiate between "self," "non-self," "us" and "other than us" (cf. Lyon et al., 2021). The study of plant communication is part of broader comparative research in biocommunication, which explores how different organisms use various systems of signs and signals, following syntactic, semantic, and pragmatic rules suited to their contexts (cf. Bonato et al., 2021). Like animals, plants ignore irrelevant messages (e.g., from species with which they do not interact) and can detect and "decode" molecular combinations that hold essential "meaning" for their survival (cf. Bianchi et al., 2025a).

Plant chemical signaling has been examined in various ways and at multiple levels. For instance, allelopathy refers to processes where plants selectively secrete substances that inhibit and slow the development of neighboring plants (the reverse phenomenon is the so-called "allelobiosis," a term use to describe non-competitive, potentially cooperative chemical interactions between plants). These processes have been associated with territorial competition and collaboration, particularly at the root level (cf. Mancuso, 2006).

In plant signaling, a distinction is often made between internal communication within the plant and external chemical signaling, which can influence both the functioning of the plant itself and that of other plants or organisms. Internal signaling between different tissues coordinates responses to environmental stimuli, shaped by physical, structural, and physiological constraints. For example, vascular communication occurs between leaves connected by the vascular system (cf. Orians & Jones, 2001), while volatile communication can occur between unconnected leaves. However, volatile signals are effective only over short distances, as they quickly degrade due to wind and other climatic factors. Unlike internal or intercellular signals, volatile signals are "public," traveling through the air and potentially being intercepted by other plants or nearby organisms (cf. Heil & Karban, 2010). Long-distance communication, on the other hand, is facilitated through mycorrhizal networks (cf. Karban, 2017).

In recent years, research on volatile communication, particularly under the label of volatile organic compounds (VOCs), has become extensive (cf. Bianchi et al., 2025a). Thousands of these compounds have been identified, spanning more than 20 different molecular groups, including alcohols, aldehydes, esters, large classes of terpenoids, aromatic compounds, and various nitrogen- or sulfur-containing molecules (cf. Raguso & Kessler, 2017). Many VOCs remain unidentified, and their roles in biological communication are not yet fully understood. The diversity of these signals—such as the distinct VOC "signature" released during herbivore attacks compared to that following other types of wounds (cf. Arimura et al., 2005)—likely arises from the vast array of interactions, not only between plants of the same or different species but also with soil organisms (particularly fungi and microbes) and various insect species (cf. Bais et al., 2003; Witzany, 2008). The release of VOCs is primarily triggered by the need to attract pollinators (cf. War et al., 2012), respond to attacks from different predators or pathogens, and adapt to environmental stressors (cf. Baldwin et al., 2006).[10]

Plant communication involves not only chemical signals related to VOCs but also changes in shape and color, which are often directed at non-plant organisms, such as pollinators. One example is *Desmodium setigerum*, which changes the color of its flowers from lilac-purple (more attractive to bees) to white or blue (less attractive to insects) after pollination. This color change is accompanied by a shift in the shape of the flowers, where the petals fold over the reproductive parts once pollination has occurred. It is interesting to note that *D. setigerum* has the ability to reverse the change in flower color and shape if pollination has been insufficient, allowing for a second opportunity for pollination (cf. Willmer et al., 2009).

These visual signaling processes extend beyond flowers to include leaves as well. In addition to emitting volatile compounds, green leaves produce visual cues—such as radiometric signals and reflectance profiles—that herbivores and other animals use to assess a plant's overall health as a potential food source or oviposition site. These signals, which are "readable" by other species, are invisible to the human eye without special equipment. However, they can be detected using tools like infrared cameras, imaging technologies, and spectrometers, which are valuable for assessing a plant's stress level (cf. Nansen, 2017).

Plant communication occurs not only above ground, but also below-ground through root and mycorrhizal networks (cf. Simard et al., 2012; Johnson & Gilbert, 2015; Simard, 2018). Since the research by Simard et al. (1997) on the relationship between paper birches and Douglas firs, it has been understood that the exchange of nutrients and resources along mycorrhizal networks can be interpreted as a form of "cooperation." In this intricate network, several signaling processes critical for plant survival have been identified (cf. de Kroon et al., 2009), including root exudates, composed of thousands of secondary metabolites, plant hormones (cf. Wang et al., 2009), and maybe electrical signals within the network (cf. Thomas & Cooper, 2022). These signals have even been compared by some researchers to a rudimentary nervous system (cf. Rhodes et al., 1996; Baluška & Mancuso, 2007). On the other hand, the role of mycorrhizae in plant communication has recently been called into question by some researchers (cf. Karst et al., 2023; Henriksson et al., 2023).

[10] For instance, certain herbaceous plants, such as *Epipactis helleborine*, emit a spectrum of compounds similar to those released by other plant species when attacked by herbivores, effectively signaling for the aid of predators that target the herbivores (cf. Brodmann et al., 2008). Even more interestingly, some orchid species have been observed to release these compounds even in the absence of herbivore attacks, using them instead to attract insects for pollination (cf. Bonato et al., 2021).

In general, numerous studies have shown that information exchange at the root level, particularly between roots, mycorrhizal fungi, and certain bacteria species, influences gene regulation. This, in turn, impacts processes such as photosynthesis, nitrogen fixation and conversion into ammonia and nitrates by nitrogen-fixing bacteria, and overall plant defenses and stress responses (cf. cf. Falik et al., 2012; Pellegrino & Di Paola, 2019, pp. 95–96).[11]

A frequently asked question is whether plants, by emitting and detecting these informational signals, are actively and "intentionally" communicating (as many animals do) or whether they are merely engaging in signal detection, akin to "eavesdropping" on environmental information (cf. Chamovitz, 2012; Karban, 2017; Raguso & Kessler, 2017). Regardless of whether communication is initiated with "intention" (understood here as a targeted or goal-directed behavior rather than conscious will), the key point is that plants can perceive, differentiate, and appropriately respond to these signals. From an ecological perspective, plant communication systems are highly efficient for their ecological needs and the functions they must perform, which often involve collective protection. Some have even suggested this aligns with group selection theories, where plants "warn" their conspecifics to defend against predators or environmental stressors (cf. Pellegrino & Di Paola, 2019, p. 98).

Beyond the debate over intention in plant communication, there are indications of specificity and greater communicative efficiency among plants within the same population. This has led some (cf. Moreira et al., 2016) to liken plant communication to the linguistic phenomenon of dialects, coining the term "phytodialect." Additionally, variations in communication may be influenced by kinship, with some findings suggesting that plants could modify their signaling patterns, potentially even developing distinct signal variants, to minimize the risk of interception by antagonists. These geographical "phytodialects" of VOCs may reflect differences in pollination strategies and competitive or collaborative interactions within local communities compared to distant populations of the same species. Supporting this idea, plants with greater genetic similarity should respond more strongly to VOCs released during herbivore attacks than to signals from more genetically distant individuals, suggesting a "private" communication channel shared among relatives (cf. Raguso & Kessler, 2017). Geographic variation in plant communication has been observed in several cases. For instance, fig species such as *Ficus racemosa* and *Ficus hispida* exhibit divergence in their terpenoid profiles, likely due to geographic isolation (cf. Soler et al., 2011). Additionally, studies on sage plants have shown that the effectiveness of VOC signals varies depending on the plants' origin. Local plants respond more strongly to damage

[11] An example showing that plants may be able to enhance their preparedness for potential future osmotic and drought stress has been observed in *Pisum sativum*, *Cynodon dactylon*, *Digitaria sanguinalis*, and *Stenotaphrum secundatum*, which were found to be able of communicating the imminent arrival of severe drought conditions through their root network (Falik et al., 2012). This is evidenced by the closure of stomata in neighboring plants that have not yet experienced water stress, thereby preventing potential water loss. This type of information is critical to the plant community, as dry soil is generally not conducive to plant survival.

signals than plants of the same species growing several hundred kilometers away. Furthermore, sage plants with leaf damage release VOCs that are characterized by two heritable chemotypes, which enhance communication compared to different chemotypes (cf. Bonato et al., 2021). These dynamics are also being investigated within the frameworks of biosemiotics and phytosemiotics, which examine the interpretative and communicative dimensions of biological interactions (cf. Witzany, 2006).

In addition to genetic links and geographical factors, plant communication is also influenced by context and dosage (the amount of substance emitted). Both abiotic and biotic factors, such as the types of species present, or stage of development of the plant, constrain and shape the content of the information. This means that the same compound can mediate a variety of interactions over time and in different contexts (cf. Witzany, 2016; Raguso & Kessler, 2017, pp. 27–30).

Another important aspect, often compared to animal communication, is the combinatorial nature of plant signaling, which allows for differentiation in meaning (cf. Bonato et al., 2021). Terpenoids, one of the most diverse classes of compounds involved in plant communication, may play a key role in this process (cf. Penuelas et al., 1995). These chemical "units" appear to combine into larger "messages" whose meaning depends on context, including concentration. For example, at lower concentrations certain terpenoids may attract pollinators, wheras at higher concentrations they can act as repellents (cf. Harborne, 1991).

Within the broader debate on biocommunication, as recently noted by Bianchi et al. (2025a), a promising direction for further investigation—given the multisensoriality of plants, i.e., their possession of multiple sensory and perceptual abilities to respond to environmental stimuli and stressors (cf. Lamers et al., 2020)—could concern the possibility that plants engage in multimodal communication. This would involve the use of multiple communication channels, such as chemical signaling, visual cues, and movement patterns, allowing plants to flexibly regulate their behavior across different environmental contexts. This raises the question of whether plants communicate exclusively through "multicomponent" modalities, involving multiple signal elements within a single mode of communication (cf. Higham & Hebets, 2013), or whether they truly engage in "multimodal" communication (cf. Partan, 2013).

Both multisensoriality and multimodality could offer evolutionary advantages in terms of communication and survival. For example, they may enhance signal detection in "noisy" environments, where certain communication channels may be diminished or disrupted by external conditions (cf. Munoz & Blumstein, 2012; Otto et al., 2013). Multimodal signaling, as a form of phenotypic plasticity, could help to compensate for weak signals, enhance communication, and facilitate plant adaptation in dynamic environments. At the same time, different types of signals (e.g., chemical, movement-related, color- or shape-based) exhibit distinct properties, such as different transmission ranges and temporal persistence (cf. Colemann, 2009). Consequently, it remains unclear to what extent these different channels are integrated, and how they interact in non-neural organisms such as plants.

Recent studies have shown that certain flowering species communicate with pollinators using composite signals, including visual, olfactory, tactile, electrical and thermal cues, which significantly improve the efficiency of pollen localization (cf. Rands et al., 2023).

Key questions for future research include whether plants use "fixed" signals—i.e., signals that are always combined across modalities (e.g., a visual cue that is always paired with a chemical or acoustic cue)—or "free" signals, in which elements from different modalities occur independently and are not necessarily coupled (cf. Partan & Marler, 2005). Thus, it remains to be explored whether plants have fixed or free combinations, or both, and to what extent they exhibit flexibility in their signaling strategies. Another central issue concerns sensory dominance: do plants primarily rely on certain modalities in specific contexts, or are they capable of switching from one communicative channel to another in response to ecological demands? Additionally, it is necessary to investigate how sensory and communicative capacities evolve throughout the individual development of the plant—i.e., how the type and effectiveness of communication change across life stages—and whether such dynamics differ between species (cf. Bianchi et al., 2025a).

A deeper investigation of these dimensions of plant communication will enhance our understanding of behavioral flexibility in plants and their interaction with the environment. In turn, this will help clarify what is specific and unique to plants and what is shared with animals and other organisms, ultimately expanding our ecological knowledge and contributing to a more integrated perspective on the communication capabilities of living systems across biological kingdoms.

3.4 Processes of Memory, Learning, Decision-Making, Anticipation, and Attention

In this section, I examine the relevance of understanding how cognitive behaviors in plants are expressed through processes such as memory formation, learning, decision-making, anticipatory responses, and possibly even rudimentary forms of attention. Without these capacities, it would be difficult for an organism to behave effectively; moreover, from an explanatory standpoint, it would be challenging to observe or substantiate claims of cognitive behavior.

3.4.1 Types of Memory

To begin, I consider the concept of memory. At a general level, within the study of psychobiological faculties, memorization capacity has been conceptualized in multiple ways. Initially, when the underlying mechanisms of different memory forms were not well understood, it was metaphorically perceived as a physical space for storing information. Over time, memory has been reconceived as an active process, representing the neurocognitive capacity to acquire and retrieve various kinds of information. From the perspective of encoded information, memories are structured configurations that allow past experiences—once present events—to shape current behavior when reactivated through specific procedural mechanisms. As bioecological knowledge has advanced, the reifying conceptions of memory, which focus less on its multifactorial and processual nature, have proven limited from several perspectives. These conceptions have gradually been either refined or replaced by approaches better suited to address the multilevel complexity of the systems and processes involved (cf. Bianchi, 2021).

In other cases, the presence of the process that is referred to as "memory" has been completely criticized. For example, the traditional cognitive science notion of memory as the retrieval and processing of information has faced opposition within the ecological approach and related theoretical frameworks. In fact, memory, as understood in cognitivism, is considered extraneous to the ecological approach to perception. Instead, this approach focuses on mechanisms that support environmental interaction and learning by examining the intrinsic link between individual development (ontogeny) and the evolutionary development of a species (phylogeny). This connection is essential to an organism's capacity to "know" its environment by discerning its relevant features (cf. Michaels & Carello, 1981, p. 76). While this perspective offers valuable insights—suggesting that many of an organism's behavioral possibilities are already available for use, and that many aspects of its environment are simply recognized without requiring complex information processing—the terms "memory" and "memorization processes" are used here to conceptually frame an organism's capacity to track variations in environmental interactions over time. A renewed conception of memory, which is more nuanced, empirically grounded, and aligned with the bio-ecological realities of organisms, is particularly valuable today.

One might wonder why memorization processes, operating at different levels, are so crucial to the organization of living (cognitive) systems. One possible answer is that memory capacity is fundamental to the agentive identity of organisms, forming the basis for diverse historical pathways in individual experience. This perspective allows these processes to be viewed in terms of a system's history of interactions rather than as mere stereotyped responses to stimuli. Even in cases of stereotyped responses, could these still reflect a form of species-specific memory or information storage? In a sense, this perspective can connect to the ecological view of the link between species, individual and environmental characteristics. A related aspect concerns how the processes through which information is retained are transmitted across generations, a form of intergenerational memory.

From the perspective of the organs involved in memory processes—a localized yet widespread capacity—in the context of classical neurobiology, research has predominantly focused on the formation of new connections and the strengthening of existing ones within neuronal networks in animal brains (cf. Baker, 2017, p. 151), which can channel information flows in complex ways (cf. Trewavas, 2014, p. 272), as observed across simpler and more complex systems (cf. Kandel, 2001). While memorization dynamics in nervous systems have been the primary focus, recent years have seen growing attention to non-neural "biological memory" processes (cf. Baluška & Levin, 2016). Memory and learning do not strictly require a nervous system, though the complexity and flexibility of these processes may vary greatly.[12]

Memory capacity can be studied at multiple levels within an organism, starting with cellular functions relevant to both unicellular and multicellular organisms (LeDoux, 2020, p. 71). Beyond synaptic memory, which involves specialized cells like neurons, memory has been approached systemically, in terms of "memories of the body" or "bodily memories", that arise from chemical and electrical interactions among various cell types and tissues (cf. Shomrat & Levin, 2013). Biological memory is a process that can integrate the function of an entire organ—and, in some cases, the whole organism—in continuous interaction with its environment (cf. Morabito, 2020, p. 131).

A range of processes involving aspects of memorization do not rely directly on the nervous system. For example, the immune system exhibits memory-like functions through the actions of white blood cells and antibodies. In certain cases, forms of long-term immune memory support cell replication, forming the basis for the initiation of defense responses (cf. Trewavas, 2014, p. 273).

Overall, these broadened perspectives on memory have led to operational definitions that are applicable to non-neural organisms as well.

As reported by Galviz et al. (2020), here are some wide-ranging definitions from the Merriam-Webster dictionary: (i) "*the power or process of reproducing or recalling what has been learned and retained especially through associative mechanisms*," (ii) "*the store of things learned and retained from an organism's activity or experience as evidenced by modification of structure or behavior or by recall and recognition*" or, simply, (iii) "*capacity for storing information.*" The study of plant memory is likely to deepen our understanding of the cellular basis of memory in living things and to support the development of a concept of "systemic memory," a set of processes characteristic of biological networks involved in signaling and storing information across various levels.

Regarding types of memory attributed to humans and, in some cases, other animals, psychology and cognitive studies distinguish several categories: working memory; procedural memory (also referred to as implicit or skill memory); conceptual memory (or semantic/factual memory); and episodic memory (cf. Marraffa & Paternoster, 2013). Working memory involves the capacity to hold information while it is actively being used by the system; procedural memory, which does not provide a representation of how something is or was, refers to the ability to remember how to perform tasks (concerns the ability do something, such as riding a bike); conceptual memory, often overlapping with what cognitive psychology refers to as semantic memory, covers the recall of specific or general facts (e.g., the release year of a film or the name of the capital of Norway); finally, episodic memory involves specific personal experiences (such as recalling a concert attended) (cf. Maher, 2017, pp. 81–82). An interesting aspect emerged from brain imaging techniques is

[12] In single-celled protists, evidence shows the presence of learning and memory abilities, even in the absence of a brain (cf. LeDoux, 2020, p. 71).

that brain regions associated with episodic memory activate even when a person imagines a possible future event. This finding suggests that such abilities are part of a broader capacity for offline processing, applicable both to past experiences and to the imagination of hypothetical or future events (cf. Godfrey-Smith, 2021, pp. 288–291). These distinctions are traditionally used to differentiate forms of human memory, yet researchers have attempted, with necessary modifications, to extend these concepts to other species as well.

In the study of animal memory, an important consideration raised by LeDoux (2020, p. 395) concerned whether animals form unified (and possibly conscious) "what-where-when" memories, or instead rely on multiple discrete memory traces or simple associative mechanisms. A possible example of unified memory is found in studies by Nicky Clayton and colleagues (2001, 2003) on Western scrub-jays (*Aphelocoma californica*), which can hide worms (what) in various locations (where) and later retrieve them (when), feeding first on older food.

A type of procedural memory could also be relevant in describing certain abilities in plants. For instance, Chamovitz (2012) suggested that, based on current understanding, plants appear to possess an anoetic, nonconceptual capacity (not related to self-conscious or metacognitive representations) that enables them to remember "how" to do something. This would be a case of knowing "how" rather than knowing "what" (cf. Patterson, 2020). Nevertheless, it may not be necessary to align plant memory abilities with types of memory identified in humans or other animals. If this were possible, it would certainly facilitate our understanding. However, regardless of the specific type of memory under consideration, it remains that plant behavior involves some capacity to keep track of information. Indeed, plant behavior manifests the ability to organize and utilize information in ways that support adaptive responses to their environment (cf. Maher, 2017, pp. 83–85; Bianchi & Castiello, 2023a, 2023b).

3.4.2 Memory Processes in Plants

To date, numerous mechanisms for storing information in plants have been identified, enabling both metabolic and anatomical adaptations (cf. Thellier & Lüttge, 2013; Cvrčková et al., 2016). Memory functions contribute to various physiological and behavioral processes in plants, including predatory behavior in certain carnivorous species, thigmotropic and phototropic responses, and adaptations like stem thickening in response to wind (cf. Maher, 2017). Depending on the level and scope of analysis—ranging from more localized to systemic perspectives—various memory-related processes can be identified. These include the highly dynamic and flexible nature of metabolic memory, which involves biological networks that modulate, store, and reconfigure biochemical states; the more conservative and stable domain of genetic memory; and epigenetic memory, which occupies an intermediate position between the two (cf. De la Fuente, 2015).

Therefore, it is essential to clarify the appropriate approach for studying a specific process, in this case, plant memory, whether focusing on an isolated mechanism or exploring a broader systemic organization. For instance, even plants, which lack neurons, exhibit a form of "synaptic" memory connected to chemical-electrical signaling between non-neural cells (cf. Calvo et al., 2016).

Research on plant memory brings forth fascinating questions about the type and extent of memory integration in modular organisms, grounded in the analysis of signaling networks (cf. Bianchi & Castiello, 2023a). Here, memory is perceived as a decentralized phenomenon affecting various types of cells and tissues, not limited to neural structures, and influencing organismal responses at multiple levels. Investigations into plant memory processes, understood as signaling variations at the cellular level, can thus deepen our understanding of memory itself, revealing its functions across diverse organizational scales (cf. Castiello, 2019, pp. 99–100; Bianchi & Castiello, 2023a, 2023b).

With regard to the physiological processes underlying information storage in plants, much remains to be clarified (cf. Gagliano et al., 2014). Several studies have investigated calcium signaling pathways—including calcium ion channels and calmodulin-mediated mechanisms—which are known to regulate gene expression in plant cells (cf. Yellen, 1998; Halling et al., 2005; Tuteja & Mahajan, 2007; Kim et al., 2009). Similar to processes observed in animal brains, bioelectrical gradients between non-neural plant cells can accumulate and transmit information via electrical signaling pathways (cf. Volkov et al., 2008).

Epigenetics constitutes a crucial level of regulation for molecular memory, a process common across taxa (cf. Ginsburg & Jablonka, 2009) and particularly pronounced in plants. It enables the retention of chromatin modifications, such as DNA methylation and histone acetylation, which modulate gene expression. In some cases, these modifications can even be transmitted to the next generation as a form of "soft inheritance" (cf. Michmizos & Hilioti, 2019).

Epigenetic changes may be induced by external stimuli (cf. Yaish et al., 2011), particularly by environmental stressors (cf. Ding et al., 2012). However, it is important not to assume that these mechanisms are universal or constantly active. Some researchers have argued that epigenetic memory may occur less frequently than previously thought, with stress recovery and epigenetic resetting often representing the predominant responses (cf. Crisp et al., 2016). A well-documented example of somatic epigenetic memory is vernalization, the requirement for cold exposure in certain plant species to initiate flowering. In *Arabidopsis thaliana*, the gene *Flowering Locus C* (FLC) encodes a protein that inhibits flowering. Together with other regulatory genes, FLC responds to prolonged cold exposure by becoming epigenetically silenced, thereby lifting the repression on flowering after a sufficient period of vernalization. Chromatin modifications associated with this silencing are maintained through cell division, enabling subsequent generations of cells, even those not directly exposed to cold, to flower at the appropriate time (cf. Karban, 2008, p. 734; Castiello, 2019, pp. 91–92).

Compared to environmental stress memory (cf. Conrath et al., 2001)—involving stressors such as extreme frost, drought, wind, herbivory, and pathogen attacks—plants adapt through priming, which enhances resistance and response intensity to future exposures (cf. Haider et al., 2021). Since stressors often recur (cf. Hilker & Schmülling, 2019), this adaptation is advantageous. However, in certain cases, it may be more adaptive for organisms to "forget" stress-related memories, as the retention of such information can sometimes impede recovery and growth. Striking a balance between memory formation and physiological restoration is therefore essential for optimal resilience (cf. Crisp et al., 2016). Moreover, plants exposed to specific stressors can produce offspring with greater resistance to the same stress, as observed with UV exposure, which can alter gene expression variability and be inherited by all progeny. If this had resulted from a genetic mutation, the sequence would be present in only a subset of the offspring (cf. Molinier et al., 2006).

Beyond epigenetic memory, researchers have proposed that plants may possess an "immunological" memory (cf. Yakura, 2020) to defend against pathogens, including fungi, bacteria, and viruses. This memory allows defense signals to be swiftly transmitted to unaffected parts of the plant and recalled during subsequent attacks (see Systemic Acquired Resistance and Induced Systemic Resistance).

In terms of the duration of plant memory phenomena, a distinction exists between short-term memory (lasting seconds to minutes) and long-term memory (spanning weeks to months) (cf. Hilker & Schmülling, 2019). Instances of short-term memory in plants include, for example, the coiling of tendrils in response to touch after a period of darkness and re-exposure to blue light, which provides the energy for movement and is stored as information in plant cells (cf. Jaffe & Shotwell, 1980). Another example is the rapid trap closure in *Dionaea muscipula* (or *Drosera capensis L.*), which requires the temporary storage of bioelectrical information for about 20 seconds to complete the movement (cf. Hedrich, 2012). In contrast, an example of longer-term memory, lasting approximately a month, can be observed in the habituation response of *Mimosa pudica* L., in which the plant gradually ceases to respond to a repeated stimulus that has proven to be non-threatening, provided the environmental conditions remain stable (cf. Gagliano et al., 2014).

Many memory processes occur at the cellular and intercellular levels, while those involving electrical signaling can also extend to the entire plant body. Going beyond the physical, porous, boundaries of the plant itself, some authors have proposed the concept of "ecological memory," an "extended" memory that refers to the ability of plants to leave chemical traces in the environment (cf. Simard, 2018). These traces influence the growth of future generations of plants, as well as animals, fungi, and bacteria sharing the same ecological niche, thereby establishing a loop of reciprocal influence that shapes the development of all involved.

3.4.3 *Learning Processes*

Studies on memory are significant due to their close connection with the capacity for learning and anticipatory behavior in organisms, an indicator of cognitive activity often considered fundamental to biological intelligence (cf. Trewavas, 2014,

p. 229). While memory is distinct from learning, it is essential for the acquisition and retention of information over time, both at the species level (for innate aspects of behavior and all that is inherited) and the individual level (for elements that have been experienced). Learing, therefore, requires the capacity to monitor and retain information within the organism's functional processes, enabling its reconstruction, organization, and subsequent reuse (cf. Maher, 2017, p. 82).

On a broader level, one may ask: What is learning? Various definitions exist, including the inclusive conception of learning as a biological and ecological process through which organisms associate and acquire new information, or "knowledge," about their environment (cf. Kandel et al., 2014). Learning has been described as the ability to adjust behavior (in general terms, a form of adaptive plasticity) by modifying the intensity or nature of responses to significant environmental stimuli, based on past experiences (cf. Ginsburg & Jablonka, 2009; Lyon et al., 2021). In this sense, learning can be described as a historical process in which an organism's behavior transforms through individual experience, supporting the maintenance of its organization (cf. Maturana & Varela, 1980; Nucara, 2014, p. 98).

Learning implies some form of change; in fact, all natural processes undergo change (cf. Bateson, 1976, pp. 328–329), enabling quick and efficient responses to numerous environmental challenges, critical for survival (cf. Darwin, 1859). Considering the interconnected structure of an organism and its environment, learning does not involve simply acquiring something from the environment. Rather, it expresses the compatibility of interactions between an organism's behavior and the environment in which it exists (cf. Maturana & Varela, 1987).

In developing an ecological approach to understanding perceptual activity in organisms—an area that seeks to dismantle forms of epistemic dualism, as was similarly noted regarding the conpect of memory—the goal is to understand how an animal (or more broadly, any organism) becomes capable of detecting structured distributions of energy that hold ecological value for it. From this perspective, learning becomes an "education of attention" to invariants that guide a living system's activity. Organisms understand their environments through evolutionary processes, which prevent them from starting from scratch and enable them to discern and learn details of their specific environments. Both evolutionary history and individual experience contribute not to a mere accumulation of knowledge, but to an enhancement in understanding and an improved capacity to interact with the environment (cf. Michaels & Carello, 1981, pp. 77–82).

Various approaches have been used to describe natural learning processes, each emphasizing different aspects of the phenomenon. Generally, learning ability has been studied in organisms with nervous systems (both simple and complex). However, recent studies have shown that learning also occurs in organisms without brains (cf. Baluška & Levin, 2016). The adaptive advantage provided to animals by neurons, which can be modified through interaction with the environment due to synaptic plasticity, forms the basis of their learning capabilities and is indeed relevant. Although all organisms, even those without nervous systems, are capable of simple learning according to their unique ecological demands, the nervous system allows for more flexible and variable learning (cf. LeDoux, 2020, p. 177). In some species, there is also evidence of the capacity to "learn to learn," a higher-order skill under investigation (cf. Bateson, 1976, p. 395). For instance, recent studies in the field of animal cognition have focused on aspects of cognitive flexibility, as well as on the ability to adapt learning strategies based on previous experiences (cf. Pearce, 2008; Gupta et al., 2021).

Regarding the study of learning in non-neural systems, such as plants, a recurring and relevant question concerns whether learning occurs solely at the cellular level or, at the systemic level, across the entire organism (cf. Cvrčková et al., 2009). This distinction is particularly important for us, "unitary" organisms (organisms with a relatively unitary body plan), where there is a strong interest in understanding learning as an integrated process that might involve a "felt" experience at the organismal level (cf. Affifi, 2013).[13]

Another important point to remember is that learning should not be seen as a linear, cumulative, and continuous acquisition of new traits or behaviors. Much of an organism's biological organization is maintained rather than altered, and learning capacity is always constrained; in some species, learning is significantly more limited than in others.

Moreover, historically, there has been no clear consensus in research on the mechanisms underlying learning processes. For example, the Danish immunologist Niels Jerne (1911–1994) proposed in 1967 that learning in the brain might function analogously to the immune system. According to Jerne, the environment selects from preexisting neural circuits, activating those most relevant to a given situation. In this view, learning time corresponds to the time required for the brain to retrieve or activate the neural circuit best suited to address a particular challenge from its existing repertoire (cf. Gazzaniga, 2019, p. 272). Despite these differing perspectives, organisms, through memory and learning processes, are able to navigate their environments, select among options shaped by both species-level and individual experiences, and refine their responses within dynamic yet recurrent interaction contexts.

Plants also appear to engage in this kind of adaptive dynamic.

3.4.4 *Ability to Learn in Plants*

At a general level, plant learning abilities are grounded in forms of intracellular epigenetic and intercellular memory associated with the exchange of bioelectrical and hormonal signals, and are shaped through the formation or modification of pathways for information flow (cf. Trewavas, 2005; Heras-Escribano & Calvo, 2019; Ginsburg & Jablonka, 2021). At the physiological level, plant learning is understood to occur through complex networks of chemical and electrical exchanges

[13] Based on recent evidence, the possibility of learning on an organism-wide scale, tailored to the specific organism, is highly probable across all organisms. However, it is more challenging to assert that this capacity also entails a "felt" experience at the level of the entire organism (cf. Affifi, 2013).

that can be modified over the course of development (cf. Trewavas, 1999). Due to fundamental anatomical differences between plants and animals, learning in plants does not involve the reinforcement of neural connections, but rather the modulation of plasmodesmal connectivity and tissue organization. Depending on the type of environmental stimulus and the task at hand, plants adapt by forming new structural and signaling networks, rather than altering dendritic connections (cf. Trewavas, 2003, p. 14; Trewavas, 2014). Specific stimuli activate signal transduction pathways that lead to changes in plant behavior (cf. Ginsburg & Jablonka, 2009; Trewavas, 2014, p. 170). These behavioral modifications are informed by prior interactions (cf. Michmizos & Hilioti, 2019, p. 211) and may include feedback regulation mechanisms that align with the organism's functional goals (cf. Trewavas, 2003; Clark, 2013).

From a practical standpoint, an effective way to determine if an organism has the capacity to learn is to place it in novel situations and observe its responses (this has typically been carried out using animal organisms).

Within the context of plant learning, a general critique has emerged. Building on Firn's (2004) arguments, Adams (2018), for example, expressed skepticism by suggesting that what some researchers interpret as plant learning might instead reflect genetically predetermined responses, rather than an ability to "learn" over the organism's lifetime in the traditional sense. However, recent findings indicate that plants may enhance their chances of survival during development by regulating their behavior more actively than previously assumed (cf. Segundo-Ortin & Calvo, 2019). One compelling issue raised by these critiques is the distinction between individual and intergenerational forms of learning.

In this case what makes the issue particularly interesting—and why we should remain open to exploring and testing new hypotheses—is the possibility, over time, of identifying both common traits and species-specific differences that could enrich our understanding of biological functioning and the ecological relationships of different living systems. Achieving this, however, requires patience and a willingness to adjust initial expectations, which are often shaped by analyses based on human or other animal characteristics.

In studying the behavior of living things, various types of learning have been identified, with some of the simplest being forms of nonassociative learning. One example is habituation, a process enabling organisms to ignore nonharmful or irrelevant environmental information and concentrate on stimuli essential for survival (cf. Eisenstein et al., 2001; Castiello, 2019, p. 93). The opposite of habituation is sensitization, which involves heightened perception or an intensified response to a stimulus that was previously unperceived (cf. Bianchi, 2021).

A central concept in behaviorism that illustrates the close relationship between learning ability and survival potential is operant conditioning. First explored by Edward Thorndike in the late nineteenth century through his law of effect experiments, and later formalized by B. F. Skinner (1954), operant conditioning refers to a form of learning in which the likelihood of a behavior (e.g., the ability to learn to perform a task, such as pressing a lever to escape a cage) is modified by its consequences, enabling animals to acquire or suppress actions based on positive reinforcement (e.g., food rewards) or negative reinforcement (e.g., discouraging certain actions). Repeated exposure to positive reinforcement increases the behavior being repeated in future situations. Through goal-directed, instrumental learning, outcomes learned by trial and error directly shape current behavior. However, for creative problem-solving, the ability to anticipate future outcomes becomes essential (cf. LeDoux, 2020, p. 281). In the analysis of behavior, past, present, and future are recognized as inseparably intertwined within dynamic interactions. Memory enables learning, learning fosters anticipation, and anticipation is made possible by the processes of previous interactions, creating a continuous cycle of adaptation and response.

A well-known form of associative learning is classical conditioning, a process by which an organism learns to associate two stimuli: an unconditioned stimulus (naturally elicits a response or instinctive behavior) and a conditioned, initially neutral stimulus (which at first evokes no spontaneous response) (cf. Castiello, 2019, p. 96; Bianchi, 2021).[14] This type of conditioning has been observed across nearly all animal classes. Functional analogues have also been reported in protozoa and bacteria, aneural organisms that cannot rely on synaptic plasticity, but instead use intra- or intercellular molecular changes to establish stimulus-response associations (LeDoux, 2020, p. 264).

Forms of learning have also been investigated in plants, and interest in plant learning has grown significantly in recent years. From a behavioral psychology perspective, studies have examined various processes. For instance, Calvo and Keijzer (2011) discussed how plants might exhibit trial-and-error-like strategies, adjusting inter-organ signaling pathways in response to environmental stimuli. This mechanism allows plants, as observed by Darwin (1875) and von Sachs (1887) in climbing species, to disregard unsuitable support structures. In such cases, tendrils initiate coiling, stop, unwind, and search for another target. Similarly, if a support is moved, vines detect the change and adjust their growth direction, seemingly following a trial-and-error approach rather than simple reflex movement (cf. Pellegrino & Di Paola, 2019).

Research on learning capacity in plants has focused on two main types of behavior: habituation and classical conditioning. Habituation studies have shown positive results, while classical conditioning requires further investigation and consistent experimental setups in both controlled environments and natural growth contexts (cf. Ponkshe et al. 2024).

One of the earliest documented examples of plant habituation was described by Darwin (1875), who observed that the tendrils of *Passiflora gracilis* could respond to mechanical stimulation approximately 21 times over a period of about 54 hours before ceasing to react (cf. Trewavas, 2014, p. 86; Castiello, 2019, p. 95). Another

[14] One of the most widely known examples of classical conditioning is Pavlov's (1849–1936) experiment with dogs. Initially, the dogs salivated at the sight of food (unconditioned stimulus). After repeated pairings of the sound of a bell (conditioned stimulus) with the presentation of food, the dogs began to salivate at the sound of the bell alone. In summary, they were conditioned to associate the bell with food.

widely cited case concerns the behavior of *Mimosa pudica*, which rapidly closes its leaflets in response to touch (presumably as an anti-predatory defense mechanism). The first study on habituation in *Mimosa pudica* in response to non-harmful stimuli, evidenced by the absence of leaflet folding, was conducted by Wilhelm F.P. Pfeffer (1845–1920) in 1873. In recent years, Gagliano et al. (2014) conducted a systematic analysis by dropping *Mimosa* plants in plastic pots from a height of 15 cm, securely mounted on steel rails. After a brief period, the plants habituated to the repeated drop stimulus, perceiving it as harmless, yet continued to close their leaves in response to other stimuli, such as lateral shaking or touch. These control tests confirmed that the plants' reduced response was due to habituation rather than "fatigue" (depletion of energy resources) or sensory adaptation (cf. Rankin et al., 2009). Remarkably, this habituation lasted up to about one month, indicating long-term memory. The study validated Pfeffer's findings and revealed individual variations in plant responses based on context and condition. Further observations showed that plants grown under energetically demanding conditions, such as low-light environments, habituated more quickly to the dropping stimulus, ceasing to fold their leaves sooner, likely due to the higher energetic cost of leaf closure (cf. Reed-Guy et al., 2017).

Researchers have also attempted to study plant learning through classical conditioning, aiming not to reinforce an existing behavior but to introduce a new behavior by creating a response association triggered by a stimulus that initially elicits no automatic response from the plant (cf. Castiello, 2019, pp. 94–96). Trewavas (2014, p. 250) pointed out the specific challenges in studying this form of learning in plants (and plant learning in general), largely due to the limited range of stimuli that prompt a growth response, primarily sunlight, the main energy source. An additional complication is that plant behavior involves growth and metabolic changes that continuously reshape the organism's structure. As a result, it becomes difficult to assess the same individual consistently over time (you won't have to deal with the same plant twice). While animals also undergo constant changes, these transformations are less apparent in the short term and easier to control in behavioral studies.

Despite these challenges, often conditioned by the limitations of experimental designs adapted from animal studies, one of the first attempts to study associative learning in plants was conducted by Holmes and Gruenberg (1965). They attempted to condition a plant to associate a curling response to touch (unconditioned stimulus) with an electric discharge (neutral stimulus). However, after four days of training, the experiment was unsuccessful (cf. Castiello, 2019, pp. 96–97).

In a later, well-known study, Gagliano et al. (2016) investigated associative learning in *Pisum sativum* (pea plants) using a Y-shaped structure. They paired the plants' innate response to light with the conditioned stimulus of an air current from a small fan, which indicated the future appearance and location of the light. The results suggested that conditioned plants—those repeatedly exposed to the fan in conjunction with the light—tended to grow toward the arm of the maze where the fan had been located, even in the absence of light. This behavior indicated that the plants might encode spatial and temporal information regarding light, associating the two stimuli and adjusting their growth accordingly. However, subsequent attempts to replicate the experiment have so far not succeeded in producing the same results. For example, Markel (2020) found no conclusive evidence to support associative learning in plants. Instead of consistently bending toward the arm where light was last presented, the pea seedlings appeared to grow randomly in both maze arms, suggesting no anticipation of the light source's return. It is necessary to clarify the importance of the interactions between phototropism, gravitropism, and circumnutation (which, for example, increases during the dark phase in some species). It is also important to determine whether the tips of growing shoots can remotely perceive the physical constraint of the maze, thereby altering their growth direction. Moreover, monitoring the internal dynamics occurring within the maze through time-lapse imaging is required. Further studies are needed to refine the understanding of this response mechanism. Research at the MINT Lab has addressed these and other related aspects (cf. Ponkshe et al., 2024). In any case, this line of research has sparked a stimulating debate and laid the groundwork for future investigations.

The analysis of this interesting case underscores the importance of experimental reproducibility. Many findings (including those in animal cognition studies) used widely in the literature lack replication. Therefore, to assess the associative learning abilities of plants, more evidence is required (cf. Abramson & Chicas-Mosier, 2016), by adjusting test configurations to align with the bio-ecological characteristics of plants, and by monitoring plant responses over time.

However, even if plants were unable to engage in associative learning, this would not negate their potential for other forms of learning or, more broadly, (cognitive) behavior.

Moreover, a goal for future research is to verify plant learning capabilities in natural contexts, where growth and interaction occur spontaneously, considering the full range of ecological factors involved.

3.4.5 Anticipatory Behavior

Closely linked to studies on memory and learning abilities in plants is the exploration of anticipatory behavior, one of the traits currently considered most indicative of cognitive activity in organisms. From a sensory-behavioral perspective, anticipation is the capacity to predict (or estimate) what is likely to occur based on a prior stimulus (cf. Lyon et al., 2021). Organisms maintain their overall organization partly by using this predictive ability, adapting themselves to anticipated future conditions. It seems that only systems that effectively anticipate their environment, rather than merely reacting to it, can survive (cf. Hohwy, 2013; Clark, 2016). The basic logic of this survival dynamic has been observed even in simple life forms. For instance, *Escherichia coli* uses environmental sensors to process information, thereby guiding its responses in ways that increase the likelihood of acquiring bioenergetic resources (cf. Trewavas, 2014, p. 208).

Predictive processing, central to research on biological cognition and life, has been examined more systematically in recent years through statistical models in accordance with the Free Energy Principle (FEP) (cf. Friston, 2010, 2013; Kiverstein & Sims, 2021). In this framework, free energy, is a theoretical quantity that approximates surprisal, the degree to which incoming sensory data deviate from what is predicted by the system's internal model. Biological systems are thought to maintain their organization by minimizing such free energy over time, thereby reducing environmental uncertainty. Predictive processing is widely regarded as the mechanism through which the FEP is implemented in cognitive systems: it generates predictions about sensory inputs, compares them with actual inputs, and updates internal models to reduce prediction error, the mismatch between expected and observed signals. Minimizing such errors is essential to maintaining systemic integrity and preventing disintegration (cf. Friston & Kiebel, 2009; Allen & Friston, 2018).

Anticipation relies on estimating the probability that a given sensory state corresponds to an actual environmental source affecting the system. To reduce prediction error, organisms adopt various strategies, including the development of hierarchically structured generative models of their environments, the use of perceptual inferences, and the use of active inferences that enable direct engagement with the environment (cf. Calvo & Friston, 2017). In this framework, even organisms without nervous systems appear capable of selecting contextually appropriate behaviors that maximize practical epistemic value, that is, actions that are accurate, adaptive, and conducive to sustaining future viability.

From this standpoint, in the specific case of plants (see plant predictive processing [PPP]), Calvo and Friston (2017) proposed analyzing certain plant behaviors through active inference, a process in which plants predict the environmental sources of sensory stimuli. This form of anticipatory perception allows plants to minimize prediction errors by updating "expectations" through plastic changes in their phenotype. These adjustments combine forms of "perceptual" inference (adapting internal states) with "active" inference (selectively engaging with the environment). Anatomically, this predictive processing may involve the plasmodesmata conduction pathways and the highly connected hierarchical structure of vascular bundles, through which electrical and chemical information flows. Like other living organisms, plants can endogenously shape environmental variation.

A significant theoretical issue recently discussed (cf. Bianchi & Castiello, 2023a, 2023b) is whether plant anticipatory behavior necessarily involves computational and representational elements (i.e., whether any form of coding and inference must rely on representational models) (cf. Calvo et al., 2016; Gallagher, 2017). This question is especially pertinent when considering the potential to explain these processes through non-representational post-cognitivist approaches, particularly certain forms of enactivism (cf. Kirchhoff & Robertson, 2018) and ecological approaches to cognition (cf. Heras-Escribano & Calvo, 2019). In this context, Calvo and Friston (2017) hypothesized that, although inferential processes may be involved, plant behavior could function as a system of dynamic coupling between plant and stimulus. Importantly, terms like "inference" or "model" do not necessarily imply "representation" in the traditional sense, a notion whose meaning, as will be discussed in Chap. 4, is broad and interpretation-dependent. For instance, in climbing plants, anticipatory behavior is evident in how they perceive and plan movement to grasp a support. The plant perceives environmental affordances directly, without requiring internal reconstructive representations; instead, the plant itself, in relational terms, constitutes the model that realizes the possibility of action directly (cf. Souza et al., 2018). This issue deserves further investigation, particularly regarding the mechanisms underlying these processes.

It is now possible to briefly recall some examples of plant behavior where anticipatory ability seems to be involved. In general, as in other organisms, predictive capacity is essential for plants to sustain and regulate their internal condition. Anticipatory behaviors, which relate to potential future interactions with environmental elements, are visible both above and below ground. For instance, shoots can branch, flower, and initiate tropisms depending on the sources and availability of light; many plants can anticipate future shading by detecting far-red light reflected from nearby vegetation, triggering phenotypic changes in response (Trewavas, 2014, p. 88). Similarly, at the root level, plants adjust their growth in response to the availability and type of nutrients, and the presence of obstacles in the soil, as well as through whole-plant modifications in response to environmental stress, preparing them to mitigate or better respond to future stressors (cf. Ding et al., 2012).

Starting with the aerial parts of the plant, some deciduous species appear capable of anticipating future environmental conditions, such as the onset of winter. In response to the shortening photoperiod in autumn, these plants shed their leaves in preparation for the colder temperatures that could otherwise damage leaf tissues (cf. Karban, 2008, p. 733). Additionally, studies have shown that chloroplasts can exhibit predictive responses to anticipated changes in light, adjusting the photosynthetic apparatus according to current conditions through a form of physiological computation (cf. Latzel et al., 2016).

Another example often cited as anticipatory behavior is the nocturnal reorientation of *Lavatera cretica* L. leaves toward the direction of sunrise (cf. Kreps & Kay, 1997; García Rodríguez & Calvo Garzón, 2010). This heliotropic movement persists for several days even in the absence of sunlight (cf. Schwartz & Koller, 1986), suggesting an endogenous mechanism that incorporates essential environmental rhythms into the plant's daily functioning. According to García Rodríguez and Calvo Garzón (2010), such temporal regulation does not necessarily require representational explanatory models, but can instead be explained through causal covariation, a type of explanation often applied in dynamic or oscillatory systems (cf. Heras-Escribano & Calvo, 2019).

Precisely because the capacity for anticipation is widely regarded as a "hallmark of cognitive activity," some authors have argued that interpreting the nocturnal reorientation of *Lavatera cretica* L. as indicative of an offline cognitive response is misleading (cf. Adams, 2018). But this does not so much clarify why the example should not be considered anticipatory behavior with cognitive relevance; rather, as Segundo-Ortin and Calvo (2019) have pointed out, it shifts the focus to the underlying causes of the behavior, rather than the behavioral dynamics themselves. In doing so, it risks conflating the explanandum with the explanans, implicitly reaffirming the notion that anticipation qualifies as cognitive only when it involves future-oriented representations. Here, the debate pivots on the level of cognition and the nature of representation itself. However, the questions and critiques raised in this area are instrumental in advancing research, ultimately helping to clarify the processes involved and the most accurate ways to describe these phenomena.

Instead, an example of goal-directed behavior involving the entire plant body, even though it concerns a very special kind of plant, is found in the parasitic, non-photosynthesizing genus *Cuscuta* (previously mentioned), which seeks nutrients from host plants. This search begins even before reaching the host and obtaining nutrients, suggesting a perception of how much energy to invest relative to the anticipated nutrient gain (cf. Calvo & Friston, 2017).

Fascinating studies have also explored root behavior that appears to involve anticipatory abilities. In terms of future resource acquisition, plants exhibit the ability to allocate root biomass according to resource availability and variability. Specifically, in *Pisum sativum* L. specimens whose roots of the same plant were separated and grown in pots under different nutrient regimes, varying over time in both quantity and resource availability, root growth was observed to increase under stable conditions of nutrient availability. In contrast, if nutrient concentration was insufficient to support immediate plant development, biomass increased in pots with varying levels of concentration, even when absolute resource availability was lower (cf. Dener et al., 2016). This and similar behaviors indicate that plants do not respond indiscriminately to absolute resource availability; rather, they can sense and track fluctuations in resource dynamics over time (increases and decreases), enabling them to anticipate growth and maximize long-term performance (cf. Novoplansky, 2016).

3.4.6 Decision-Making

Closely related to analyzing memory, learning, and anticipation, is the study of an organism's decision-making ability and prioritize responses based on pressing survival demands. The need to make a "choice"—not necessarily dependent on self-reflective or self-conscious capacities—stems from life unfolding in ever-changing environments characterized by uncertainty, where, on the other hand, it is still possible to estimate the likely outcomes of various behaviors. Decision-making involves integrating and organizing diverse types of information from multiple stimuli to select actions aimed at achieving a specific goal (cf. Lyon et al., 2021). When multiple needs arise, behavioural strategies must exist to initiate the most urgent response (cf. Trewavas, 2014, p. 75). Without the capacity to perceive, distinguish the various elements, and act appropriately, an organism would quickly perish. Responding in one way rather than another, or not responding at all, may play a role in preserving or improving an organism's condition.

From an observer's perspective, particularly in animals, a difficult choice or behavioral option made under stressful conditions is often evident in the increase in the slowness with which it is made. In some species, it has been hypothesized that goal achievement or resource acquisition relies on the use of spatial maps constructed from information about the relationships between environmental landmarks. These maps enable animals to compare options or identify the most efficient, shortest, or safest routes. Such behaviors have been observed in mammals, birds, and bees as they navigate threats, forage for food, or move (cf. LeDoux, 2020, p. 283).

3.4.7 Response Selection Processes in Plants

As with other organisms, the environment in which plants live is complex, variable, and competitive. The most effective decision-making strategy depends not only on the context—such as the presence of competing species or resource scarcity (response selection is a relational process)—but also on the specific traits of the plant species, including its size, developmental stage, and vigor (cf. Bianchi, 2022).

Similar to the dynamic interactions among cells that can modulate the frequency of transduction signals based on stimulus strength, plants also appear capable of adjusting the type and intensity of their responses according to a functional hierarchy of priorities (cf. Trewavas, 2014).

As observed in animal life, and likely in all forms of life,[15] the environment is cognitively parsed through a basic dichotomy, with gradations in between, into zones of immediate importance (those that, for various reasons, elicit greater attention) and more distant or less salient zones (those perceived as less relevant or attractive) (cf. Marraffa & Paternoster, 2013, p. 183). Similarly, plants appear capable of distinguishing between beneficial and harmful elements, thus discriminating what is useful and relevant from what is not. These categorizations of the environment—ways of structuring life contexts—lead organisms to differentiate responses based on utility/relevance/familiarity versus non-utility/irrelevance/unfamiliarity. This consideration becomes especially significant when studying plant cognition, as it underscores the idea that every living organism "knows" (recognizes) only what it needs to know (recognize) (cf. De Waal, 2016).

Some examples of processes in plants that resemble decision-making can be observed in light-acquisition strategies. These strategies include outcompeting neighboring plants, initiating structural modifications, such as increasing leaf surface area under shaded conditions to enhance light capture, shifting into

[15] Even a bacterium, equipped with sensors on its cell surface that respond to various environmental factors such as temperature, osmotic pressure, salinity, light, oxygen, and nutrients, can send signals to its genome that trigger responses essential for survival. Through this mechanism, it can distinguish between two types of environments: one containing beneficial substances and one lacking them or containing harmful substances.

energy-conserving modes, or adopting alternative growth trajectories (cf. Castiello, 2019, pp. 120–121). For example, shade tolerance is typically observed when a plant is faced with structurally taller competitors, whereas vertical growth strategies, such as those seen in *Potentilla reptans* L., are more likely to occur when the plant is densely surrounded by shorter neighbors (Gruntman et al., 2017).

Additional examples include the ability of climbing plants to distinguish and select supports based on specific characteristics. Regarding the duration of support-seeking behavior, Darwin (1875) reported that *Humulus lupulus L.* ceases searching after approximately five consecutive days of unsuccessful attempts (cf. Castiello, 2019, p. 129; Wang et al., 2023). Another example is *Berberis vulgaris* L., which, when its seeds are infested by parasites, may abort developing seeds to prevent the spread of infestation, effectively killing the larva by depriving it of nourishment (cf. Castiello & Guerra, 2020).

Even at the root level, processes have been observed that appear to indicate some degree of response selection capacity (cf. Hodge, 2009). One example is salt-avoidance behavior in *A. thaliana*, where salt exposure, which disrupts root cellular biochemistry, prompts roots to grow upwards, reversing their natural gravitropic tendency. A different type of response, involving the activity of the entire plant (cf. Trewavas, 2003, p. 5), is seen in parasitic, non-photosynthetic plants such as *Cuscuta pentagona*, which choose their host based on its nutritional composition. This species relies on both chemical cues (emitted substances) and light radiation from nearby plants to assess potential hosts (cf. Segundo-Ortin & Calvo, 2022). Host selection can be influenced by the availability of key nutrients; for instance, if a potential host lacks sufficient nitrogen, the parasite may abandon it in search of a more suitable alternative.

A more debated case that raises the question of whether it represents a form of decision-making or a simple reaction to environmental conditions is the timing of seed germination. Seeds should germinate neither too early (e.g., risking damage from lingering cold) nor too late, when competing plants may have already claimed available resources. As mentioned in Chap. 2, *A. thaliana* seeds contain a control mechanism involving two distinct groups of cells (cf. Topham et al., 2017). One cell group keeps the seed in a static state, while the other initiates metabolic activation for germination. Communication between these cell groups, relaying environmental information, occurs via hormonal signaling (cf. Castiello, 2019, pp. 123–124).

In this case, as perhaps in all cases discussed, one might argue that these are not true "decision-making" processes or "choices." Instead, plants may be in physiological states where a particular need arises (e.g., a deficiency of a specific mineral), prompting a specific response, perhaps the only feasible one. Thus, behavior is shaped not only by the external context but also by the plant's state in any given moment, where a particular need triggers a certain response. Constraints on the system may mean the response is almost determined, as it is the only viable option (cf. Bateson, 1976, p. 435). However, this could arguably apply to all organisms, including humans.

One possible way to confirm truly decision-making behaviors is to observe instances where, despite a pressing need, an organism "chooses" to delay a response in order to secure a potentially larger or safer future reward, rather than simply utilizing immediately available resources, even though future availability remains uncertain (cf. Kacelnik & Bateson, 1997). Noteworthy examples interpreted this way come from studies on sensitivity (inclination/aversion) to risk, particularly concerning the quantity and variability of resources available to roots. Generally, an organism will take risks to survive when resources are limited (but it depends on the context, presence, and what kind of species). In line with many animals, plants also appear to "play it safe" under stable, favorable conditions but tend to embrace risks in uncertain environments or "hard times" (cf. Castiello, 2019, pp. 125–126). A study by Dener et al. (2016) provided evidence of this behavior, emphasizing anticipatory abilities. As previously mentioned in the context of research on anticipatory behavior, the experiment involved dividing roots between two pots with different nutrient regimes: one with a constant nutrient concentration and the other with a variable concentration. The findings revealed that when the constant nutrient concentration was sufficient for development, *Pisum sativum L.* allocated more roots in this pot. Conversely, when the constant concentration was inadequate to support metabolic needs, more roots were directed toward the pot with variable nutrient concentrations. This behavior suggests that plants can shift between risk-averse and risk-prone behavior depending on their detection of resource availability. Schmid (2016) interpreted this behavior, exhibited in response to changing environmental conditions, as a form of decision-making, through which the plant selects the option that best ensures its future viability.

3.4.8 Attentional Processes

A capacity that appears to be closely associated with decision-making and the effective regulation of responses in organisms is attention, which generally enables the identification of salient elements in the environment, allowing for selective engagement with certain aspects of a situation while filtering out others (cf. Lyon et al., 2021).

In a broad sense, attention has been linked to various sensory-cognitive processes and abilities for identifying specific objects of interest. It can be described as a targeted selection process that shapes interactions with the surrounding environment, a cognitive function enabling the selection and processing of information (or sets of information) relevant to a specific context and time (cf. Grondin, 2016). Attentional processes can vary by intensity and selectivity.

The definition of "attention" varies slightly depending on the theoretical framework. Within ecological psychology, Gibson argued that the source of experiential variation lies not in the perceiver's mind, but in the diversity of environmental stimulation, which the organism actively selects. In this view, attention is understood as the process of engaging with stimuli, regarded as the control of detection of information relevant to ongoing activities and situated goals, rather than as an internal filtering mechanism (cf. Michaels & Carello, 1981, pp. 69–70). In contrast, dominant traditional accounts conceive of attention as the flexible control of limited cognitive resources, a mechanism for shielding the organism from cognitive overload by reducing the amount of information that must be managed to complete tasks essential for survival (cf. Kahneman, 1973; Castiello & Umiltà, 1990).

From a cognitive efficiency perspective, in integrated and dynamic biological systems, attentional processes, interacting with perception and memory processes, help reduce the complexity of the diverse, multilevel analysis of the reality. While much of the research on attention has focused on its relationship with perception and memory, studies have also demonstrated that this capacity can be operationalized through action. This is exemplified in the premotor theory of attention (cf. Rizzolatti et al., 1987), which conceives of attention as "selection for action" (cf. Allport, 1987), effectively preparing for the action. For example, when selecting a single object among many from a container, only the target object influences the type and direction of the movement (cf. Morabito, 2020, p. 105).

How can attention span be assessed? In humans and across the animal kingdom, attentional processes—their presence, increase, or decrease—have been studied throughout various phases of the sleep-wake cycle or during task execution following the administration of sedatives, with observable changes in behavior. Even after tranquilizers are administered, attention, responsiveness, and performance levels can be boosted by making activities more engaging. This can be achieved by introducing novel elements, adding irregularities, increasing stress levels, or offering rewards for task completion (cf. Oken et al., 2006; Parise et al., 2022).

Notably, understanding attention does not always necessitate the presence of a nervous system. A foundational basis for attentional or attention-like processes could lie in the capacity to distinguish and select relevant elements while disregarding others, moment by moment, even in aneural systems.

3.4.9 *Attention in Plants?*

At present, discussions of attention on non-animal organisms, particularly plants, remain rare, as research in this area is still in its early stages. An attention span in these organisms could enable them to identify relevant stimuli for performing specific tasks as required, depending on the context and the plant's particular physiological state. Even regarding this specific capacity, a relationship could exist with different levels of bodily memory and with the potential for learning and decision-making (cf. Thellier & Lüttge, 2013; Calvo et al., 2016; Galviz et al., 2020; Bianchi & Castiello, 2023a). In many species, distinguishing, recognizing and retrieving information seems to be linked to levels of attention (cf. Lozito & Mulligan, 2006).

Parise et al. (2021) have examined this possibility, exploring ways to detect and experimentally investigate plant attention variability in response to diverse environmental signals. This interdisciplinary approach includes research from psychology and philosophy—especially with a phenomenological emphasis (cf. Arvidson, 2006; Marder, 2009, 2011, 2013)—as well as plant physiology, notably electrophysiology, which is particularly relevant since electrical signaling is one of the primary channels for rapidly integrating body information to mount coordinated responses to environmental challenges (cf. Brenner et al., 2006; Fromm & Lautner, 2007; Baluska & Mancuso, 2009; de Toledo et al., 2019; Miguel-Tomé & Llinás, 2021).

How can such phenomena be investigated? In the human nervous system, bioelectrical activity—particularly gamma wave oscillations (typically ranging from 30 to 100 Hz) detectable via electroencephalography—has been associated with high-level information processing (cf. Berthouze et al., 2010). Such activity is thought to correlate with attention, as well as capacities for learning and problem-solving.

Parise et al. (2021, 2022) proposed the use of electrophysiological analysis techniques to investigate changes in the overall bioelectrical activity of plants in the presence of different ecological stimuli. Specifically, depending on the environmental stimulus, variations in the plant electrome can be analyzed (cf. Souza et al., 2017; de Toledo et al., 2019). The term 'electrome,' coined by De Loof (2016), denotes the totality of electrical activity—such as cell membrane depolarization processes, ion fluxes, and other bioelectrical phenomena—within an organism or its parts over time, measurable through emerging patterns. In humans, for example, this is captured through electroencephalograms (EEGs). In studies on the parasitic plant *Cuscuta racemosa* Mart., electrophysiological analyses revealed that the plant's electrome—recorded two hours prior to and two hours following exposure to potential hosts—varied depending on the type and viability of the host (bean or wheat, either viable or nonviable). The findings indicated that, even from a distance, the electrome exhibited increased energy, regularity, and predictability, particularly in the presence of a live, suitable host. These changes have been interpreted as indicative of a possible form of selective attention in plants (cf. Parise et al., 2021). This case is particularly intriguing because the task, which leads to electrome variation, requires coordinated behavior (and possibly an attention span) across the entire plant body, a phenomenon that seems rare in modular organisms, which consist of multiple semi-autonomous units (cf. Lüttge, 2021). For instance, a leaf typically does not respond to the same conditions as a root, which grows in a separate local environment. It would appear, therefore, that the extent and distribution of cognitive or attentional processes may vary accordingly, ranging from localized responses to organism-wide coordination. Besides the type of stimulus and context, changes in a plant's electrome—linked to possible attention processes and their duration—also depend on the plant species, its condition, and developmental stage (cf. Parise et al., 2022).

Other examples of processes that appear to involve a form of attention include, more broadly, goal-directed behavior, which entails the ability to distinguish relevant factors from irrelevant ones in order to initiate specific responses. A salient case is the range of movement behaviors in climbing plants—such as *Pisum sativum* L.—that are modulated according to stimulus features and demonstrate error-correction capabilities (cf. Elhakeem et al., 2018; Guerra et al., 2019; Ceccarini et al., 2020). In such instances, attentional processes may be inferred from the modulation of movement dynamics, including the speed of approach toward stimuli and variations in tendril aperture depending on the diameter of available support (cf. Castiello & Umiltà, 1992). Similarly, in certain forms of learning previously mentioned, such as habituation (e.g., the cessation of leaf-closing behavior in *Mimosa pudica* L. [cf. Gagliano et al., 2014]) and sensitization (cf. Conrath, 2006), attention processes may also be at play. These are observable as a decrease in responsiveness to insignificant stimuli and, conversely, an increased responsiveness to previously ignored stimuli. Even at the root level, cases of identifying different types of soil obstacles suggest some capacity for attention and discernment.

One of the main reasons for studying attention in plants is to understand how organisms lacking a nervous system and unable to locomote (though capable of movement and interaction, including chemical modulation) can execute accurate responses to address survival challenges. In this context, accuracy of action can be understood as a form of attention (cf. Alegiani & Marraffa, 2023; Bianchi & Castiello, 2023b).

What, then, makes this line of inquiry particularly relevant? Probably because continued research on the connections between memory, learning, anticipation and attention processes in plants could clarify the physical and processual basis of the flexibility of plant behavior, contributing to a more coherent picture of plant cognition and communication (cf. Bianchi & Castiello, 2023a, 2023b). This research may further inform a view of cognition that is not "hermeneutic" or cognitive in the classical sense of involving thought and concepts, but rather "pragmatic" (cf. Dominey et al., 2016) and action-oriented (cf. Morabito, 2023), closer to a "know-how," relating to the "where" and "how" rather than a "knowing of knowing," which relates to the "what" (cf. Bianchi, 2021, 2022, 2024).

Finally, as Parise et al. (2022) noted, while being cautious not to anthropomorphize (or zoomorphize) the issue, investigating potential attentional processes in plants may shed light (and does not seem entirely implausible) on aspects of plant "consciousness" or minimal awareness. This, in turn, could help to understand how plants coordinate the response behavior of a set of modules in reaction to a signal (a stimulus or group of stimuli). Might it be said, then, that in attending to a stimulus, a plant is, in some sense, "aware" of it?

3.5 Individuality and Consciousness

Among the most controversial and widely debated topics in recent studies of plant cognition is the question of whether plants possess some form of "consciousness."

The issue of consciousness is inherently complex, and even in animal studies, there is no single interpretation. Some researchers argue that only certain species possess consciousness, while others suggest that consciousness, in various forms and degrees, could extend to all life systems, including plants. Research in this area is ongoing, and the debate remains open. Here, I attempt to clarify some of these ideas, distinguishing between different types of consciousness (or "awareness") and, further, between consciousness and self-consciousness, terms that are often used interchangeably.

A goal here is also to adjust and rescale expectations, inviting us into a naturalistic perspective focused on understanding the traits of processes that are bio-ecologically relevant for plants, rather than for us, in relation to our constitution and our world of life, meanings, and interactions. If something is less complex or simply different, this does not mean that it is trivial or less interesting.

3.5.1 *Individuality in Divisible Organisms: The Problem of Integration and Coordinated Responses*

As mentioned at the beginning of the chapter, plants are modular organisms (divisible and dissectable) and this very characteristic, which relates to the limited integration of signals across the whole organism, lies at the heart of ongoing discussions about plant "individuality" and "consciousness." Specifically, Trewavas's (2002, 2003) description of plants as intelligent organisms has been challenged by Firn (2004), who argued that the concept of individuality cannot apply to plants due to their modularity, which supposedly prevents unitary responses. In response, Trewavas (2004) contended that, in general, individual plant parts cannot survive independently before they develop specific organs, and plants are indeed capable of coordinated responses (e.g., root behavior influences the aerial part through various signaling mechanisms, and vice versa) (cf. Latzel et al., 2016). Moreover, the vascular system, as previously noted, plays a fundamental role in transmitting various types of signals.

This argument can also connect to earlier discussions on memory levels, exploring the minimum degree of integration, organization, and capacity for information retention necessary for the organism. A pertinent question that arises at this point is: What type and minimum level of memory are required to establish the identity of the "self" or, rather, the individuality of the "agent"? Does this require a form of systemic memory? While leaving this question open, it is helpful to note that the concept of "subjectivity" as commonly attributed to humans, with self-reflective and conceptual dimensions, is often confused with the simpler notion of "individuality."

Given our current knowledge, while we may not speak of "subjectivity" in plants, we can at least consider the notions of history and individual developmental trajectory (individual path of experience). To start, it is important to understand what is meant by "individuals" in the broadest sense. For instance, one might argue that organisms qualify as individuals to the extent that they possess a life history, a unique trajectory characterized by a distinct sequence of events. This sequence is "unique" because it results from different genomes, constitutions, starting conditions, random events, and contingencies (cf. Kull & Velmezova, 2019). From the perspective of interaction dynamics within an environment, if we ask what enables a system to remain "the same" over time—that is, to preserve its identity and organizational integrity by minimizing randomness—we might hypothesize that this capacity lies in the system's ability to act in a functionally coherent way within the course of its life. Such capacity stems from its agent-level organization. It is precisely this capability that grounds the possibility for an organism to have its own "perspective", an individual perspective, on its interactions with the surrounding environment (cf. Bianchi, 2024).

The concept of individuality can be examined from multiple perspectives. From a biochemical reactivity standpoint, Williams (1956) argued that each biological entity responds somewhat differently to a given drug dose, implying that uniform administration across an entire population may not be feasible. This perspective is echoed by Trewavas (2003), who emphasized the intrinsic variability in biological systems. This observation illustrates differences in constitution and response to substances, even among members of the same species. Similarly, in biological studies, individuality is often associated with genomic uniformity—that is, each organism possessing its own DNA, despite also hosting DNA from numerous endosymbionts. In humans, for example, only a portion of our cells and DNA are exclusively "our own." Thus, from a more realistic perspective, it is more appropriate to conceive of shared functions within "co-individuals," entities that are "co-inhabited" by countless other organisms (cf. Gagliasso, 2019). This perspective has, for example, contributed to the development of the concept of the "holobiont" (cf. Bonfante, 2021). Furthermore, in plants, it is also necessary to consider the widespread phenomenon of genetic mosaicism (cf. Tomimoto & Satake, 2023).

However, abandoning the concept of individuality altogether, even for practical reasons, seems problematic. While the term should be used cautiously, it remains valuable, given that each phenotype results from a specific genotype and environmental influences, in a fundamentally relational framework. The approache to individuality outlined above is applicable to plants as well. Concerning possessing a unique genome, it's important to avoid confusion when discussing forms of distributed intelligence. In fact, a clear distinction should be made between highly cooperative colonies of organisms of the same species, each with a distinct genome (e.g., ants or bees), and a single plant conceived as a modular system, whose components share a common genome (except for variations due to somatic mutations). The comparison is useful for examining different degrees of behavioral coordination and emergent processes, but taken too literally, it can mislead due to varying levels of integration and organization of signals, depending on the specific case (cf. Godfrey-Smith, 2021, p. 173; pp. 257–258).

Although it is widely considered prudent not to refer to plants as "subjects" (a term distinct from "agents," which is plausible), some may even hesitate to define them as "individuals" (beyond, of course, the fact that most plants are "divisible" or "dissectable" [cf. Mancuso & Viola, 2015]). To address this issue, some researchers have proposed a third option between the concepts of individuals and non-individuals, suggesting the term "pre-individuals" for plants (cf. Faucher, 2014). This perspective considers not only the factors just mentioned but also the standard concept of a unitary, hierarchically organized individual, as seen in many vertebrates. At a general level, however, a potential concern with analyzing plants through the lens of pre-individuality is that it might unintentionally frame plants as lacking something in comparison to animals, rather than as organisms structured and organized in different ways. To avoid such misunderstandings, it may be more appropriate to refrain from categorizing plants as "individuals" and instead refer to their "individual life histories" shaped by unique paths, and contingencies, a perspective that can be validly applied to any living being. In this way, plant behavior—as well as their forms of intelligence, cognition, and communication—can be explained by taking into account both the modular characteristics of their decentralized systems and the elements of "individuality" and relative integration (non-random organization of information) within their systems or subsystems, as evidenced by their ability to respond with sufficient accuracy in pursuit of specific goals.[16]

This hybrid nature, characterized by semiautonomy and a non-random behavior, represents one of the most compelling aspects for advancing our understanding of these organisms. It enriches our knowledge and carries practical and technological implications, starting from an exploration of the diverse ways of life in nature and the different modes of "being in the world."

Moreover, beyond the traditional concept of individuality, the study of clonal plants in relation to distributed intelligence has deepened our understanding of what might be described as a form of "modular individuality," an almost oxymoronic term. Trewavas (2003) noted that even in non-clonal plants, modularity should not be viewed as an obstacle to intelligent behavior; instead, individual ramets can be seen as processing units, each capable of handling different types of information based on distinct interactions that shape the plant's structure over time. Latzel et al. (2016) suggested that the higher degree of modularity and autonomy in clonal plants—expressed as networks of interconnected, semi-independent systems—might offer them a greater capacity for intelligent behavior compared to non-clonal plants. This capacity is evident in the flexibility and adaptability of the information they store, acquired over time in various, sometimes distant, local environments due to the prolonged development of different components. In clonal species, the genetic individual (*genet*), derived from a single zygote, does not directly correspond to its physiological manifestation (*ramet*), which constitutes a subunit within the genet capable of attaining a degree of physiological autonomy (cf. Oborny, 2019).[17]

A further level of analysis that challenges the concept of individuality is the ecological-relational perspective, which considers the survival of plants in terms of their interactions with other species in their community. Where, then, does the porous boundary of the individual lie? Each organism is part of a system larger than

[16] Whereas the decentralization of functions was once viewed as a limitation, there is now a growing appreciation for the potential of systems with distributed functional organization (cf. Bonato & Castiello, 2020), and this interest is increasingly being applied in technological development.

[17] Each *ramet* can function as an individual plant, generally capable of surviving independently if separated from the rest of the *genet*, whether through natural detachment or physical damage (cf. Latzel et al., 2016).

the physical confines of its body (even if those boundaries are distinguishable by us). For example, consider the emergent properties of a forest that are not present or are different in a single tree. These properties have led some authors to examine certain processes from a systemic perspective, and in some cases, even to describe aspects of plant life as not only interactive but also as "social" within plant communities (cf. von Humboldt & Bonpland, 1807).

There is, generally, a tendency to connect individuality with consciousness. We might ask: How could an organism—or any agent with its own individual life path—be completely unaware of its surroundings and needs? How would it function without some form of "awareness" to guide its interactions at the right time and in the appropriate manner? To fully appreciate this question, however, it's essential to distinguish among various conceptions of consciousness to assess which framework might best apply to plants, if any.

3.5.2 Conceptions of Consciousness

As a starting point, we should not assume that every behavior or process involving cognitive activity is conscious. In such cases, reference is made to non-conscious cognitive processes, that is, physiological information-processing mechanisms that are essential for conscious activity to occur at various levels of organization (cf. LeDoux, 2020, pp. 259–260).

Etymologically, the term 'consciousness' derives from the Latin *conscire*, meaning "to know together." In its more modern philosophical usage, the concept was significantly shaped in the seventeenth century by René Descartes (1596–1650), and has generally been used to indicate a form of awareness accompanying knowledge, for instance, awareness of one's own mental content. This "metacognitive" view of consciousness is widely recognized, although the term is now used across vastly different contexts to indicate a variety of phenomena (cf. Gazzaniga, 2019, p. 22). Typically, "consciousness" is employed transitively to indicate being conscious *of* something, or intransitively to describe the alert state of someone who is neither asleep, fainted, nor in a coma, and thus able to respond to environmental stimuli (cf. Marraffa & Paternoster, 2013, p. 44). In behavioral terms, being conscious, or aware of what is happening around an organism, could primarily serve to maintain a coherent focus toward a particular goal. From this perspective, consciousness may have a "defensive" function, providing a framework in which different actions are justified and directed toward specific objectives (cf. Perconti, 2017, p. 91; cf. Di Francesco et al., 2017, pp. 62–63).

Historically, there has been a shift in both the subjects considered capable of consciousness and the locations (organs) and types of functions (processes involved) believed to enable it. For example, panpsychist theories propose that all matter, even inorganic, possesses some form of consciousness, though it varies depending on the degree of organizational complexity (cf. Gazzaniga, 2019, p. 94).

When it comes to identifying the presumed "seat of consciousness," countless hypotheses have been proposed over the centuries, making it impossible to address them all here.

Notably, some approaches viewed consciousness as a condition emerging from a single organ or produced by a singular mechanism or process. For instance, neo-mechanistic theories suggest that consciousness arises from a specific neural network or circuit (ibid., p. 106). Until recently, it was widely believed that consciousness was localized in a single brain region, the cerebral cortex.[18] However, clinical studies on children with anencephaly or hydranencephaly (cf. Merker, 2007) have demonstrated that, even in the absence of a cerebral cortex, these individuals can orient themselves in space and in some way interact with their environment, responding to external stimuli with coherent emotional reactions (cf. Solms & Panksepp, 2012). This suggests that the cortex contributes to higher-order processing of cognitive activity, while the potential for consciousness also relies on the activity of subcortical structures, particularly the midbrain.[19] In summary, multiple brain areas and processes, including subcortical structures and other bodily functions, appear to contribute to the emergence of consciousness (cf. Gazzaniga, 2019, pp. 190–192; p. 283).

Over time, localizationist hypotheses have been significantly scaled back, if not entirely abandoned, though this shift has not resolved the debate. Some authors, such as Llinás (1998), have proposed that consciousness, similarly to the sleep state, arises from large-scale brain dynamics, specifically, from rhythms and resonances emerging within brain activity. This approach focuses on the self-organizing processes of the nervous system. Conversely, other authors argued that the significance attributed to certain rhythmic patterns in consciousness processes has been overestimated. They suggested, more realistically, that these patterns may simply serve to maintain stable brain function by preventing the disorganization of brain activity (cf. Godfrey-Smith, 2021, p. 231).

According to proponents of emergentism, consciousness arises when the organization of certain biological processes reaches a sufficient level of complexity. Gerald M. Edelman (1929–2014) proposed that conscious activity emerges from reentrant brain circuits, neural activity that loops back on itself, creating self-resonance. Similar ideas have been suggested by Dewey, Varela, and Bernstein in their studies on intelligent behavior (cf. Edelman et al., 2011; Morabito, 2020, p. 126).

In all these perspectives, which tend to move beyond rigid localization theories, consciousness is not confined to a specific brain area but is seen as an intrinsic feature of the brain's diverse capacities (cf. Gazzaniga, 2019, p. 300). Some

[18] Today, this view is no longer sustainable. For instance, the 2012 Cambridge Declaration on Consciousness states that the absence of a neocortex does not appear to prevent certain organisms from displaying intentional behaviors and, in some cases, experiencing "affective states" (cf. Calvo, 2017).

[19] The capacity for spatial orientation has also been observed in the central complex structures of insect brains, which is analogous to the vertebrate midbrain (cf. Barron & Klein, 2016).

researchers even suggest that consciousness might not be limited to a particular organ or system (such as the nervous system) but could involve widespread intercellular signaling activities (cf. Baluška et al., 2021).

An example of a contemporary non-localizationist perspective, at least at the cerebral level, is Gazzaniga's (2019) view, which conceptualizes consciousness as "an instinct." Although the term instinct is somewhat vague, it is interestingly understood here as an innate capacity that does not need to be "learned" and enables those who possess it to navigate their environment effectively (cf. Gazzaniga, 2019, pp. 301–307). From this perspective, consciousness (at least in humans) is not tied to a specific neural structure, network, or circuit but is instead distributed across various cognitive activities. Support for this comes from observations of patients with advanced Alzheimer's disease or severe brain injuries. When certain brain circuits cease to function, these patients remain unaware of the loss of that specific activity, as though it had never existed.[20]

As the author noted, one of the earliest indications that consciousness is not confined to a specific neural network comes from studies on split-brain patients who underwent resection of the corpus callosum. In these individuals, both hemispheres continue to function, but under certain experimental conditions—such as when different inputs are presented to each visual field—they may operate as separate conscious systems, each processing information independently (ibid., pp. 15–16; 113; 137). Although these patients typically behave normally, specific tests reveal that they can exhibit dual responses, suggesting the presence of "two minds" within a single body. Furthermore, even when one hemisphere lacks the capacity to verbalize information, emotional stimuli presented to it can still influence the individual's overall mood (cf. Godfrey-Smith, 2021, pp. 175–182).

While fundamental studies continue on the mechanisms, processes, and dynamics—not only within the brain but also throughout the entire body in relation to environment—that contribute to conscious activities, and despite the challenges in quantitatively measuring systemic phenomena, theoretical research has proposed various levels of consciousness. These range from simpler forms present in all living things to more complex forms unique to humans, with intermediate forms seen in other animal species. For instance, among the more restrictive perspectives is that of conceptualists, who raise the so-called intellectualistic objection: can consciousness exist without concepts, that is, as a form of pre-conceptual or pre-reflective experience? A separate question is whether reflection itself is possible without concepts as traditionally understood in discursive, "linguistic" human thought. As Marraffa and Paternoster (2013, pp. 85–90) explained, this objection argues that it is indeed possible to experience an object without categorizing or conceptually recognizing it. There are diverse philosophical perspectives

[20] Furthermore, like bubbles rising in a boiling pot, the processes generated by certain neural circuits—through a competitive dynamic shaped by various needs and environmental stimuli—can reach the level of consciousness, while others do not. Some may emerge later, when they become the most active circuits, with their associated processes defining the individual's conscious state at that particular moment (cf. Gazzaniga, 2019, p. 140).

on the role of concepts in experience in general, and in conscious experience in particular. Gareth Evans (1946–1980) proposed that, at least in humans, perceptual experience involves a progression from nonconceptual content—characterized by coarser granularity—to conceptual content, which enables finer-grained discrimination and a more detailed representation of complex features. By contrast, Fred Dretske (1932–2013) emphasized the epistemic function of conceptualization, arguing that it serves to filter out irrelevant details and direct attention toward what is most significant for information acquisition.

Then, there is the stance of "anti-conceptualists" or those open to the idea that perceptual experiences in various forms of life may not necessarily rely on concepts. For instance, a perceiver might be able to locate a nearby object without possessing a specific concept for it. This view aligns with findings from empirical psychology, particularly studies in Gestalt and phenomenological perceptual psychology, which suggest that perception and conceptual thought function as two relatively distinct systems. This distinction is further illustrated by optical illusions—such as the well-known Ponzo and Müller-Lyer illusions—which demonstrate that certain perceptual processes can operate independently of higher-level cognitive influences (e.g., expectations or beliefs).

A compelling point is that if perceiving an object (or being aware of its presence) required having the corresponding concept, then newborns and many animals—as well as plants and other aneural organisms, based on current knowledge—would be unable to perceive key objects in their environment, a notion that is intuitively hard to accept and has been experimentally disproven. However, the discussion does not end there. The point is to understand the boundary between perception-action processes and self-consciousness phenomena.

In formulating inclusive views on consciousness, scholars have sought to identify fundamental levels of consciousness that may extend across various non-human organisms, including those without conceptual thought. For instance, in *The Feeling of What Happens* (1999), Antonio Damasio distinguished between "core" consciousness—immediate, non-reflective awareness of bodily and environmental states—and "extended" consciousness, which involves autobiographical memory and self-reflection, typically associated with humans. Core consciousness pertains to the direct sensation of one's current context, a kind of "feeling without knowing," experienced in the "here and now" (cf. Marraffa & Paternoster, 2013). This form of consciousness is likely common to many animals. Similarly, Mark H. Bickhard (2005) described "primary" consciousness as the basic capacity of a system to interact with its environment, suggesting that such consciousness does not require conceptual content. While an organism's experience may develop to distinguish its surroundings with increasing clarity and detail, it is possible for an organism to exist as "a self" without necessarily possessing a conceptual sense of itself (cf. Frazier, 2021).

From the perspective of practical action, a "conscious" organism does not need to know "how" or "why" it perceives, but simply "what" it perceives (cf. Bateson, 1976, p. 175). A capacity for consciousness understood in this way enables the organism to navigate and act safely within its environment without requiring reflective or conceptual awareness, although it must still have some ability to distinguish itself from everything else; otherwise, it would be unable to orient itself and would risk making fatal errors (cf. Gazzaniga, 2019, p. 283). This is one of the main reasons why the concept of consciousness has been expanded.

The most inclusive approaches to consciousness stem from an expanded understanding of the factors and processes involved in conscious activity and a broadening of the boundaries considered in these dynamics. For instance, Brentano, diverging from the common intuition that consciousness is a private, entirely internal phenomenon, conceived of it in relational terms, as an active relationship in which the subject constructs representations of the world (cf. Marraffa & Paternoster, 2013, p. 44).

More recent approaches, which have further explored and problematized the relationship with the representational aspect, suggest that the concept of a "bodily self" has also shaped the understanding of consciousness. Consciousness is thought to depend not only on an animal's neural activity but also on how it is integrated into the agent's sensory-motor interactions within its environment (cf. Palmiero & Borsellino, 2018, p. 125).

This expanded framework has enabled increasingly rigorous scientific investigations into animal consciousness, a field now so advanced that it includes studies on self-awareness and metacognitive abilities in non-human organisms (cf. De Waal, 2016). While self-consciousness is traditionally considered a human trait, and perhaps present in some primates and mammals, the possibility of conscious experience in animals is no longer generally denied. Although such experiences would likely differ greatly from human consciousness, attributing conscious states solely based on human experience to differently constituted organisms is misguided. However, this should not lead us to dismiss other forms of consciousness altogether (cf. LeDoux, 2020, pp. 27, 399). We must be patient and wait for future research results in this area. Anyway, even if not fully understood today, it seems difficult to deny any trace of consciousness in animals simply because they, unlike humans, lack verbal language to express what they feel and perceive. Among other things, verbal language cannot be considered a decisive factor, as it is absent not only in non-human animals—who nonetheless possess their own forms of communication—but also in prelinguistic infants and in adults with speech disorders.

Taking the analysis to an even more fundamental level, one that can extend to forms traditionally considered incapable of consciousness, Lynn Margulis (1938–2011) and Carl E. Sagan (1934–1996) argued that consciousness, in its simplest meaning, can be understood as "awareness of the external world." In this sense, not only are animals conscious (i.e., aware of what is happening around them, of what to seek out, such as beneficial substances, and what to avoid, such as harmful ones), but all living systems may possess some level of "awareness," including plants and, according to these authors, potentially even individual cells (cf. Trewavas, 2014, p. 255). Alternatively, a more moderate position suggests that if it seems too daring to consider individual cells as conscious in this way, they may nonetheless engage in processes that parallel, or at least resemble, those underlying conscious activity (cf. Gazzaniga, 2019, p. 259).

3.5.3 Consciousness and Self-Consciousness

A fundamental distinction to keep in mind—especially when addressing consciousness from a behavioral perspective in non-human organisms—is that between "simple" consciousness and self-consciousness. According to the most inclusive approaches, consciousness does not equate to self-consciousness. The former corresponds to what is often described in philosophy as the capacity for experience (or for environmental interaction), namely, the ability to engage in perceptual and agentive contact with the world and, in doing so, to undergo sensations. Self- consciousness, on the other hand, involves the capacity to objectify a particular internal state resulting from various interactions, making it itself the object of reflection. This level requires introspective and metacognitive abilities. It is likely, therefore, that more manifestations of the former case exist in the living world, as understood in this sense (cf. Marraffa & Paternoster, 2013, pp. 44–45; 143; cf. Maher, 2017, p. 57).

Some authors, such as José L. Bermúdez, have expressed similar ideas but use the term 'self-consciousness' to describe a minimal form of proprioceptive or somatic awareness (a "bodily" or "ecological self," not based on conceptual content). This minimal self- consciousness relates to the sensorimotor systems' capacity to tacitly distinguish bodily stimuli from non-bodily ones. In this minimal sense, many living beings could possess a pre-reflective self-consciousness. In any case, to avoid confusion, it is useful to distinguish between consciousness (or awareness) and self-consciousness. Thus, there is a difference between stating that an organism is a subject, or more precisely, an agent, and saying that it perceives itself as such, an assertion that would mark its entry into the sphere of self-consciousness (cf. Marraffa & Paternoster, 2013, pp. 102–103).

On a practical level, various tests have been developed to determine whether an organism is self-conscious, specifically if it can recognize itself. The best-known example is the mirror test, which has been administered to several animal species. This involves marking the animal, placing it in front of a mirror, and observing whether it identifies and touches the mark, thus indicating consciousness that the mark is on its own body (cf. De Waal, 2016, pp. 30–33). Different species exhibit a range of reactions to mirror exposure. Certain mammals and birds, such as primates, dolphins, elephants, and magpies, have demonstrated behaviors consistent with self-recognition. Others, including dogs and cats, often respond to their reflection as if encountering another individual, displaying social or exploratory behaviors. Some species, like snails, may not exhibit any notable response, treating the mirror as a neutral object. These varied reactions suggest a spectrum of self-awareness, ranging from basic visual-motor coordination to genuine self-recognition, and, in humans, extending to reflective self-consciousness and considerations of appearance (cf. Perconti, 2017, p. 96). However, the mirror test is not suitable for all species, as self-recognition may occur through modes less reliant on vision and more, for instance, on tactile perception or chemical recognition, factors that could be relevant for plants. As emphasized throughout these pages, recognizing this diversity highlights the importance of understanding a species' biology and ecology to select the most appropriate test for investigating a particular capability (cf. Bianchi, 2021, pp. 122–123).

3.5.4 Awareness in Plants?

Regarding plants, several critiques have challenged the possibility of plant consciousness. These critiques often present arguments that assume plants lack any form of consciousness, using criteria based on typically human standards rather than directly demonstrating the absence of consciousness, or something akin to it, in plants (cf. Taiz et al., 2019; Mallatt et al., 2021). However, focusing on a particular level of consciousness does not automatically imply that organisms lacking the characteristics to meet this standard are devoid of an analogous capacity or of any other property related to consciousness. As Maher (2020) observed, focusing on a specific definition of consciousness does little to answer whether some version of the concept might apply to plants, let alone whether they might possess a form of environmental "awareness."

For example, assuming that a nervous system is a prerequisite for consciousness inevitably leads to the exclusion of any evidence of conscious-like activity in brainless organisms, despite their perceptual and chemo-electrical signaling capabilities.

For example, Mallatt et al. (2021) critically assessed the applicability of Integrated Information Theory (IIT) (cf. Tononi & Koch, 2015; Tononi et al., 2016) in supporting the notion of minimal consciousness in plants (cf. Trewavas et al., 2020). The critics argued that IIT's premises are overly broad, potentially extending the attribution of consciousness to non-living systems based solely on information processing and integration. This, they contend, risks trivializing the concept of consciousness by making it excessively inclusive, thereby undermining a robust foundation for the idea of plant consciousness. Furthermore, Mallatt et al. (2021) suggested that proponents of plant consciousness, and more broadly, researchers in plant cognition and intelligence, have extended terms such as 'cognition,' 'intelligence,' 'consciousness,' and 'sentience' to encompass all living systems. According to the authors, this broad application would necessitate the use of different terminology to describe these capacities in neural, multicellular organisms, which they argued possess internal representations of their experiences (cf. LeDoux, 2020). On the other hand, adopting an inclusive, empirically informed, and comparative approach to the study of living beings' capacities allows for the identification of characteristics shared across taxa, while also making it possible to account for distinct biological and ecological organizations shaped by different strategies and possibilities. Obviously, not every ability must be present in all forms of life or expressed in the same way, but as long as the presence of a particular capacity cannot be entirely ruled out, continued investigation is essential (cf. Bianchi & Castiello, 2023b).

Another criticism raised by Mallatt et al. (2021) involves the nature of electrical activity in plants, which some have linked to the potential for conscious activity. Although plants exhibit various types of electrical potentials, their electrical responses are generally considered too slow to process information in the same way as neuronal activity. At this stage of research, pending further findings, it can simply be noted that each living system employs strategies and mechanisms suited to its own possibilities and needs, operating on species-specific time scales and rhythms of growth and ecological interaction (cf. Bianchi et al., 2025b). The slower pace of plant life and behavior—especially those involving growth (though cellular reactivity, certain forms of chemical signaling, and exudate emissions can occur rapidly)—is well-suited to their "mode of life" within their respective ecological niches (cf. Bianchi et al., 2025a). Thus, a slower response may not imply less effective responsiveness for plants within their context of interaction, nor does it exclude the possibility of a form of sensitivity and environmental awareness. In any case, the question remains open.[21]

Finally, according to Mallatt et al. (2021), the articles addressing the topic of plant consciousness are overly speculative. Even if that were the case, it is important to bear in mind that we are only at the beginning of these stimulating lines of inquiry, which ought to be pursued with flexibility, avoiding premature conclusions or overly rigid claims, while remaining mindful of the many aspects of "mental life" that were initially denied and only later recognized in various animal species. In any case, a particular model for studying a specific capacity can help illuminate certain aspects of a behavior or ability, even if it may not provide a complete explanation. A comprehensive approach should aim to leverage the strengths of various models, each contributing to our understanding of different facets of the issue (cf. Bianchi, 2024).

Beyond explicit criticisms of the possibility of plant consciousness, there are more cautious positions (cf. Godfrey-Smith, 2021, p. 262) that express reasonable doubt but remain open to future revision should evidence suggest that plants possess a form of consciousness (or "sentience"). This form of sentience is understood as "felt experience," associated with the development of a self with its own perspective, generally arising from the integrated activity of a nervous system. From this perspective, while plants exhibit a high degree of sensitivity and can signal and actively respond to many environmental stimuli, this may not be sufficient for them to have a true "experience" in the subjective sense (ibid., p. 257). The reason may be that plants have followed an evolutionary path that does not encompass this form of experience.

This view seems to imply a conception of consciousness that involves a certain level of self-reflective capacity, which is difficult to apply to plants. In this sense, rather than possessing consciousness "in the strong sense" associated with complex cognition or inner, "subjective" life (cf. Hiernaux, 2021), it may be more fitting, as Trewavas (2021) suggested, to speak of a form of "awareness" of the surroundings,

[21] Could plants operate so slowly because they "know" the chemical composition of their surroundings in detail and therefore do not need to rush to react to avoid being caught off guard? (cf. Bianchi et al., 2025).

allowing plants to engage in diverse interactions with significant aspects of their ecological niche. Such a form of awareness could be seen as a characteristic of all living systems, including plants.

As is true for many animal species (cf. LeDoux, 2020, pp. 401–402), which may not be reflexively aware of themselves through a representation of their present, past, and future conditions, an organism must still be aware of essentials such as distinguishing edible from poisonous food, identifying harmless versus threatening species, recognizing cooperative partners, discerning potential enemies, and detecting possible mates. Similarly, it is hypothesized that plants may possess a basic awareness of their surroundings—differentiating themselves, their bodies, and their activities from all that is not themselves—and recognizing what to seek out and what to avoid (cf. Trewavas, 2021). This line of research is relevant because, regardless of the degree of awareness, understanding the forms of awareness that plants may have could offer a more comprehensive view of the evolution of consciousness in living systems (cf. Calvo, 2017).

Recently, specific approaches have been proposed to study potential forms of consciousness in plants. One example is the predictive processing analysis framework, based on the assumption that information necessary for conscious experience could be embedded in the generative models of organisms. From this perspective, consciousness is seen as the process of perceptual inference about states of the world that produce sensations (cf. Hobson et al., 2014). Consequently, one approach to exploring plant consciousness would involve investigating, in line with the work initiated by Calvo and Friston (2017), how plants probabilistically predict sensory variations to guide their behavior. Certain assumptions within predictive theories help to develop the basis for an embodied and grounded theory of consciousness, understood as a form of awareness arising from active inferences that emerge through organism-environment interactions (cf. Kirchhoff & Robertson, 2018). Within this framework, predictive models, particularly in relation to plant sensitivity, can be effectively integrated into a more ecological approach that takes into account interspecific and contextual dynamics. This integration can help determine whether, and to what extent, representational and inferential aspects may be relevant in explaining these processes (cf. Bianchi & Castiello, 2023b).

From a practical perspective, however, the question arises as to how the presence of a minimal form of awareness in plants might be verified.

One of the most effective methods involves administering anesthetics typically used in animals, which can also induce a reversible loss of sensitivity in plants. When plants are damaged or under significant stress, they produce ethylene, a substance that acts as an anesthetic in animals, causing a temporary suppression of responsiveness (cf. Pollan, 2001; Grémiaux et al. 2014). A concrete example is the effect of ether, an anesthetic once used in humans, in dampening thigmotropism in plants (cf. Maher, 2017, p. 56). Another example is the cessation of the trap-closing behavior in *Dionaea muscipula* after treatment with anesthetic substances, or, similarly, the failure of *Mimosa pudica* leaves to fold in response to touch (cf. Yokawa et al., 2018). A working hypothesis is that anesthetics, as in animals, act on action potentials that involve chemical substances like glutamate and GABA, thereby inhibiting the production of electrical signals and diminishing the plants' capacity to produce behavioral responses (cf. Baluška & Yokawa, 2021).

More specifically, plant awareness appears to be closely linked to the capacity to distinguish between self and non-self, a distinction that echoes early descriptions of immune system function (cf. Burnet, 1971). In this context, "self" refers to the components and activities of the organism's own body, while "non-self" encompasses everything external that does not elicit an immune response (cf. Castiello, 2019, p. 134). On the basis of this analogy, it could be argued that plants possess a capacity to discriminate between what pertains to their own body and activity and what lies outside of it. However, current understandings of immune system dynamics reveal a far more complex and nuanced picture, one that is considerably less binary than initially assumed.

Self-recognition at the molecular level is widespread across living organisms. In plants, for example, mechanisms exist for chemically discriminating the type of pollen entering the ovary, including anti-selfing systems that prevent self-fertilization. These systems are mediated by proteins encoded at a specific genetic region known as the S-locus (cf. Trewavas, 2014, pp. 181–183).

The recognition of self versus non-self is an ecologically indispensable aspect for plants, manifesting not only at the genetic-molecular level but also in various physiological processes. Recent studies have begun to explore this phenomenon in roots and tendrils of certain climbing plants, suggesting a more integrated capacity for physiological coordination that may involve proprioceptive mechanisms (cf. Gruntman & Novoplansky, 2004). This line of research originated from observations of different parts of a single plant's root system, where physically separated sections, despite sharing the same genome, gradually began to interact as if they were from distinct individuals. Such behavior suggests that recognition and associated responses involve a level of physiological coordination (cf. Falik et al., 2006; Thompson, 2018, pp. 42–45). Comparable research has also been conducted on tendrils. For instance, beyond the observation that tendrils from the same plant recognize one another and avoid intertwining (cf. Trewavas, 2014, p. 86), in a manner analogous to how roots distinguish self from non-self to avoid competition (cf. Karban, 2017, p. 11), it has been observed that *Cayratia japonica* can respond not only to tendrils from genetically distinct plants but also to genetically identical tendrils that were previously part of the same plant, before being divided (Fukano & Yamawo, 2015).

In addition to self/non-self discrimination, plants are also capable of recognizing genetically related individuals, or "kin," as evidenced by behavioral differences observed both in root systems and in above-ground structures. These forms of kin recognition appear to be mediated by chemical exudates, surface-bound enzymes, and electrical signals, all of which can trigger adaptive behavioral responses (cf. Castiello, 2019, p. 138; cf. Novoplansky, 2009; Biedrzycki et al., 2010; Semchenko et al., 2014). Among the earliest studies focused on these aspects is that by Dudley and File (2007), who found that *Cakile edentula* plants allocated more biomass to root development when growing alongside unrelated individuals than when growing with kin. Similar evidence of kin recognition and corresponding resource allocation strategies has been discussed by Anten and Chen (2021). Supporting this line of inquiry, Biedrzycki et al. (2010) demonstrated that *Arabidopsis thaliana* produced more lateral roots in response to exudates from non-kin plants compared to those released by genetically related individuals.

Kin recognition also manifests in modifications to the aerial parts of plants, such as branches and leaves. Photoreceptors play a crucial role in this process, enabling plants to detect the presence of neighboring individuals by sensing variations in blue and far-red light (cf. Crepy & Casal, 2015; Bais, 2015). For instance, *Arabidopsis thaliana* adjusts its leaf orientation depending on the genetic relatedness of nearby plants: when growing alongside kin, it directs its leaves away from neighboring individuals to reduce competition for light, a behavior not observed in the absence of kinship (cf. Castiello, 2019, p. 42). Similar kin-dependent responses have been documented in species of the genus *Balsamina*, which produce more branches and elongate their stems when growing near genetically related individuals, thereby reducing mutual shading from overlapping foliage (cf. Murphy & Dudley, 2009; Dudley et al., 2013).

In biological studies, researchers have investigated the possible causes of reduced competition or even cooperative behavior in the presence of relative plants. Hamilton (1964), for instance, introduced the concept of "kin selection," suggesting that if one individual assists another with whom it shares genetic proximity there is a greater chance that some of those alleles will be passed to the next generation. Overall, the effects of kin recognition have ecological implications for the structure and dynamics of plant populations within communities, affecting resource and information exchange (cf. Pickles et al., 2017; Anten & Chen, 2021).

In discussions of plant awareness, particular attention has been given to the ability of plants to distinguish and recognize a range of elements, including the self, kin, non-kin, as well as beneficial and harmful substances. As Nancy E. Baker (2017, p. 146) pointed out, the term "recognition" covers slightly different aspects depending on the criteria used and the increasing complexity of possible interactions. Thus, although "recognition" may imply something different at each stage of development, these usages are linked by a Wittgensteinian "family resemblance," explaining why the same term can be applied to different behavioral levels. This approach is more effective than using the expression "as if," which implies that plants behave as though they recognize something, rather than acknowledging that they are actively identifying and responding to ecologically significant elements for their survival.

3.6 Intelligence and Cognition in Plants

In this section, I focus more specifically on the concepts of plant intelligence and cognition, two increasingly debated ideas that, despite initial criticisms and understandable perplexities, are gaining greater acceptance in current discourse and conceptual analysis related to emerging research on plant behavioral activities.

While "intelligence" and "cognition" are often used interchangeably to describe certain interpretations of plant behavior, a distinction can generally be made. In a broad, non-definitional sense, intelligence can be understood as the capacity of living systems to solve survival-related problems by selecting appropriate behaviors for various situations and by gradually varying and refining their responses to specific, recurrent stimuli. Cognition, on the other hand, refers to a species-specific set of abilities—such as the recognition of differences, memory, learning, anticipation, and communication processes—that manifest in diverse behavioral strategies aimed at achieving specific goals and effectively adapting to changing environments.

From a practical standpoint, research on plant intelligence and cognition is currently being conducted primarily under controlled laboratory conditions. However, to gain a more comprehensive understanding of these behavioral phenomena, it is also essential to investigate plant growth and interactions within natural environments over extended timeframes. Such an approach emphasizes generational changes within plant populations and their adaptive responses over time (cf. Calvo et al., 2020). Both ongoing research directions are relevant and warrant further exploration.

3.6.1 On Intelligence

With regard to intelligence, the term "intelligence" is heavily laden with connotations, particularly in Western culture, where it has traditionally been associated with distinctly human abilities, primarily logical and mathematical skills, often measured through reductive IQ tests. However, this view of intelligence represents only one facet of the abilities possessed by organisms with complex brains. There has been a persistent tendency, especially in human theoretical practices, to categorize aspects of reality according to hierarchical importance. This approach often places human-specific traits, considered unique or highly developed in our species, at the pinnacle of such hierarchies (cf. Bianchi, 2021).

A different attempt to explain intelligence, albeit sometimes criticized by mainstream psychology for its limited empirical backing, was introduced by Howard Gardner (1983), who emphasized the cultural relativity of intelligence, proposing that behaviors considered "intelligent" in one society may be interpreted or valued differently in another. Cultural factors form an intrinsic part of the human ecological niche, one that blurs the line between nature and culture, making culture an integral element of our nature and a foundation for many behavioral dynamics. From this perspective, Gardner broadened the concept of human intelligence to include bodily-kinesthetic, spatial, musical, interpersonal, and intrapersonal intelligences, thereby acknowledging a wide range of human abilities and distinct behavioral strengths (cf. Trewavas, 2014, p. 194).

In philosophical reflection, there has been a gradual shift from highly intellectualist perspectives to more embodied, agent-centered views that connect intelligence to practical action within the framework of biological motivation and survival behavior. Until the early twentieth century, dominant theories about human cognition and intelligence largely assumed that ideas and reflective consciousness played central roles in goal-directed behavior. However, certain philosophers, such as John Dewey (1859–1952), challenged these predominantly “intellectualist” views, highlighting the adaptive, reactive flexibility of the body as a key component of human intelligence (cf. Maher, 2017, pp. 113–114). Henri-Louis Bergson (1859–1941), influenced in part by the pragmatist approach of William James (1842–1910), proposed that intelligence is rooted in the capacity to carry out actions with adaptive value. He argued that, prior to the development of philosophical and scientific thought, and indeed in everyday reflective processes, intelligence, which enables us to direct bodily action, serves as a practical tool for navigating. These perspectives highlight the biological, functional, and adaptive dimensions of intelligent behavior, which are fundamental to the scientific understanding of survival and self-preservation in living beings (cf. Morabito, 2020, pp. 32; 64–65).

Even within the field of psychology, various characteristics of intelligence have been identified. When Robert J. Sternberg and Douglas K. Detterman (1986) compiled responses from 24 psychologists regarding the nature of intelligence, the most commonly cited attributes included problem-solving ability, decision-making skills, and abstract reasoning, reflecting a general emphasis on adaptive cognitive functioning. While the latter may be limited to relatively few species, the other two, especially problem-solving ability, appear to be widespread, expressed in diverse forms across species, and variable among individual organisms (cf. Trewavas, 2014, pp. 74; 195). The notion of intelligence as a problem-solving capacity, especially in the context of survival challenges, has been adopted by various authors to describe plant intelligence as well, as will be further discussed (cf. Trewavas, 2005; Mancuso & Viola, 2015; Pollan, 2013). Defining intelligent behavior in this way suggests that organisms must have threshold mechanisms to identify specific problems and then respond coherently (cf. Trewavas, 2014, p. 91).

The problem of reducing intelligence to a single form arises when we consider it as a unique or exclusive trait of our species, despite current understanding that intelligence is a property of the phenotype and manifests through a range of processes and behaviors across different contexts. Since the capacity for intelligent behavior is closely tied to the sensory modalities through which an organism perceives its world—with varying perceptual and agentive possibilities (cf. Warwick, 2000)—comparisons between species are possible. However, such comparisons should be made with careful attention to both shared features and species-specific differences, acknowledging that they often emphasize traits that are highly developed in humans and thus considered important according to human standards of value (cf. Trewavas, 2014, pp. 193–197). Moreover, many characteristics traditionally associated with human intelligence—such as complex language and tool-making abilities—have been identified, in various forms, within the behavior of other species as research progresses. For instance, cephalopods, particularly certain species of octopus, have demonstrated notable tool-use abilities, such as employing coconut shells or mollusk shells as portable shelters for protection against predators (cf. Godfrey-Smith, 2021, p. 167).

Thus, it is important to note that early reflections on intelligence emerged when most knowledge about behavior was focused solely on human behavior (cf. Trewavas, 2014, p. 197). As studies on the abilities and life contexts of other organisms have advanced, our understanding has evolved, enabling a broader and deeper analysis of the diverse forms of intelligence present in nature. In a more inclusive, naturalistic framework, intelligence can be seen as a fundamental biological trait essential for an organism's survival, comparable to other traits that facilitate the possibility of survival and evolution over time. In line with the biogenic approach to cognition (cf. Lyon, 2006), intelligence can be studied from a bioecological perspective across the evolutionary tree of life. This approach encourages moving beyond anthropocentric assumptions and facilitates the exploration of cognitive capacities in both simpler and more complex organisms. It is, of course, from a human perspective that any analysis, observation, and interpretation begins (cf. Bianchi, 2024).

Following Trewavas' (2014) inclusive interpretation of intelligence—as a concept that, like other abilities, can be defined in relation to different biological structures and ecological demands—the Latin root of *intelligentia*, derived from *interlegere*, can be considered. The term 'intelligence' originates from the Latin *inter* (between) and *legere* (to pick, to collect, to read), conveying the meaning "to choose between." This broader definition encompasses the ability to discern and choose between different, sometimes conflicting, options, making it applicable to a wide range of organisms (ibid., p. 192). On a general level, intelligence can be recognized in an organism's ability to make appropriate choices in specific situations and to adjust its responses to changing circumstances. For instance, while Darwin (1871) did not offer a systematic definition of intelligence, he discussed intelligent behaviors within the framework of evolution and natural selection, suggesting that intelligence lies in a species' ability to effectively undertake actions necessary for survival, a process that inherently requires adaptability (cf. Trewavas, 2014, p. 10). The reason intelligent behaviors are observable in living systems is because they are essential for survival; without the ability to discern and selectively adjust the type and intensity of responses, which can be refined to become more effective over time, an organism would not survive for long. George J. Romanes (1848–1894), building on Darwin's insights into intelligence, defined it as the faculty enabling the intentional adaptation of means to ends, that is, the ability to modify behavior purposefully to achieve specific goals. Similarly, Jennings (1908, 1923, p. 334) described intelligence as the capacity to alter behavior freely, without reliance on automatic reflexes, based on experience and directed toward a goal (cf. Calvo et al., 2020). Intelligence has also been interpreted as a teleonomic concept, which does not imply a providential or perfection-oriented goal or a fixed purpose. Instead, it refers to an organism's ability to achieve biological goals within changing environments (ibid., p. 192).

In recent years, particularly in the context of plant intelligence, David Stenhouse's (1932–2013) conception of animal intelligence—defined as the adaptive behavioural variability within the lifetime of the individual (1974)—has been revisited and expanded within an evolutionary framework. According to this view, the greater an organism's adaptive behavioral variability, the greater its intelligence (cf.

Trewavas, 2003). Similarly, Sternberg (1986) described intelligence as the adaptation to real-world environments, enhancing the ability to deal effectively with novel situations. A related perspective defines intelligence as the species-specific capacity to apply information, structures, and strategies to particular contexts.

In short, in a basic sense, intelligence can be understood as an organism's capacity to apply available information to the conditions it encounters. This process entails evaluating the relative costs and benefits of different behavioral options within specific environmental contexts, enabling the selection of responses that enhance functional effectiveness (cf. Castiello & Guerra, 2020).

In contemporary terms, intelligence is understood as the capacity for problem-solving, involving behavioral flexibility, adaptive plasticity, and the anticipatory ability to respond to environmental changes. This form of adaptation refers to an organism's ability to dynamically adjust its behavior in response to environmental conditions (cf. Trewavas, 2014, p. 194). As such, it also emphasize the relational dimension of intelligence, emphasizing the ongoing interaction between the organism and its environment.

Instead of asking *what* intelligence is or its nature, it might be more productive to ask *why* organisms need it and what function it serves. At a general level, and without presuming to fully define it, intelligence can be understood as arising from the interaction between an organism and its environment. Through various behavioral strategies, intelligence enables an organism to respond effectively to environmental challenges or solve problems related to survival, achieving success or gaining an advantage toward a specific goal. According to some interpretations, intelligence is greater the more flexibly an organism can adapt to different contexts and situations (cf. Trewavas, 2014, pp. 192–195).

With regard to plants, strong criticism and the view that developing a concept of plant intelligence is scientifically unhelpful remain widespread. For example, according to some authors, as Pollan (2013) noted, while they do not believe that plants *possess* intelligence, they are not entirely opposed to the idea that plants can exhibit intelligent behavior. However, this perspective loses much of its force if it is based on an overly reified notion of intelligence, tied exclusively to a specific organ or trait. Additionally, a Cartesian-derived position, often cited as a reason for denying intelligence in plants, emphasizes their lack of "interiority," as if this trait alone could define intelligence in all its complexity. Yet, even in humans, interiority (or self-consciousness) is not always present when displaying behaviors that can be interpreted as "intelligent," let alone other animals (cf. Baker, 2017, p. 150). In any case, regardless of the type of criticism, if discussing plant intelligence is generally deemed inappropriate or as an overextension, it may simply indicate that the traditional concept of intelligence does not align well with plant characteristics. However, if the inquiry is approached in a way that respects the unique biological and ecological structures and organization of plants, it becomes possible to reveal how non-neural organisms like plants exhibit intelligent behaviors through their specific structures and functional systems (cf. Trewavas, 2014, p. 278). This, once again, invites a reassessment of our assumptions and underscores the need to contextualize the analysis within specific cases.

For example, considering the unique structural and functional organization of plants, some interpretations suggested that root systems display neural-like activity, driven by multiple forms of signaling, pointing toward a form of distributed and modular intelligence. This implies an ability to coordinate behavior in response to diverse stimuli. To substantiate this hypothesis, recent studies frequently compare the root tips' capacity for information coordination with the behavioral patterns found in bird flocks or insect swarms, which involve forms of communication and cooperation (cf. Baluška & Mancuso, 2013). On the other hand, some caution is warranted against overextending this analogy. Unlike bird flocks or insect swarms, where each individual is a distinct member with a unique genome and independent locomotor abilities, the root tips belong to the same plant and share the same genome, thus minimizing genetic conflict. These differences, however, do not preclude the possibility of making cautious and critical comparisons. While it's essential to approach different systems with care, the analogy between colonies and modular plants can stimulate research into understanding how behaviors are coordinated in variable environments involving multiple, diverse, elements. Each case may involve varying levels of signaling and integration (cf. Bianchi & Castiello, 2023a).

In support of the "root-brain hypothesis," research has highlighted the presence of synchronous electrical activity within the elongation zone of root tips near meristematic regions (cf. Masi et al., 2009). This activity may enable the integration of information from multiple sensory inputs, which is crucial for root development as roots monitor environmental parameters and respond adaptively. On this basis, Baluška and Mancuso (2013) proposed that this region of the root tip functions as a "command center" for information processing, comparable to a simple animal brain, in its capacity to integrate and relay signals to surrounding root tissues and the rest of the plant. However, observations by Rehm and Gradmann (2010) suggested that the detected synchronized oscillations might be artifacts introduced by the electrodes used in these studies (cf. Taiz et al., 2019). While further experimental validation is needed, these concerns do not preclude the possibility of employing enhanced or alternative techniques to better understand how roots integrate chemical and electrical cues from the environment, potentially exhibiting behavior that, in a broad sense, could be interpreted as "intelligent."

3.6.2 Plant Intelligence?

I now discuss in what sense we might refer to the intelligence, or intelligent behavior, of plants. As noted in the previous chapter, I prefer to talk in terms of behavior or cognition rather than in terms of "intelligence." However, once the most appropriate conception for the case has been clarified, there should be no misunderstandings. Researchers should remain open-minded and not be intimidated by unconventional proposals. Future findings are expected to shed light on these issues.

The most effective approach, applicable across different living systems, appears to rely on observing behavioral variability throughout an organism's development and growth, in line with Stenhouse's (1974) definition. From this perspective, plants address local challenges through multiple levels of behavioral interaction with their environment, relying on forms of phenotypic plasticity associated with meristematic and metabolic activity (cf. Trewavas, 2014, p. 196). Even in the absence of a brain—a condition explored through the concept of "Mindless Mastery" (cf. Trewavas, 2002)—certain behaviors are proposed to emerge from intercellular chemical and electrical interactions, with information flows varying in response to environmental contexts of interaction (cf. Trewavas, 2003). One objective will be to understand the degree of flexibility in individual behavior and the extent to which transformations occur at the intergenerational level.

Calvo et al. (2020) build on conceptualizations proposed by Legg and Hunter (2007) to describe general traits of intelligence—traits potentially observable across different forms of life—arguing that these accounts can also provide a useful framework for understanding plant intelligence. According to this perspective, the intelligence of an agent or living system (such as a plant or "plant-agent") is a diachronic property, one that develops and changes over time in response to various challenges and ecological demands. This adaptability is reflected in the expression, and sometimes the modification, of a set of behaviors that the plant-agent exhibits in interaction with its environment. Bio-ecological intelligence, therefore, depends on the organism's capacity to pursue goals and adjust to changing environmental conditions, and it is related to its ability to succeed or benefit in relation to a goal (including the preservation of itself). These descriptions are not intended as rigid definitions, but rather as conceptual distinctions that help clarify and make sense of the phenomenon being investigated.

One might wonder why we should consider plants as intelligent agents or systems; in other words, what benefits come from recognizing plants as organisms capable of intelligent behavior? Building on and extending the analysis by Calvo et al. (2020), it is possible to list several advantages of approaching plant activities through the lens of intelligence.

As a first point, this perspective encourages a multidisciplinary, comprehensive, and multilevel understanding of what a plant truly is. It enables a more nuanced study of plant behavior, focusing on experimentally verifiable processes such as decision-making, error perception, action anticipation, and the ability to respond adaptively to various environmental factors. These processes were once either unrecognized, misinterpreted, or seen as unexplained phenomena, far from being associated with behavioral flexibility.

Examining the diverse adaptive strategies employed by plants can contribute to establishing a comparative framework for analyzing the various forms of intelligence found in nature. Such an analysis may uncover shared patterns and highlight distinct behavioral traits used to address similar challenges (or similar strategies in varied contexts). Despite differences in physical structures and ecological niches, these insights could point to a common evolutionary basis of intelligence across taxa. This approach also holds potential in systems biology by enabling analysis across multiple organizational levels, from cellular processes involving active, nonrandom information processing (cf. McClintock, 1984), to intercellular signaling in various tissues, to organ and system-level interactions, and ultimately to the entire organism in relation to the environment (cf. Bianchi, 2024).

Depending on the task under investigation, a systemic-relational approach makes it possible to contextualize the goal-directed activities of plants within various forms of ecological interaction. This broadens the scope of analysis—from studies on clonal plants (cf. Latzel et al., 2016)[22] to mycorrhizal networks and other modes of interspecific exchange—allowing for an assessment of how, and to what extent, intra- and interspecific complexity contributes to understanding the interdependent nature of life. This approach also supports a more in-depth examination of cooperative or "social" dynamics among plants (cf. Dudley, 2015). Moreover, investigating the behavioral flexibility of plants may contribute to potential technological applications. It enables the observation of diverse processes involved in plant behavior across multiple levels, from molecular and electrical signaling within interactomes to broader, often hidden and spatially extended root activities (cf. Mazzolai, 2021).

Understanding the molecular and genetic basis of plant adaptability, expressed through nonrandom interactions, can improve crop management and bolster plant adaptability in varied agricultural environments. Decoding plant signaling and behavior will also contribute to reducing pesticide use, promoting more sustainable agricultural practices. Furthermore, a comprehensive grasp of plant interactions, both intra- and interspecific, under varying environmental conditions, supports ecological restoration efforts. This includes facilitate the remediation of polluted areas, prevent soil deterioration and slow the myriads of problems related to the ecological crisis caused by anthropogenic impacts and related climate change (cf. Bianchi, 2021).

In summary, the diverse means and strategies that plants employ to solve distinct challenges are the results of evolutionary adaptations to specific ecological niches. Fully understanding these behaviors requires setting aside biases that may limit or distort the analysis. All organisms, in one way or another, "act intelligently" within their environments, in the sense of effectively responding to challenges and opportunities, aiming to preserve, and, when possible, enhance their condition. This is also true of plants, although their different temporal scale and a greater emphasis on molecular aspects have long prevented recognition of their behavioral abilities (cf. Trewavas, 2014, p. 197). Now the ground is fertile enough to explore them further.

[22] As suggested by Latzel, Rendina González and Rosenthal (2016) intelligent plant behavior could increase in clonal plant networks, especially thanks to the possibility of modification linked to epigenetic memory, which allows the accumulation of different types of long-term information related to growth in microcontexts with different local conditions. According to the authors, the ability of plants to engage in intelligent behaviors, characterized by adaptive flexibility, increases with the increase in the level of complexity, provided by clonality, which offers eco-evolutionary advantages, the basis of their successful diffusion and longevity.

3.6.3 Cognitive Behaviors in Plants

In line with approaches adopted in the study of intelligence, interpreting plant behavior through a cognitive lens offers a framework for exploring diverse modes of interaction, including perception-action, memory, learning, and communication processes, topics that were examined throughout this chapter. From this perspective, plant behavior may not be entirely preprogrammed or automatic; rather, it can be seen as displaying a degree of flexibility, goal-directedness, and adaptability, depending on the type of stimulus, the plant's physiological state, and the specific context of interaction (cf. Bianchi et al., 2025a).

To briefly summarize, nervous systems appear to have significantly expanded and enhanced cognitive abilities, yet they do not seem to have originated these capacities, nor are they strictly required for their most basic manifestations (cf. Calvo & Keijzer, 2009). Consequently, inclusive approaches to cognition, such as biogenic perspectives, the heterogeneous 4E cognition research program, systemic models, and predictive models, have begun to explore behavior emerging from biological signaling networks in non-neural tissues in relation to environment (cf. Lyon et al., 2021). These approaches can also help guide research toward a more comprehensive exploration of the distinctive physical and interactional characteristics of plants.

Even among proponents of plant cognition, much remains to be understood about how plants use these abilities and the full set of processes that make them possible. The concept of plants as organisms capable of basal cognition (or conceived as "cognitive agents") is still a minority view and is not without criticism. For instance, some have contended that plant cognition is overly simplistic, implying that when a skill becomes too elementary, it ceases to meet the criteria for being considered "cognitive" (cf. Adams, 2018). However, given the gaps in our understanding, we should be cautious in assuming the simplicity of all plant behaviors, as it may be premature to generalize the capabilities of plants to a single level of complexity (cf. Bianchi, 2024).

Depending on the processes under consideration, what may be perceived as "simple" or "rudimentary" could, in fact, reflect a form of "diversity" when viewed against the models and characteristics typically regarded as standards. Plant responses and interactions may demonstrate complexity in unique ways, for instance, through their forms of communication, particularly their chemical abilities to "know" and chemically "recognize" their environment. Moreover, even if these capabilities were simpler than those of other living things, simplicity is not a flaw, on the contrary. The level of complexity attributed to a behavioral trait depends on the factors considered and the standard of comparison. The point is not to view plant capabilities hierarchically against those of other organisms, but rather to appreciate the ways in which each organism efficiently responds to environmental challenges and interacts with other living systems in its ecosystem (cf. Bianchi, 2021).

Another line of argument, viewed through a critical lens, seeks to emphasize that what is often interpreted as a form of cognition may, in many cases, be more accurately described as mere perceptual capacity. For instance, *seeing* something as sky blue and *believing* it to be sky blue are thought to be two distinct processes. Basic visual perception is typically classified as a perceptual, rather than a cognitive, ability, as it does not necessarily involve conceptual representations (cf. Adams, 2018). However, two questions arise: first, is it truly possible to draw a clear boundary between perceptual and cognitive processes, even at the most basic levels? As might be expected, the answer depends on how perception and cognition are defined and interpreted. Moreover, the example itself may be somewhat misleading: the cognitive component arguably lies less in the immediate perception of color and more in the ability to discriminate among different colors, an ability that likely involves detection/recognition, classification, and memory processes.

A second consideration concerns the question of whether genuine non-representational cognitive forms exist or whether cognitive processes can be entirely divorced from the standard concept of "representation." This issue, which will be examined further in Chap. 4, points to the debate surrounding non-representational cognition. Some theories propose non-representational conceptions (cf. Linson & Calvo, 2020) or reinterpret representations as the ability of a state of information to initiate or guide behavior (cf. Bianchi & Castiello, 2023a). These perspectives suggest a view of basal cognition tied to the coherent modulation of responses aimed at achieving specific goals, something that lies between pure biological information processing and the traditional notion of mental representation. At the level of the organism, this translates to an ability to organize different types of information in a nonrandom, manner. Such organization allows the organism to orient itself within its environment, regulating its states and interactions effectively (cf. Bianchi & Castiello, 2023b).

Another point of criticism concerns the claim that cognition in plants is entirely innate and unconscious. A misleading example in this context is the capacity of seeds to germinate at the appropriate time, often explained by presupposing a form of action program not tied to self-reflective consciousness or to any form of "knowing that one knows." Critics argue that it is more accurate and sufficient to describe such innate programming as genetic information acquired through the evolutionary process via natural selection. They maintain that this is fundamentally different from cognition as it is typically understood in humans (cf. Taiz et al., 2019). Of course, this perspective does not cast doubt on the fact that every trait an organism possesses is the result of its evolutionary history, which, undoubtedly, it is. However, acknowledging the evolutionary origins of a behavior does not preclude an analysis of the potential it holds for interaction and problem-solving across the organism's lifespan and throughout successive plant generations. Moreover, while it may be reasonable and correct to conclude that certain processes, such as those occurring within seeds, should not be classified as "cognitive," particularly when considered in isolation, this does not mean that all aspects of cognition must necessarily involve representational, conceptual, propositional, or self-conscious elements as they appear in humans when considering organism-level behaviors.

Provided that plants do not have capacities superior to those currently hypothesized, it would seem restrictive to consider their anoetic level (in the sense of being unrelated to thought) as entirely non-cognitive. One aspect that warrants further investigation is the relationship between decision-making, anticipatory behaviors, and ways of conceiving inferential abilities in nonconceptual terms, through forms of sensory inference (cf. Calvo & Friston, 2017). In any case, if cognitive analysis is restricted to reflective mental or representational abilities, as typically conceived in humans, plants will be unfairly excluded from consideration. Instead, it is worth exploring the possibility that plants engage with their environment through alternative relational strategies (cf. Bianchi & Castiello, 2023a).

Tying cognition exclusively to reflective abilities and self-awareness may be overly restrictive, particularly in light of increasing indications suggesting that plants might exhibit a form of environmental awareness that supports and guides their behavior (cf. Trewavas et al., 2020). While much of plant activity is genetically, and in many cases epigenetically, regulated, and many behaviors should indeed be understood through the lens of intergenerational transmission and modification, plants also demonstrate the ability, over the course of their lifespan, to differentiate between stimuli, modulate the type and intensity of their responses according to contextual relevance, retain information, and exhibit forms of learning (cf. Baluška et al., 2018). These capacities may cautiously be interpreted in cognitive terms, aligning more closely with goal-directed behavior than with entirely conditioned or automatic responses.

Furthermore, even behaviors that appear relatively inflexible can evolve or disappear over time in response to species-level evolutionary dynamics. This highlights the importance of analyzing every process, transformation, and behavior within an appropriate context, whether situated in a specific phase of an organism's life, across its entire developmental trajectory, or within population-level successions throughout the evolutionary history of a species.

These are just some of the (useful) criticisms raised within the debate on plant cognition, along with possible responses to them. As discussed in the previous chapter, even if the cognitive interpretation of plant activities, either in general or in relation to specific capacities, proves to be inadequate, uninformative, or incorrect as research evolves, exploring this possibility could still help clarify the modalities and processes underlying plant behavior. Thus, this type of analysis appears neither superfluous nor unjustified (cf. Bianchi, 2024).

Considering the behavioral and cognitive abilities of plants does not, naturally, imply that these organisms should be viewed as substitutes for animals. Once a shared ground has been established, it will be crucial to specify the differences and unique characteristics when describing plant behavior (cf. Bianchi et al., 2025a).

Building on advancements in research regarding the perceptual and information-signaling abilities of these organisms, alongside studies on other aneural organisms (such as protists and bacteria), the term "minimal" or "basal" cognition has been introduced (cf. Calvo & Keijzer, 2009; Lyon et al., 2021). These frameworks describe how these living systems interact with their environment, including their capacity to retain past information and anticipate future events (cf. Novoplansky, 2016). From this perspective, plants can be said to "behave" in a non-random way, meaning their responses go beyond mere physical reactions (cf. Karban, 2008;

Cvrčková et al., 2016; Calvo et al., 2020). For example, the behavior of roots adjusting their growth direction to avoid obstacles differs from a purely physical process such as the dissolution of salt. While both involve a change of state, the dissolution of salt cannot meaningfully be described as a behavior or as a response to an environmental challenge, whereas the plant's adjustment can, since it is driven by changing conditions related to survival and self-organization. At the same time, it is particularly important to emphasize that the behavior of any living organism, including humans, could be described in purely physical terms. Yet such a description would represent only one specific level of analysis among the multiple perspectives needed to account for the full complexity of biological phenomena.

Bio-ecological cognition is thus closely tied to the dynamics of life and suggests that organisms operate from an "organism-relative perspective" shaped by their internal organization and environmental context (cf. Bianchi, 2021; Godfrey-Smith, 2021).

More specifically, within the biogenic approach to cognition, a necessary component of cognitive activity appears to be the organism's capacity to influence, or modulate, its environment in ways that support and sustain its metabolic organization. This capacity spans a wide range of processes and levels of interaction, from molecular regulation to behavior and ecological engagement (cf. Thompson, 2007).

In recent years, the growing body of evidence concerning plant capacities such as memory, learning, and anticipation has underscored the need for a more unified and coherent theoretical framework within plant sciences. This has prompted efforts to integrate these capabilities into the broader domain of cognitive science. Promising directions include embodied, situated, enactive, and extended approaches to cognition, as well as systemic-relational frameworks that investigate behavior across various biological systems. Moreover, predictive modeling has gained traction as a valuable tool for exploring anticipatory capacities, which are increasingly regarded as key indicators of cognitive activity.

In the study of plant cognition, the absence of a brain does not necessitate identifying a functional analogue, such as a singular organ or mechanism unrelated to neurons. A similar limitation applies to approaches that study cognition exclusively through the functioning of the nervous system, in isolation from the rest of the body and the surrounding environment. By contrast, plant cognition is understood to involve a network of multilevel processes distributed across the plant body, where both material composition and functional organization are essential to understanding plant-environment interactions (cf. Calvo & Friston, 2017).

Despite having a sensorimotor organization that differs from that of many animals (cf. Calvo & Keijzer, 2009), a core principle of embodied and enactive approaches to cognition (cf. O'Regan & Noe, 2001) can also be applied to the study of plant behavior. In plants, sensorimotor contingencies are evident in their ability to develop and regulate the accuracy of their responses to environmental stimuli (cf. Alegiani & Marraffa, 2023). Enactive-ecological perspectives (cf. Thompson, 2007; Di Paolo et al., 2017; Hutto & Myin, 2017) suggest that even non-neural organisms may enact the capacities required to maintain integrity, autonomy, and survival within dynamic environments (cf. Wang et al., 2021). This includes interpreting plant activity as the expression of open systems that sustain themselves within ecological niches by modulating their sensorimotor coupling with the environment in response to varying needs and circumstances (cf. Bianchi et al., 2025a).

Thus, a broader and more nuanced exploration of a processual and relational concept of cognition—one that carefully considers the complex interplay of various dynamics—could have profound implications for our understanding of plant ecology and behavior (cf. Bianchi et al., 2025b).

More generally, such research could reshape and expand our conception of cognition itself. It may contribute to a deeper understanding of what cognition is, the roles it plays, its origins, and its evolution, particularly in relation to how different bodily organizations interact with and adapt to their environment, understood as a collective network of systems.

References

Abramson, C. I., & Chicas-Mosier, A. M. (2016). Learning in plants: Lessons from Mimosa pudica. *Frontiers in Psychology, 7*, 417.

Adams, F. (2018). Cognition wars. *Studies in History and Philosophy of Science Part A, 68*, 20–30.

Affifi, R. (2013). Learning plants: Semiosis between the parts and the whole. *Biosemiotics, 6*(3), 547–559.

Alegiani, A., & Marraffa, M. (2023). Cognizione vegetale ed elaborazione predittiva. *Giornale Italiano di Psicologia, 50*(2), 381–386.

Allen, M., & Friston, K. J. (2018). From cognitivism to autopoiesis: Towards a computational framework for the embodied mind. *Synthese, 195*(6), 2459–2482.

Allport, A. (1987). Selection for action: Some behavioral and neurophysiological considerations of attention and action. In H. Heuer & A. F. Sanders (Eds.), *Perspectives on perception and action* (pp. 395–419). Erlbaum.

Anten, N. P., & Chen, B. J. (2021). Detect thy family: Mechanisms, ecology and agricultural aspects of kin recognition in plants. *Plant, Cell & Environment, 44*(4), 1059–1071.

Appel, H. M., & Cocroft, R. B. (2014). Plants respond to leaf vibrations caused by insect herbivore chewing. *Oecologia, 175*, 1257–1266.

Arimura, G. I., Kost, C., & Boland, W. (2005). Herbivore-induced, indirect plant defences. *Biochimica et Biophysica Acta (BBA)-Molecular and Cell Biology of Lipids, 1734*(2), 91–111.

Arvidson, P. S. (2006). *The sphere of attention: Context and margin*. Springer.

Bais, H. P. (2015). Shedding light on kin recognition response in plants. *New Phytologist, 205*(1), 4–6.

Bais, H. P., Vepachedu, R., Gilroy, S., Callaway, R. M., & Vivanco, J. M. (2003). Allelopathy and exotic plant invasion: From molecules and genes to species interactions. *Science, 301*(5638), 1377–1380.

Baker, E. N. (2017). The intelligence of plants and the problem of language: A Wittgensteinian approach. In M. Gagliano, J. C. Ryan, & P. Vieira (Eds.), *The language of plants: Science, philosophy, literature* (pp. 136–154). University of Minnesota Press.

Baldwin, I. T., Halitschke, R., Paschold, A., Von Dahl, C. C., & Preston, C. A. (2006). Volatile signaling in plant-plant interactions «talking trees» in the genomics era. *Science, 311*(5762), 812–815.

Baluška, F. (2010). Recent surprising similarities between plant cells and neurons. *Plant Signaling & Behavior, 5*(2), 87–89.

Baluška, F., Gagliano, M., & Witzany, G. (Eds.). (2018). *Memory and learning in plants*. Springer International Publishing.

Baluška, F., & Levin, M. (2016). On having no head: Cognition throughout biological systems. *Frontiers in Psychology, 902.*
Baluška, F., & Mancuso, S. (2007). Plant neurobiology as a paradigm shift not only in the plant sciences. *Plant Signaling & Behavior, 2*(4), 205–207.
Baluška, F., & Mancuso, S. (2009). Plant neurobiology: From sensory biology, via plant communication, to social plant behavior. *Cognitive Processing, 10*, 3–7.
Baluška, F., & Mancuso, S. (2013). Root apex transition zone as oscillatory zone. *Frontiers in Plant Science, 4*, 354.
Baluška, F., & Mancuso, S. (2016). Vision in plants via plant-specific ocelli? *Trends in Plant Science, 21*(9), 727–730.
Baluška, F., Miller, W. B., Jr., & Reber, A. S. (2021). Biomolecular basis of cellular consciousness via subcellular nanobrains. *International Journal of Molecular Sciences, 22*(5), 2545.
Baluška, F., & Yokawa, K. (2021). Anaesthetics and plants: From sensory systems to cognition-based adaptive behaviour. *Protoplasma, 258*, 449–454.
Barron, A. B., & Klein, C. (2016). What insects can tell us about the origins of consciousness. *Proceedings of the National Academy of Sciences, 113*(18), 4900–4908.
Bateson, G. (1976). *Verso un'ecologia della mente*, Longo, G., & Trautteur, G. (Trans.). Adelphi (Original work published 1972, *Steps to an ecology of mind: Collected essays in anthropology, psychiatry, evolution, and epistemology*. Chandler Publishing Company).
Berthouze, L., James, L. M., & Farmer, S. F. (2010). Human EEG shows long-range temporal correlations of oscillation amplitude in Theta, Alpha and Beta bands across a wide age range. *Clinical Neurophysiology, 121*(8), 1187–1197.
Betti, S., Castiello, U., & Begliomini, C. (2021). Reach-to-grasp: A multisensory experience. *Frontiers in Psychology, 12*, 614471.
Bianchi, M. (2021). *La vita ramificata.* Cognizione e comportamento nelle piante fra scienza e filosofia. Mimesis.
Bianchi, M. (2022). Networks and ramifications: Relational perspectives in plant cognition. *Rivista Internazionale di Filosofia e Psicologia, 13*(2), 157–168.
Bianchi, M. (2024). Plant cognition. Challenges and horizons in philosophical reflection. *Paradigmi, 1*, 21–46.
Bianchi, M. (2025). Oltre il Mondo Animale: i Darwin e la Botanica. In E. Gagliasso, S. Pollo & E. Severini (Eds.), *Origine e attualità dell'Origine dell'uomo. Scienza, Etica, Letteratura* (pp. 33–54). Edizioni ETS.
Bianchi, M., Avesani, S., Bonato, B., Dadda, M., Guerra, S., Ravazzolo, L., Simonetti, V., & Castiello, U. (2025a). Plant behavior: Theoretical and technological advances. *Current Opinion in Psychology, 64*, 102026.
Bianchi, M., Guerra, S., Bonato, B., Avesani, S., Ravazzolo, L., Simonetti, V., Dadda, M., & Castiello, U. (2025b). Temporal Dynamics With and Without a Nervous System: Plant Physiology, Communication, and Movement. *Cognitive Science, 49*, e70079.
Bianchi, M., & Castiello, U. (2023a). Il concetto di rappresentazione nelle scienze cognitive classiche e post-classiche: la sfida della cognizione vegetale. *Giornale Italiano di Psicologia, 50*(2), 349–385.
Bianchi, M., & Castiello, U. (2023b). Riflessioni corali nell'analisi interdisciplinare sul concetto di rappresentazione nella cognizione vegetale. *Giornale Italiano di Psicologia, 50*(2), 419–452.
Bickhard, M. H. (2005). Consciousness and reflective consciousness. *Philosophical Psychology, 18*(2), 205–218.
Biedrzycki, M. L., Jilany, T. A., Dudley, S. A., & Bais, H. P. (2010). Root exudates mediate kin recognition in plants. *Communicative & Integrative Biology, 3*(1), 28–35.
Bonato, B., & Castiello, U. (2020). Dagli abeti ai piselli. Storie di ordinaria resilienza. *Psiche, 7*(1), 113–121.
Bonato, B., Peressotti, F., Guerra, S., Wang, Q., & Castiello, U. (2021). Cracking the code: A comparative approach to plant communication. *Communicative & Integrative Biology, 14*(1), 176–185.

Bonato, B., Simonetti, V., Bulgheroni, M., Wang, Q., Guerra, S., Quaggiotti, S., et al. (2023). Evidence of motor intentions in plants: A kinematical study. *Journal of Comparative Psychology, 137*(4), 228–237.

Bonato, B., Wang, Q., Guerra, S., Simonetti, V., Bulgheroni, M., Quaggiotti, S., & Castiello, U. (2024). 'United we stand, divided we fall': Intertwining as evidence of joint actions in pea plants. *AoB Plants, 16*(1), plad088.

Bonfante, P., (2021). *Una pianta non è un'isola. Alla scoperta di un mondo visibile*. Il Mulino.

Bose, J. C. (1907). *Plant response as a means of physiological investigation*. Longman, Green and Company.

Bose, J. C. (1913). *Researches on the irritability of plants*. Longman, Green and Co..

Bose, J. C. (1926). *The nervous mechanism of plants*. Longman, Green and Co.

Bose, J. C., & Guha, S. C. (1922). The dia-heliotropic attitude of leaves as determined by transmitted nervous excitation. *Proceedings of the Royal Society of London. Series B, Containing Papers of a Biological Character, 93*(651), 153–178.

Braam, J. (2005). In touch: Plant responses to mechanical stimuli. *New Phytologist, 165*(2), 373–389.

Bradshaw, A. D. (1965). Evolutionary significance of phenotypic plasticity in plants. Advances in Genetics, 13, 115–155, Elsevier.

Brenner, E. D., Stahlberg, R., Mancuso, S., Vivanco, J., Baluška, F., & Van Volkenburgh, E. (2006). Plant neurobiology: An integrated view of plant signaling. *Trends in Plant Science, 11*(8), 413–419.

Brodmann, J., Twele, R., Francke, W., Hölzler, G., Zhang, Q. H., & Ayasse, M. (2008). Orchids mimic green-leaf volatiles to attract prey-hunting wasps for pollination. *Current Biology, 18*(10), 740–744.

Burdon-Sanderson, J. S. (1873). Note on the electrical phenomena which accompany irritation of the leaf of Dionæa muscipula. *Proceedings of the Royal Society of London, 21*, 495–496.

Burdon-Sanderson, J. S. (1899). On the relation of motion in animals and plants to the electrical phenomena which are associated with it. *Proceedings of the Royal Society of London, 65*, 37–64.

Burnet, F. M. (1971). "Self-recognition" in colonial marine forms and flowering plants in relation to the evolution of immunity. *Nature, 232*(5308), 230–235.

Cahill, J. F., McNickle, G. G., Haag, J. J., Lamb, E. G., Nyanumba, S. M., & St. Clair, C. C. (2010). Plants integrate information about nutrients and neighbors. *Science, 328*(5986), 1657.

Calvo, P. (2017). What is it like to be a plant? *Journal of Consciousness Studies, 24*(9–10), 205–227.

Calvo, P., & Friston, K. (2017). Predicting green: Really radical (plant) predictive processing. *Journal of the Royal Society Interface, 14*(131), 20170096.

Calvo, P., & Keijzer, F. (2009). Cognition in plants. In *Plant-environment interactions: From sensory plant biology to active plant behavior* (pp. 247–266). Springer.

Calvo, P., & Keijzer, F. (2011). Plants: Adaptive behavior, root brains and minimal cognition. *Adaptive Behavior, 19*(3), 155–171.

Calvo, P., & Segundo-Ortin, M. (2023). Plant sentience revisited: Sifting through the thicket of perspectives. *Animal Sentience, 8*(33), 32.

Calvo, P., Baluška, F., & Sims, A. (2016). «feature detection» vs. «predictive coding» models of plant behavior. *Frontiers in Psychology, 7*, 1505.

Calvo, P., Gagliano, M., Souza, G. M., & Trewavas, A. (2020). Plants are intelligent, here's how. *Annals of Botany, 125*(1), 11–28.

Calvo, P., Sahi, V. P., & Trewavas, A. (2017). Are plants sentient? *Plant, Cell & Environment, 40*(11), 2858–2869.

Campbell, N. A., & Reece, J. B. (2009). *Biologia. 6. Ecologia e comportamento*. Pearson Italia.

Campbell, C., Yang, S., Albert, R., & Shea, K. (2015). Plant–pollinator community network response to species invasion depends on both invader and community characteristics. *Oikos, 124*(4), 406–413.

Castiello, U. (2019). *La mente delle piante. Introduzione alla psicologia vegetale*. Il Mulino.
Castiello, U. (2021). (Re) claiming plants in comparative psychology. *Journal of Comparative Psychology, 135*(1), 127.
Castiello, U., & Guerra, S. (2020). La psicologia vegetale: alle radici della cognizione. *Giornale Italiano di Psicologia, 47*(3–4), 683–708.
Castiello, U., & Umiltà, C. (1990). Size of the attentional focus and efficiency of processing. *Acta Psychologica, 73*(3), 195–209.
Castiello, U., & Umiltà, C. (1992). Splitting focal attention. *Journal of Experimental Psychology: Human Perception and Performance, 18*(3), 837–848.
Ceccarini, F., & Castiello, U. (2018). The grasping side of post-error slowing. *Cognition, 179*, 1–13.
Ceccarini, F., Guerra, S., Peressotti, A., Peressotti, F., Bulgheroni, M., Baccinelli, W., & Castiello, U. (2020). Speed-accuracy trade-off in plants. *Psychonomic Bulletin & Review, 27*, 966–973.
Chamovitz, D. (2012). *What a plant knows: A field guide to the senses*. Farrar, Straus and Giroux.
Chamovitz, D. A. (2018). Plants are intelligent; now what? *Nature Plants, 4*(9), 622–623.
Chen, R., Huangfu, L., Lu, Y., Fang, H., Xu, Y., Li, P., et al. (2021). Adaptive innovation of green plants by horizontal gene transfer. *Biotechnology Advances, 46*, 107671.
Clark, A. (2013). Whatever next? Predictive brains, situated agents, and the future of cognitive science. *Behavioral and Brain Sciences, 36*(3), 181–204.
Clark, A. (2016). Attention alters predictive processing. *Behavioral and Brain Sciences, 39*.
Clayton, N. S., Yu, K. S., & Dickinson, A. (2001). Scrub jays (Aphelocoma coerulescens) form integrated memories of the multiple features of caching episodes. *Journal of Experimental Psychology: Animal Behavior Processes, 27*, 17–29.
Clayton, N. S., Yu, K. S., & Dickinson, A. (2003). Interacting cache memories: Evidence for flexible memory use by Western scrub-jays (Aphelocoma californica). *Journal of Experimental Psychology: Animal Behavior Processes, 29*, 14–22.
Coleman, S. W. (2009). Taxonomic and sensory biases in the mate-choice literature: There are far too few studies of chemical and multimodal communication. *Acta Ethologica, 12*, 45–48.
Conrath, U. (2006). Systemic acquired resistance. *Plant Signaling & Behavior, 1*(4), 179–184.
Conrath, U. (2011). Molecular aspects of defence priming. *Trends in Plant Science, 16*(10), 524–531.
Conrath, U., Thulke, O., Katz, V., Schwindling, S., & Kohler, A. (2001). Priming as a mechanism in induced systemic resistance of plants. *European Journal of Plant Pathology, 107*, 113–119.
Creath, K., & Schwartz, G. E. (2004). Measuring effects of music, noise, and healing energy using a seed germination bioassay. *The Journal of Alternative & Complementary Medicine, 10*(1), 113–122.
Crepy, M. A., & Casal, J. J. (2015). Photoreceptor-mediated kin recognition in plants. *New Phytologist, 205*(1), 329–338.
Crisp, P. A., Ganguly, D., Eichten, S. R., Borevitz, J. O., & Pogson, B. J. (2016). Reconsidering plant memory: Intersections between stress recovery, RNA turnover, and epigenetics. *Science Advances, 2*(2), e1501340.
Cvrčková, F., Lipavská, H., & Žárský, V. (2009). Plant intelligence, why, why not or where? *Plant Signaling & Behavior, 4*(5), 394–399.
Cvrčková, F., Žárský, V., & Markoš, A. (2016). Plant studies may lead us to rethink the concept of behavior. *Frontiers in Psychology, 7*, 622.
Darwin, C. R. (1859). *Origin of species by means of natural selection, or the preservation of favoured races in the struggle for life*. John Murray.
Darwin, C. R. (1871). *The descent of man, and selection in relation to sex*. John Murray.
Darwin, C. R. (1875). *The movements and habits of climbing plants* (2nd ed.). John Murray.
Darwin, C., & Darwin, F. (1880). *The power of movement in plants*. John Murray.
De Kroon, H., Visser, E. J., Huber, H., Mommer, L., & Hutchings, M. J. (2009). A modular concept of plant foraging behaviour: The interplay between local responses and systemic control. *Plant, Cell & Environment, 32*(6), 704–712.

De la Fuente, I. M. (2015). Elements of the cellular metabolic structure. *Frontiers in Molecular Biosciences, 2*, 16.

De Loof, A. (2016). The cell's self-generated "electrome": The biophysical essence of the immaterial dimension of life? *Communicative & Integrative Biology, 9*(5), e1197446.

De Melo, H. C. (2023). Plants detect and respond to sounds. *Planta, 257*(3), 55.

De Toledo, G. R., Parise, A. G., Simmi, F. Z., Costa, A. V., Senko, L. G., Debono, M. W., & Souza, G. M. (2019). Plant electrome: The electrical dimension of plant life. *Theoretical and Experimental Plant Physiology, 31*, 21–46.

De Waal, F. B. M. (2016). *Are we smart enough to know how smart animals are*? Granta Books.

Demey, M. L., Mishra, R. C., & Van Der Straeten, D. (2023). Sound perception in plants: From ecological significance to molecular understanding. *Trends in Plant Science, 28*(7), 825–840.

Dener, E., Kacelnik, A., & Shemesh, H. (2016). Pea plants show risk sensitivity. *Current Biology, 26*(13), 1763–1767.

Di Francesco, M., Marraffa, M., & Tomasetta, A. (2017). *Filosofia della Mente: Corpo, Coscienza, Pensiero*. Carocci.

Di Paolo, E., Buhrmann, T., & Barandiaran, X. (2017). *Sensorimotor life: An enactive proposal*. Oxford University Press.

Ding, Y., Fromm, M., & Avramova, Z. (2012). Multiple exposures to drought'train'transcriptional responses in Arabidopsis. *Nature Communications, 3*(1), 740.

Dipietro, L., Poizner, H., & Krebs, H. I. (2011). EEG correlates of submovements. In *2011 annual international conference of the IEEE engineering in medicine and biology society* (pp. 7429–7432). IEEE.

Dominey, P. F., Prescott, T. J., Bohg, J., Engel, A. K., Gallagher, S., Heed, T., et al. (2016). Implications of action-oriented paradigm shifts in cognitive science. In A. K. Engel, K. J. Friston, & D. Kragic (Eds.), *The pragmatic turn—Toward action-oriented views in cognitive science, 18* (pp. 333–356). MIT Press.

Dudley, S. A. (2015). Plant cooperation. *AoB Plants, 7*, plv113.

Dudley, S. A., & File, A. L. (2007). Kin recognition in an annual plant. *Biology Letters, 3*(4), 435–438.

Dudley, S. A., Murphy, G. P., & File, A. L. (2013). Kin recognition and competition in plants. *Functional Ecology, 27*, 898–906.

Edelman, G. M., Gally, J. A., & Baars, B. J. (2011). Biology of consciousness. *Frontiers in Psychology, 2*, 4.

Eisenstein, E. M., Eisenstein, D., & Smith, J. C. (2001). The evolutionary significance of habituation and sensitization across phylogeny: A behavioral homeostasis model. *Integrative Physiological & Behavioral Science, 36*, 251–265.

Elhakeem, A., Markovic, D., Broberg, A., Anten, N. P., & Ninkovic, V. (2018). Aboveground mechanical stimuli affect belowground plant-plant communication. *PLoS One, 13*(5), e0195646.

Falik, O., de Kroon, H., & Novoplansky, A. (2006). Physiologically-mediated self/non-self root discrimination in Trifolium repens has mixed effects on plant performance. *Plant Signaling & Behavior, 1*(3), 116–121.

Falik, O., Mordoch, Y., Ben-Natan, D., Vanunu, M., Goldstein, O., & Novoplansky, A. (2012). Plant responsiveness to root–root communication of stress cues. *Annals of Botany, 110*(2), 271–280.

Faucher, K. (2014). Phytosemiotics revisited: Botanical behavior and sign transduction. *Semiotica, 202*, 673–688.

Firn, R. (2004). Plant intelligence: An alternative point of view. *Annals of Botany, 93*(4), 345–351.

Fitts, P. M. (1954). The information capacity of the human motor system in controlling the amplitude of movement. *Journal of Experimental Psychology, 47*(6), 381–391.

Fradet, L., Lee, G., & Dounskaia, N. (2008). Origins of submovements during pointing movements. *Acta Psychologica, 129*(1), 91–100.

Frazier, P. A. (2021). On the possibility of plant consciousness: A view from ecointeractivism. *Mind and Matter, 19*(2), 229–259.

Friston, K. J. (2010). The free-energy principle: A unified brain theory? *Nature Reviews Neuroscience, 11*(2), 127–138.

Friston, K. (2013). Active inference and free energy. *Behavioral and Brain Sciences, 36*(3), 212.

Friston, K., & Kiebel, S. (2009). Predictive coding under the free-energy principle. *Philosophical Transactions of the Royal Society B: Biological Sciences, 364*(1521), 1211–1221.

Fromm, J., & Lautner, S. (2007). Electrical signals and their physiological significance in plants. *Plant, Cell & Environment, 30*(3), 249–257.

Fukano, Y., & Yamawo, A. (2015). Self-discrimination in the tendrils of the vine Cayratia japonica is mediated by physiological connection. *Proceedings of the Royal Society B: Biological Sciences, 282*(1814), 20151379.

Gagliano, M., Mancuso, S., & Robert, D. (2012). Towards understanding plant bioacoustics. *Trends in Plant Science, 17*(6), 323–325.

Gagliano, M., & Renton, M. (2013). Love thy neighbour: Facilitation through an alternative signalling modality in plants. *BMC Ecology, 13*(1), 1–6.

Gagliano, M., Renton, M., Depczynski, M., & Mancuso, S. (2014). Experience teaches plants to learn faster and forget slower in environments where it matters. *Oecologia, 175*(1), 63–72.

Gagliano, M., Ryan, J. C., & Vieira, P. (Eds.). (2017). *The language of plants. Science, philosophy, literature*. Minneapolis, University of Minnesota Press.

Gagliano, M., Vyazovskiy, V. V., Borbély, A. A., Grimonprez, M., & Depczynski, M. (2016). Learning by association in plants. *Scientific Reports, 6*(1), 38427.

Gagliasso, E. (2019). Condividui in evoluzione: quale filosofia?. In M. Monti, & C. Redi (Eds.), *CON-dividuo. Cellule e Genomi* (pp. 81–96). Collegio Ghislieri.

Gallagher, S. (2017). *Enactivist interventions: Rethinking the mind*. Oxford University Press.

Galviz, Y. C., Ribeiro, R. V., & Souza, G. M. (2020). Yes, plants do have memory. *Theoretical and Experimental Plant Physiology, 32*(3), 195–202.

Garcia Rodriguez, A., & Calvo Garzon, P. (2010). Is cognition a matter of representations? Emulation, teleology, and time-keeping in biological systems. *Adaptive Behavior, 18*(5), 400–415.

Gardner, H. (1983). *Frames of mind: The theory of multiple intelligences*. Basic Books.

Gazzaniga, M. S. (2019). *La coscienza è un istinto. Il legame misterioso tra il cervello e la mente* (Trans. It.). Cortina Editore (Original work published 2018, *The consciousness instinct: Unraveling the mystery of how the brain makes the mind*. Farrar, Straus and Giroux).

Getter, K. L., & Rowe, D. B. (2006). The role of extensive green roofs in sustainable development. *HortScience, 41*(5), 1276–1285.

Gianoli, E. (2015). The behavioural ecology of climbing plants. *AoB Plants, 7*, plv013.

Gianoli, E. (2017). Eyes in the chameleon vine? *Trends in Plant Science, 22*(1), 4–5.

Gianoli, E., & Carrasco-Urra, F. (2014). Leaf mimicry in a climbing plant protects against herbivory. *Current Biology, 24*(9), 984–987.

Ginsburg, S., & Jablonka, E. (2009). Epigenetic learning in non-neural organisms. *Journal of Biosciences, 34*(4), 633–646.

Ginsburg, S., & Jablonka, E. (2021). Evolutionary transitions in learning and cognition. *Philosophical Transactions of the Royal Society B, 376*(1821), 20190766.

Godfrey-Smith, P. (2021). *Metazoa* (I. C. Blum, Trans.). Adelphi. (Original work published 2020, *Metazoa: Animal life and the birth of the mind*. Farrar, Straus and Giroux).

Grémiaux, A., Yokawa, K., Mancuso, S., & Baluška, F. (2014). Plant anesthesia supports similarities between animals and plants. *Plant Signaling & Behavior, 9*(1), e27886.

Grondin, S. (2016). Perception and attention. In *Id., psychology of perception* (pp. 123–135). Springer.

Gruntman, M., Groß, D., Májeková, M., & Tielbörger, K. (2017). Decision-making in plants under competition. *Nature Communications, 8*(1), 2235.

Gruntman, M., & Novoplansky, A. (2004). Physiologically mediated self/nonself discrimination in roots. *Proceedings of the National Academy of Sciences, 101*(11), 3863–3867.

Guerra, S., Peressotti, A., Peressotti, F., Bulgheroni, M., Baccinelli, W., et al. (2019). Flexible control of movement in plants. *Scientific Reports, 9*(1), 16570.
Guerra, S., Bonato, B., Wang, Q., Peressotti, A., Peressotti, F., Baccinelli, W., et al. (2022). Kinematic evidence of root-to-shoot signaling for the coding of support thickness in pea plants. *Biology, 11*(3), 405.
Guerra, S., Bruno, G., Spoto, A., Panzeri, A., Wang, Q., Bonato, B., Simonetti, V., & Castiello, U. (2024). Ascent and attachment in pea plants: A matter of iteration. *Plants, 13*(10), 1389.
Gupta, A., Savarese, S., Ganguli, S., & Fei-Fei, L. (2021). Embodied intelligence via learning and evolution. *Nature Communications, 12*(1), 5721.
Haberlandt, G. (1905). *Die Lichtsinnesorgane der Laubblätter*. W. Engelmann.
Haider, S., Iqbal, J., Shaukat, M., Naseer, S., & Mahmood, T. (2021). The epigenetic chromatin-based regulation of somatic heat stress memory in plants. *Plant Gene, 27*, 100318.
Halling, D. B., Aracena-Parks, P., & Hamilton, S. L. (2005). Regulation of voltage-gated Ca2+ channels by calmodulin. *Science's STKE, 2005*(315), re15-re15.
Hamant, O., & Moulia, B. (2016). How do plants read their own shapes? *New Phytologist, 212*(2), 333–337.
Hamilton, W. D. (1964). The genetical evolution of social behaviour, I and II. *Journal of Theoretical Biology, 7*, 1–52.
Harborne, J. B. (1991). The chemical basis of plant defense. In R. T. Palo & C. T. Robbins (Eds.), *Plant defenses against mammalian herbivorey* (pp. 45–60). CRC Press.
Hedrich, R. (2012). Ion channels in plants. *Physiological Reviews, 92*(4), 1777–1811.
Heil, M., & Karban, R. (2010). Explaining evolution of plant communication by airborne signals. *Trends in Ecology & Evolution, 25*(3), 137–144.
Heitz, R. P., & Schall, J. D. (2012). Neural mechanisms of speed-accuracy tradeoff. *Neuron, 76*(3), 616–628.
Henriksson, N., Marshall, J., Högberg, M. N., Högberg, P., Polle, A., Franklin, O., & Näsholm, T. (2023). Re-examining the evidence for the mother tree hypothesis—Resource sharing among trees via ectomycorrhizal networks. *New Phytologist, 239*(1), 19–28.
Heras-Escribano, M., & Calvo, P. (2019). The philosophy of plant neurobiology. In S. Robins, J. Symons, & P. Calvo (Eds.), *The Routledge companion of philosophy of psychology* (pp. 529–548). Routledge.
Hiernaux, Q. (2021). Differentiating behaviour, cognition, and consciousness in plants. *Journal of Consciousness Studies, 28*(1–2), 106–135.
Higham, J. P., & Hebets, E. A. (2013). An introduction to multimodal communication. *Behavioral Ecology and Sociobiology, 67*, 1381–1388.
Hilker, M., & Schmülling, T. (2019). Stress priming, memory, and signalling in plants. *Plant, Cell & Environment, 42*(3), 753–761.
Hobson, J. A., Hong, C. C. H., & Friston, K. J. (2014). Virtual reality and consciousness inference in dreaming. *Frontiers in Psychology, 5*, 1133.
Hodge, A. (2009). Root decisions. *Plant, Cell & Environment, 32*(6), 628–640.
Hohwy, J. (2013). *The predictive mind*. Oxford University Press.
Holmes, E., & Gruenberg, G. (1965). Learning in plants. *Worm Runner's Dig, 7*, 9–12.
Huber, A. E., & Bauerle, T. L. (2016). Long-distance plant signaling pathways in response to multiple stressors: The gap in knowledge. *Journal of Experimental Botany, 67*(7), 2063–2079.
Hutto, D. D., & Myin, E. (2017). *Evolving Enactivism: Basic minds meet content*. The MIT press.
Jaffe, M. J., & Shotwell, M. (1980). Physiological studies on pea tendrils: XI. Storage of tactile sensory information prior to the blue light activation effect. *Physiologia Plantarum, 50*(1), 78–82.
Jennings, H. S. (1908). The interpretation of the behavior of the lower organisms. *Science, 27*(696), 698–710.
Jennings, H. S. (1923). *Behaviour of the lower organisms*. Columbia University Press.
Johnson, D., & Gilbert, L. (2015). Interplant signalling through hyphal networks. *New Phytologist, 205*(4), 1448–1453.
Kacelnik, A., & Bateson, M. (1997). Risk-sensitivity: Crossroads for theories of decision-making. *Trends in Cognitive Sciences, 1*(8), 304–309.
Kahneman, D. (1973). *Attention and effort* (Vol. 1063, pp. 218–226). Prentice Hall.

Kandel, E. R. (2001). The molecular biology of memory storage: A dialogue between genes and synapses. *Science, 294*, 1030–1038.

Kandel, E. R., Dudai, Y., & Mayford, M. R. (2014). The molecular and systems biology of memory. *Cell, 157*(1), 163–186.

Karban, R. (2008). Plant behaviour and communication. *Ecology Letters, 11*(7), 727–739.

Karban, R. (2015). *Plant sensing and communication*. University of Chicago Press.

Karban, R. (2017). Plant communication increases heterogeneity in plant phenotypes and herbivore movement. *Functional Ecology, 31*(5), 990–991.

Karst, J., Jones, M. D., & Hoeksema, J. D. (2023). Positive citation bias and overinterpreted results lead to misinformation on common mycorrhizal networks in forests. *Nature Ecology & Evolution, 7*(4), 501–511.

Kelly, C. K. (1992). Resource choice in Cuscuta europaea. *Proceedings of the National Academy of Sciences, 89*(24), 12194–12197.

Khait, I., Lewin-Epstein, O., Sharon, R., Saban, K., Goldstein, R., Anikster, Y., et al. (2023). Sounds emitted by plants under stress are airborne and informative. *Cell, 186*(7), 1328–1336.

Kim, M. C., Chung, W. S., Yun, D. J., & Cho, M. J. (2009). Calcium and calmodulin-mediated regulation of gene expression in plants. *Molecular Plant, 2*(1), 13–21.

Kirchhoff, M. D., & Robertson, I. (2018). Enactivism and predictive processing: A non-representational view. *Philosophical Explorations, 21*(2), 264–281.

Kiverstein, J., & Sims, M. (2021). Is free-energy minimisation the mark of the cognitive? *Biology & Philosophy, 36*(2), 25.

Kreps, J. A., & Kay, S. A. (1997). Coordination of plant metabolism and development by the circadian clock. *The Plant Cell, 9*(7), 1235–1244.

Kull, K., & Velmezova, E. (2019). Jesper Hoffmeyer: Biosemiotics is a discovery. *Biosemiotics, 12*(3), 373–379.

Lamers, J., Van Der Meer, T., & Testerink, C. (2020). How plants sense and respond to stressful environments. *Plant Physiology, 182*(4), 1624–1635.

Latzel, V., & Münzbergová, Z. (2018). Anticipatory behavior of the clonal plant Fragaria vesca. *Frontiers in Plant Science, 9*, 1847.

Latzel, V., Rendina González, A. P., & Rosenthal, J. (2016). Epigenetic memory as a basis for intelligent behavior in clonal plants. *Frontiers in Plant Science, 7*, 1354.

LeDoux, J. (2020). *Lunga storia di noi stessi: Come il cervello è diventato cosciente* (G. Guerriero, Trans.). Raffaello Cortina Editore (Original work published 2019, *The deep history of ourselves: The four-billion-year story of how we got conscious brains*. Viking).

Legg, S., & Hunter, M. (2007). A collection of definitions of intelligence. In B. Goertzel & P. Wang (Eds.), *Advances in artificial general intelligence: Concepts, architectures and algorithms. Frontiers in artificial intelligence and applications* (pp. 17–24). IOS Press.

Lev-Yadun, S. (2016). *Defensive (anti-herbivory) coloration in land plants*. Springer International Publishing.

Li, S., Liu, S., Zhang, Q., Cui, M., Zhao, M., Li, N., et al. (2022). The interaction of ABA and ROS in plant growth and stress resistances. *Frontiers in Plant Science, 13*, 1050132.

Libet, B. (1985). Unconscious cerebral initiative and the role of conscious will involuntary action. *Behavioral and Brain Sciences, 8*, 529–566.

Linson, A., & Calvo, P. (2020). Zoocentrism in the weeds? Cultivating plant models for cognitive yield. *Biology & Philosophy, 35*(5), 1–27.

Llinás, R., Ribary, U., Contreras, D., & Pedroarena, C. (1998). The neuronal basis for consciousness. *Philosophical Transactions of the Royal Society of London. Series B: Biological Sciences, 353*(1377), 1841–1849.

Lozito, J. P., & Mulligan, N. W. (2006). Exploring the role of attention during memory retrieval: Effects of semantic encoding and divided attention. *Memory & Cognition, 34*, 986–998.

Lüttge, U. (2021). Integrative emergence in contrast to separating modularity in plant biology: Views on systems biology with information, signals and memory at scalar levels from molecules to the biosphere. *Theoretical and Experimental Plant Physiology, 33*(1), 1–13.

Lyon, P. (2006). The biogenic approach to cognition. *Cognitive Processing, 7*(1), 11–29.

Lyon, P., Keijzer, F., Arendt, D., & Levin, M. (2021). Reframing cognition: Getting down to biological basics. *Philosophical Transactions of the Royal Society B, 376*(1820), 20190750.
Maher, C. (2017). *Plant minds: A philosophical defense*. Routledge.
Maher, C. (2020). Experiment rather than define. *Trends in Plant Science, 25*(3), 213–214.
Mallatt, J., Blatt, M. R., Draguhn, A., Robinson, D. G., & Taiz, L. (2021). Debunking a myth: Plant consciousness. *Protoplasma, 258*(3), 459–476.
Mancuso, S. (2006). Alcuni aspetti di neurobiologia vegetale. *Silvae, 2*(4), 239–251.
Mancuso, S. (2017). *Botanica. Viaggio nell'universo vegetale*. Aboca Edizioni.
Mancuso, S., & Viola, A. (2015). *Verde brillante.* Sensibilità e intelligenza del mondo vegetale. Giunti.
Marder, M. (2009). What is living and what is dead in attention? *Research in Phenomenology, 39*, 29–51.
Marder, M. (2011). Phenomenology of distraction, or attention in the fissuring of time and space. *Research in Phenomenology, 41*(3), 396–419.
Marder, M. (2013). Plant intelligence and attention. *Plant Signaling & Behavior, 8*(5), e23902.
Margulis, L. (1992). Biodiversity: Molecular biological domains, symbiosis and kingdom origins. *Biosystems, 27*(1), 39–51.
Margulis, L. (1993). Origins of species: Acquired genomes and individuality. *Biosystems, 31*(2–3), 121–125.
Markel, K. (2020). Response to comment on lack of evidence for associative learning in pea plants. *eLife, 9*, e61689.
Martens, M. J., & Michelsen, A. (1981). Absorption of acoustic energy by plant leaves. *The Journal of the Acoustical Society of America, 69*(1), 303–306.
Masi, E., Ciszak, M., Stefano, G., Renna, L., Azzarello, E., Pandolfi, C., et al. (2009). Spatiotemporal dynamics of the electrical network activity in the root apex. *Proceedings of the National Academy of Sciences, 106*(10), 4048–4053.
Marraffa, M., & Paternoster, A. (2013). *Sentirsi esistere: Inconscio, coscienza, autocoscienza.* Laterza.
Maturana, H. R., & Varela, F. J. (1980). *Autopoiesis and cognition: The realization of the living*. D. Reidel Publishing Company.
Maturana, H. R., & Varela, F. J. (1987). *The tree of knowledge: The biological roots of human understanding* (R. Paolucci & F. Varela, Trans.). Shambhala Publications.
Mauseth, J. D. (2017). *Botanica, fondamenti di biologia delle piante*. Idelson-Gnocchi.
Mawphlang, O. I., & Kharshiing, E. V. (2017). Photoreceptor mediated plant growth responses: Implications for photoreceptor engineering toward improved performance in crops. *Frontiers in Plant Science, 8*, 1181.
Mazzolai, B. (2021). *Il futuro raccontato dalle piante. Cosa possiamo imparare dal regno vegetale e dal suo percorso sul pianeta*. Longanesi.
Metlen, K. L., Aschehoug, E. T., & Callaway, R. M. (2009). Plant behavioural ecology: Dynamic plasticity in secondary metabolites. *Plant, Cell & Environment, 32*(6), 641–653.
McClintock, B. (1984). The significance of responses of the genome to challenge. *Science, 226*(4676), 792–801.
McShea, D. W. (1996). Perspective metazoan complexity and evolution: Is there a trend? *Evolution, 50*(2), 477–492.
Merker, B. (2007). Consciousness without a cerebral cortex: A challenge for neuroscience and medicine. *Behavioral and Brain Sciences, 30*(1), 63–81.
Michaels, C. F., & Carello, C. (1981). *Direct perception*. Prentice-Hall.
Michmizos, D., & Hilioti, Z. (2019). A roadmap towards a functional paradigm for learning & memory in plants. *Journal of Plant Physiology, 232*, 209–215.
Migliaccio, F., & Piconese, S. (2001). Spiralizations and tropisms in Arabidopsis roots. *Trends in Plant Science, 6*(12), 561–565.
Miguel-Tomé, S., & Llinás, R. R. (2021). Broadening the definition of a nervous system to better understand the evolution of plants and animals. *Plant Signaling & Behavior, 16*(10), 1927562.

Mishra, R. C., Ghosh, R., & Bae, H. (2016). Plant acoustics: In the search of a sound mechanism for sound signaling in plants. *Journal of Experimental Botany, 67*(15), 4483–4494.

Mittler, R., Zandalinas, S. I., Fichman, Y., & Van Breusegem, F. (2022). Reactive oxygen species signalling in plant stress responses. *Nature Reviews Molecular Cell Biology, 23*(10), 663–679.

Molinier, J., Ries, G., Zipfel, C., & Hohn, B. (2006). Transgeneration memory of stress in plants. *Nature, 442*(7106), 1046–1049.

Monod, J. (1971). *Chance and necessity: An essay on the natural philosophy of modern biology* (A. Wainhouse, Trans.). Alfred A. Knopf (Original work published 1970 as Le hasard et la nécessité: Essai sur la philosophie naturelle de la biologie moderne).

Monshausen, G. B., & Gilroy, S. (2009). Feeling green: mechanosensing in plants. *Trends in Cell Biology, 19*(5), 228–235.

Morabito, C. (2020). *Il motore della mente: il movimento nella storia delle scienze cognitive*. Laterza.

Morabito, C. (2023). Nel mondo dei sistemi complessi: consonanza o cosilience? *Giornale Italiano di Psicologia, 50*(2), 405–412.

Moreira, X., Petry, W. K., Hernández-Cumplido, J., Morelon, S., & Benrey, B. (2016). Plant defence responses to volatile alert signals are population-specific. *Oikos, 125*(7), 950–956.

Moulia, B., Bastien, R., Chauvet-Thiry, H., & Leblanc-Fournier, N. (2019). Posture control in land plants: Growth, position sensing, proprioception, balance, and elasticity. *Journal of Experimental Botany, 70*(14), 3467–3494.

Munoz, N. E., & Blumstein, D. T. (2012). Multisensory perception in uncertain environments. *Behavioral Ecology, 23*(3), 457–462.

Murphy, G. P., & Dudley, S. A. (2009). Kin recognition: Competition and cooperation in impatiens (Balsaminaceae). *American Journal of Botany, 96*(11), 1990–1996.

Nansen, C. (2017). Unraveling the «radiometric signals» from green leaves. In M. Gagliano, J. C. Ryan, & P. Vieira (Eds.), *The language of plants: Science, philosophy, literature* (pp. 62–68). University of Minnesota Press.

Niu, Y., Sun, H., & Stevens, M. (2018). Plant camouflage: Ecology, evolution, and implications. *Trends in Ecology & Evolution, 33*(8), 608–618.

Novoplansky, A. (2009). Picking battles wisely: Plant behaviour under competition. *Plant, Cell & Environment, 32*(6), 726–741.

Novoplansky, A. (2016). Future perception in plants. In R. Dillmann, Y. Nakamura, & D. Vernon (Eds.), *Anticipation across disciplines* (pp. 57–70). Springer.

Nozue, K., Tat, A. V., Kumar Devisetty, U., Robinson, M., Mumbach, M. R., Ichihashi, Y., et al. (2015). Shade avoidance components and pathways in adult plants revealed by phenotypic profiling. *PLoS Genetics, 11*(4), e1004953.

Nucara, L. (2014). *La filosofia di Humberto Maturana*. Le Lettere.

O'Regan, J. K., & Noe, A. (2001). A sensorimotor account of vision and visual consciousness. *Behavioral and Brain Sciences, 24*(5), 939–973.

Oborny, B. (2019). The plant body as a network of semi-autonomous agents: A review. *Philosophical Transactions of the Royal Society B, 374*(1774), 20180371.

Okamoto, K., Ueda, H., Shimada, T., Tamura, K., Kato, T., Tasaka, M., et al. (2015). Regulation of organ straightening and plant posture by an actin–myosin XI cytoskeleton. *Nature Plants, 1*(4), 1–7.

Oken, B. S., Salinsky, M. C., & Elsas, S. (2006). Vigilance, alertness, or sustained attention: Physiological basis and measurement. *Clinical Neurophysiology, 117*(9), 1885–1901.

Orians, C. M., & Jones, C. G. (2001). Plants as resource mosaics: A functional model for predicting patterns of within-plant resource heterogeneity to consumers based on vascular architecture and local environmental variability. *Oikos, 94*(3), 493–504.

Otto, T. U., Dassy, B., & Mamassian, P. (2013). Principles of multisensory behavior. *Journal of Neuroscience, 33*(17), 7463–7474.

Palmiero, M., & Borsellino, M. C. (2018). *Embodied cognition, comprendere la mente incarnata*. Aras Edizioni.

Parise, A. G., Reissig, G. N., Basso, L. F., Senko, L. G. S., Oliveira, T. F. C., de Toledo, G. R. A., & Souza, G. M. (2021). Detection of different hosts from a distance alters the behaviour and bioelectrical activity of *Cuscuta racemosa*. *Frontiers in Plant Science, 12*, 594195.
Parise, A. G., de Toledo, G. R. A., de Carvalho Oliveira, T. F., Souza, G. M., Castiello, U., Gagliano, M., & Marder, M. (2022). Do plants pay attention? A possible phenomenological-empirical approach. *Progress in Biophysics and Molecular Biology, 173*, 11–23.
Partan, S. R. (2013). Ten unanswered questions in multimodal communication. *Behavioral Ecology and Sociobiology, 67*, 1523–1539.
Partan, S. R., & Marler, P. (2005). Issues in the classification of multimodal communication signals. *The American Naturalist, 166*(2), 231–245.
Patterson, R. C. (2020). Functional view of cognition without conscious representation. *Journal of Liberal Arts and Humanities, 1*(3), 67–75.
Pearce, J. M. (2008). *Animal learning and cognition: An introduction* (3rd ed.). Psychology Press.
Pellegrino, G., & Di Paola, M. (2019). *Etica e Politica delle piante*. DeriveApprodi.
Penuelas, J., Llusia, J., & Estiarte, M. (1995). Terpenoids: A plant language. *Trends in Ecology & Evolution, 10*(7), 289.
Perconti, P. (2017). Filosofia della mente. *Il Mulino.*
Pickles, B. J., Wilhelm, R., Asay, A. K., Hahn, A. S., Simard, S. W., & Mohn, W. W. (2017). Transfer of 13C between paired Douglas-fir seedlings reveals plant kinship effects and uptake of exudates by ectomycorrhizas. *New Phytologist, 214*(1), 400–411.
Pievani, T. (2012). *Introduzione a Darwin*. Laterza.
Pievani, T. (2017). *La teoria dell'evoluzione. Attualità di una rivoluzione scientifica*. Il Mulino.
Pollan, M. (2013). The intelligent plant. *The New Yorker, 93*.
Pollan, M. (2001). *The botany of desire: A plant's-eye view of the world*. Random House.
Ponkshe, A., Blancas Barroso, J., Abramson, C. I., & Calvo, P. (2024). A case study of learning in plants: Lessons learned from pea plants. *Quarterly Journal of Experimental Psychology, 77*(6), 1272–1280.
Raguso, R. A., & Kessler, A. (2017). Speaking in chemical tongues: Decoding the language of plant volatiles. In M. Gagliano, J. C. Ryan, & P. Vieira (Eds.), *The language of plants: Science, philosophy*. University of Minnesota Press.
Rands, S. A., Whitney, H. M., & de Ibarra, N. H. (2023). Multimodal floral recognition by bumblebees. *Current Opinion in Insect Science, 59*, 101086.
Rankin, C. H., Abrams, T., Barry, R. J., Bhatnagar, S., Clayton, D. F., Colombo, J., et al. (2009). Habituation revisited: An updated and revised description of the behavioral characteristics of habituation. *Neurobiology of Learning and Memory, 92*(2), 135–138.
Reece, J. B., Urry, L. A., Cain, M. L., Wasserman, S. A., Minorsky, P. V., & Jackson, R. B. (2015). *La forma e la funzione nelle piante*. Pearson Italia.
Reed-Guy, S., Gehris, C., Shi, M., & Blumstein, D. T. (2017). Sensitive plant (Mimosa pudica) hiding time depends on individual and state. *PeerJ, 5*, e3598.
Rehm, H., & Gradmann, D. (2010). Intelligent plants or stupid studies. *Lab Times, 3*(21), 30–32.
Rhodes, J. D., Thain, J. F., & Wildon, D. C. (1996). The pathway for systemic electrical signal conduction in the wounded tomato plant. *Planta, 200*(1), 50–57.
Rizzolatti, G., Riggio, L., Dascola, I., & Umilta, C. (1987). Reorienting attention across the horizontal and vertical meridians: Evidence in favor of a premotor theory of attention. *Neuropsychologia, 25*(1), 31–40.
Robert, D. (2023). Plant bioacoustics: The sound expression of stress. *Cell, 186*(7), 1307–1308.
Rodrigo-Moreno, A., Bazihizina, N., Azzarello, E., Masi, E., Tran, D., Bouteau, F., & Mancuso, S. (2017). Root phonotropism: Early signalling events following sound perception in Arabidopsis roots. *Plant Science, 264*, 9–15.
Roy, R., & Bassham, D. C. (2014). Root growth movements: Waving and skewing. *Plant Science, 221*, 42–47.
Runyon, J. B., Mescher, M. C., & De Moraes, C. M. (2006). Volatile chemical cues guide host location and host selection by parasitic plants. *Science, 313*(5795), 1964–1967.

Schlichting, C. D., & Pigliucci, M. (1998). *Phenotypic evolution: A reaction norm perspective*. Sinauer Associates.
Schmid, B. (2016). Decision-making: Are plants more rational than animals? *Current Biology, 26*(14), R675–R678.
Schwartz, A., & Koller, D. (1986). Diurnal phototropism in solar tracking leaves of Lavatera cretica. *Plant Physiology, 80*(3), 778–781.
Segundo-Ortin, M., & Calvo, P. (2019). Are plants cognitive? A reply to Adams. *Studies in History and Philosophy of Science A, 73*, 64–71.
Segundo-Ortin, M., & Calvo, P. (2022). Consciousness and cognition in plants. *Wiley Interdisciplinary Reviews: Cognitive Science, 13*(2), e1578.
Seluzicki, A., Burko, Y., & Chory, J. (2017). Dancing in the dark: Darkness as a signal in plants. *Plant, Cell & Environment, 40*(11), 2487–2501.
Semchenko, M. S., Lepik, S., & A. (2014). Plant root exudates mediate neighbour recognition and trigger complex behavioural changes. *New Phytologist, 204*(3), 631–637.
Shepherd, V. A. (2012). At the roots of plant neurobiology: A brief history of the biophysical research of JC Bose. *Science and Culture, 78*(5/6), 196–210.
Shomrat, T., & Levin, M. (2013). An automated training paradigm reveals long-term memory in planarians and its persistence through head regeneration. *Journal of Experimental Biology, 216*(20), 3799–3810.
Silvertown, J., & Gordon, D. M. (1989). A framework for plant behavior. *Annual Review of Ecology and Systematics, 20*(1), 349–366.
Simard, S. W. (2018). Mycorrhizal networks facilitate tree communication, learning, and memory. In F. Baluška, M. Gagliano, & G. Witzany (Eds.), *Memory and learning in plants* (pp. 91–213). Springer.
Simard, S. W., Beiler, K. J., Bingham, M. A., Deslippe, J. R., Philip, L. J., & Teste, F. P. (2012). Mycorrhizal networks: Mechanisms, ecology and modelling. *Fungal Biology Reviews, 26*(1), 39–60.
Simard, S. W., Perry, D. A., Jones, M. D., Myrold, D. D., Durall, D. M., & Molina, R. (1997). Net transfer of carbon between ectomycorrhizal tree species in the field. *Nature, 388*(6642), 579–582.
Simonetti, V., Bulgheroni, M., Guerra, S., Peressotti, A., Peressotti, F., Baccinelli, W., et al. (2021). Can plants move like animals? A three-dimensional stereovision analysis of movement in plants. *Animals, 11*(7), 1854.
Skinner, B. F. (1954). The science of learning and the art of teaching. *Harvard Educational Review, 24*(2), 86–97.
Smalley, G. W., & Coats, D. W. (2002). Ecology of the red-tide dinoflagellate Ceratium furca: Distribution, mixotrophy, and grazing impact on ciliate populations of Chesapeake Bay. *Journal of Eukaryotic Microbiology, 49*(1), 63–73.
Soler, C., Hossaert-McKey, M., Buatois, B., Bessière, J. M., Schatz, B., & Proffit, M. (2011). Geographic variation of floral scent in a highly specialized pollination mutualism. *Phytochemistry, 72*(1), 74–81.
Solms, M., & Panksepp, J. (2012). The "Id" knows more than the "Ego" admits: Neuropsychoanalytic and primal consciousness perspectives on the interface between affective and cognitive neuroscience. *Brain Sciences, 2*(2), 147–175.
Souza, G. M., Ferreira, A. S., Saraiva, G. F., & Toledo, G. R. (2017). Plant «electrome» can be pushed toward a self-organized critical state by external cues: Evidences from a study with soybean seedlings subject to different environmental conditions. *Plant Signaling & Behavior, 12*(3), e1290040.
Souza, G. M., Toledo, G. R., & Saraiva, G. F. (2018). Towards systemic view for plant learning: Ecophysiological perspective. In F. Baluska, M. Gagliano, & G. Witzany (Eds.), *Memory and learning in plants* (pp. 163–189). Springer.
Stahlberg, R. (2006). Historical overview on plant neurobiology. *Plant Signaling & Behavior, 1*(1), 6–8.

Stenhouse, D. (1974). *The evolution of intelligence—A general theory and some of its implications*. Barnes & Noble Books.

Sternberg, R. J. (1986). A framework for understanding conceptions of intelligence. In R. J. Sternberg & D. K. Detterman (Eds.), *What is intelligence? Contemporary viewpoints on its nature and definition* (pp. 5–19). Ablex Publishing.

Sternberg, R. J., & Detterman, D. K. (Eds.). (1986). *What is intelligence?: Contemporary viewpoints on its nature and definition*. Praeger.

Taiz, L., Alkon, D., Draguhn, A., Murphy, A., Blatt, M., Hawes, C., et al. (2019). Plants neither possess nor require consciousness. *Trends in Plant Science, 24*(8), 677–687.

Thellier, M., & Lüttge, U. (2013). Plant memory: A tentative model. *Plant Biology, 15*(1), 1–12.

Thomas, M. A., & Cooper, R. L. (2022). Building bridges: Mycelium-mediated plant–plant electrophysiological communication. *Plant Signaling & Behavior, 17*(1), 2129291.

Thompson, E. (2007). *Mind in life: Biology, phenomenology, and the sciences of mind*. Harvard University Press.

Thompson, K. (2018). *Darwin's Most wonderful plants. A tour of his botanical legacy*. University of Chicago Press.

Tomimoto, S., & Satake, A. (2023). Modelling somatic mutation accumulation and expansion in a long-lived tree with hierarchical modular architecture. *Journal of Theoretical Biology, 565*, 111465.

Tononi, G., Boly, M., Massimini, M., & Koch, C. (2016). Integrated information theory: From consciousness to its physical substrate. *Nature Reviews Neuroscience, 17*(7), 450–461.

Tononi, G., & Koch, C. (2015). Consciousness: Here, there and everywhere? *Philosophical Transactions of the Royal Society B: Biological Sciences, 370*(1668), 20140167.

Topham, A. T., Taylor, R. E., Yan, D., Nambara, E., Johnston, I. G., & Bassel, G. W. (2017). Temperature variability is integrated by a spatially embedded decision-making center to break dormancy in Arabidopsis seeds. *Proceedings of the National Academy of Sciences, 114*(25), 6629–6634.

Toyota, M., Spencer, D., Sawai-Toyota, S., Jiaqi, W., Zhang, T., Koo, A. J., et al. (2018). Glutamate triggers long-distance, calcium-based plant defense signaling. *Science, 361*(6407), 1112–1115.

Trewavas, A. (1999). How plants learn. *Proceedings of the National Academy of Sciences, 96*(8), 4216–4218.

Trewavas, A. (2002). Plant intelligence: Mindless mastery. *Nature, 415*(6874), 841–841.

Trewavas, A. (2003). Aspects of plant intelligence. *Annals of Botany, 92*(1), 1–20.

Trewavas, A. (2004). Aspects of plant intelligence: An answer to Firn. *Annals of Botany, 93*(4), 353–357.

Trewavas, A. (2005). Green plants as intelligent organisms. *Trends in Plant Science, 10*(9), 413–419.

Trewavas, A. (2014). *Plant behaviour and intelligence*. Oxford University Press.

Trewavas, A. (2017). The foundations of plant intelligence. *Interface Focus, 7*(3), 20160098.

Trewavas, A. (2021). Awareness and integrated information theory identify plant meristems as sites of conscious activity. *Protoplasma, 258*(3), 673–679.

Trewavas, A., Baluška, F., Mancuso, S., & Calvo, P. (2020). Consciousness facilitates plant behavior. *Trends in Plant Science, 25*(3), 216–217.

Tuteja, N., & Mahajan, S. (2007). Calcium signaling network in plants: An overview. *Plant Signaling & Behavior, 2*(2), 79–85.

Van Bel, A. J., Furch, A. C., Will, T., Buxa, S. V., Musetti, R., & Hafke, J. B. (2014). Spread the news: Systemic dissemination and local impact of Ca2+ signals along the phloem pathway. *Journal of Experimental Botany, 65*(7), 1761–1787.

Veits, M., Khait, I., Obolski, U., Zinger, E., Boonman, A., Goldshtein, A., et al. (2019). Flowers respond to pollinator sound within minutes by increasing nectar sugar concentration. *Ecology Letters, 22*(9), 1483–1492.

Volkov, A. G., Carrell, H., Adesina, T., Markin, V. S., & Jovanov, E. (2008). Plant electrical memory. *Plant Signaling & Behavior, 3*(7), 490–492.

Volkov, A. G., Foster, J. C., Ashby, T. A., Walker, R. K., Johnson, J. A., & Markin, V. S. (2010). Mimosa pudica: Electrical and mechanical stimulation of plant movements. *Plant, Cell & Environment, 33*(2), 163–173.

von Humboldt, A, & Bonpland, A. (1807). *Ideen zu einer Geographie der Pflanzen nebst einem Naturgemälde der Tropenländer*. Tübingen, F.G. Cotta.

von Sachs, J. (1887). *Lectures on the physiology of plants* (Harry Marshall Ward, Trans.). Clarendon Press (Original work published 1882 as Vorlesungen über Pflanzenphysiologie).

Wang, J. W., Czech, B., & Weigel, D. (2009). miR156-regulated SPL transcription factors define an endogenous flowering pathway in Arabidopsis thaliana. *Cell, 138*(4), 738–749.

Wang, Q., Guerra, S., Ceccarini, F., Bonato, B., & Castiello, U. (2021). Sowing the seeds of intentionality: Motor intentions in plants. *Plant Signaling & Behavior, 16*(11), 1949818.

Wang, Q., Guerra, S., Bonato, B., Simonetti, V., Bulgheroni, M., & Castiello, U. (2023). Decision-making underlying support-searching in pea plants. *Plants, 12*(8), 1597.

War, A. R., Paulraj, M. G., Ahmad, T., Buhroo, A. A., Hussain, B., Ignacimuthu, S., & Sharma, H. C. (2012). Mechanisms of plant defense against insect herbivores. *Plant Signaling & Behavior*, *7*(10), 1306–1320.

Warwick, K. (2000). *QI: The Quest for Intelligence*. Piatkus Books.

Waszczak, C., Carmody, M., & Kangasjärvi, J. (2018). Reactive oxygen species in plant signaling. *Annual Review of Plant Biology*, 69(1), 209–236.

Waqas, M., Van Der Straeten, D., & Geilfus, C. M. (2023). Plants' cry' for help through acoustic signals. *Trends in Plant Science, 28*(9), 984–986.

Williams, R. J. (1956). *Biochemical individuality: The basis for the genetotrophic concept.* John Wiley & Sons.

Willmer, P. G., Nuttman, C. V., Raine, N. E., Stone, G. N., Pattrick, J. G., Henson, K., … & Knudsen, J. T. (2009). Floral volatiles controlling ant behaviour. *Functional Ecology*, *23*(5), 888–900.

Witzany, G. (2006). Plant communication from biosemiotic perspective: Differences in abiotic and biotic signal perception determine content arrangement of response behavior. Context determines meaning of meta-, inter-and intraorganismic plant signaling. *Plant Signaling & Behavior*, *1*(4), 169–178.

Witzany, G. (2008). Biocommunication of unicellular and multicellular organisms. *tripleC: Communication, Capitalism & Critique. Open Access Journal for a Global Sustainable Information Society*, *6*(1), 24–53.

Witzany, G. (2016). The Biosemiotics of Plant Communication. *The American Journal of Semiotics*, *24*(1–3), 39–56.

Yaish, M. W., Colasanti, J., & Rothstein, S. J. (2011). The role of epigenetic processes in controlling flowering time in plants exposed to stress. *Journal of Experimental Botany*, *62*(11), 3727–3735.

Yakura, H. (2020). Cognitive and memory functions in plant immunity. *Vaccines*, *8*(3), 541.

Yellen, G. (1998). The moving parts of voltage-gated ion channels. *Quarterly Reviews of Biophysics, 31*(3), 239–295.

Yiyao, L., Wang, B., Xuefeng, L., Chuanren, D., & Sakanishi, A. (2002). Effects of sound field on the growth of Chrysanthemum callus. *Colloids and surfaces B: Biointerfaces*, *24*(3–4), 321–326.

Yokawa, K., Kagenishi, T., Pavlovič, A., Gall, S., Weiland, M., Mancuso, S., & Baluška, F. (2018). Anaesthetics stop diverse plant organ movements, affect endocytic vesicle recycling and ROS homeostasis, and block action potentials in Venus flytraps. *Annals of Botany*, *122*(5), 747–756.

Zhu, H., Wang, W., & Yan, Y. (2009). Effect of proline on plant growth under different stress conditions. *Journal of Northeast Forestry University*, *37*(4), 86–89.

Zinn, A., Ward, D., & Kirkman, K. (2007). Inducible defences in Acacia sieberiana in response to giraffe browsing. *African Journal of Range & Forage Science, 24*(3), 123–129.

Zweifel, R., & Zeugin, F. (2008). Ultrasonic acoustic emissions in drought-stressed trees–more than signals from cavitation? *New Phytologist, 179*(4), 1070–1079.

Chapter 4
Philosophical Areas of Impact

Contents

4.1 Ontological Issues: Processes of Categorization of Reality 207
4.1.1 Habits of Thought: Utility and Simplification 209
4.1.2 Subdivisions Methods 210
4.1.3 Cognitive Ontologies: The Hallmarks of Cognition 215
4.1.4 Process Ontologies 217
4.2 Experiential Issues: Feeling and Acting 219
4.2.1 Perception-Action 219
4.2.2 The Concept of Agency 225
4.2.3 Are the Elements of Behavior Innate, Learned, or Both? 227
4.2.4 To Be a Plant 230
4.2.5 Observing Behavior: A Behaviorist Reintroduction? 232
4.3 Epistemological Issues: Situated and Natural Practices of Knowing 235
4.3.1 Epistemological Analyses 236
4.3.2 Conceptions and Types of Knowledge 237
4.3.3 Knowledge and Action: Relational Dimensions and Practical Forms of Knowing 240
4.3.4 Eco-Evolutionary Framework and the Biological Role of Knowledge 242
4.3.5 Knowing: Constraints and Possibilities 244
4.3.6 The Logic of Knowledge Processes 247
4.3.7 A Natural Logic? 249
4.3.8 Distinction of Quantity and Difference 250
4.3.9 Social Life and Game Theories 252
4.4 Plants in Cognitive Sciences 255
4.4.1 Historical Evolution of Approaches to Cognition 256
4.4.2 Bodily, Enactive, and Extended Turn of Cognition 260
4.4.3 4E Cognition and Systemic-Relational Approach to Plant Cognition 267
4.4.4 A Trivialization of the Cognitive? 280
4.4.5 Bio-Ecological Cognition: A Processual Perspective 281
4.5 Plants in the Philosophy of Mind 284
4.5.1 Contemporary Philosophy of Mind and Historical Reconstruction of the Concept of Mind 286
4.5.2 Embodied, Ecological and Processual Conception of the Mind 290
4.5.3 Beyond the Human Mind 292
4.5.4 Representational Cognition: An Open Question 297

M. Bianchi, *Philosophical Implications of Plant Cognition*, Philosophical Studies Series 158, https://doi.org/10.1007/978-3-032-01328-6_4

4.5.5 Varieties of Conceptions of Representation 298
4.5.6 Advantages and Disadvantages of Representations 302
4.5.7 Representations in Plants? 306
4.5.8 Cognition and Temporality of Processes 312
References 318

Abstract This chapter concerns the interdisciplinary implications that ongoing research on plant behavior may have in various areas of philosophical analysis, with potential experimental applications. Numerous lines of investigation emerge, to be explored in future theoretical and practical research.

Keywords 4E cognition · Affordance · Agency · Behaviorism · Bio-ecological cognition · Biogenic approach · Biological role of knowledge · Branching processes · Categorization of reality · Cognitive ontologies · Complex networks · Conceptions of mind · Ecological mind · Ecological networks · Ecological psychology · Embodied cognition · Embodied mind · Enacted cognition · Epistemic tools · Epistemology · Experiential issues · Extended cognition · Fallacies of reasoning · Interactome · Knowledge and action · Mental representation · Meta-analysis · Natural logic · Network theory · Perception-action · Philosophy of biology · Philosophy of cognitive science · Philosophy of Mind · Plant mind · Plant representation · Predictive models · Processual perspective · Ramifications · Relational context · Representationalism and anti-representationalism · Signaling networks · Temporality of cognition · Trivialization of cognition

In this final chapter, I outline how the growing focus on the plant kingdom may enrich both current and future theoretical perspectives, especially in the philosophy of cognitive science and the philosophy of mind. Adopting a relational view of plant activity has the potential to extend research across multiple philosophical domains, offering new insights and possibly reshaping our understanding of a wide range of issues. Indeed, this unconventional starting point provides a valuable opportunity to reexamine old questions and formulate new ones, establishing previously overlooked or unconsidered connections.

I emphasize that this type of analysis—regardless of the appropriateness (and in which contexts, and to what extent) of interpreting plant behavior in cognitive terms—opens the way for a more comprehensive examination of the assumptions about what constitutes minimal or sufficient, essential or relevant, processes and conditions for cognitive behavior. It encourages us to reflect on their scope, the levels at which they emerge, and the modes and strategies of interaction they entail.

This approach also extends to our understanding of various forms of intelligence and knowledge, broadly conceived, as encompassing everything that enables organisms to regulate their activity in response to environmental conditions and to pursue survival-related goals. Exploring the cognitive behavior of plants and other aneural organisms may therefore contribute to the development of a processual, systemic, and context-sensitive conception of cognition (cf. Bianchi, 2022, 2024).

The significance of this research is bidirectional: on the one hand, it stimulates further investigations into plant capacities; on the other, it provides new insights that are relevant to various areas of theoretical and applied research, potentially leading to technological innovation. Thus, this line of research can contribute to deepening our understanding of plant activity, which, in turn, could influence multiple branches of philosophical inquiry, creating pathways of reciprocal influence.

Of course, the reflections presented in this chapter are not intended to be conclusive, given the ongoing state of research, much of which remains in a hypothetical stage. The arguments laid out here are flexible and open to future revisions and additions as new findings emerge.

4.1 Ontological Issues: Processes of Categorization of Reality

The term "ontology" has a long-standing tradition in philosophical thought. It derives from the Greek words *ὄντος* (being) and *λόγος* (discourse), meaning "discourse on being," or from *τὰ ὄντα*, meaning "the entities." Traditionally, ontology has referred to the study of being and its fundamental categories, a focus rooted in Aristotle's "first philosophy," later termed "metaphysics." In more recent contexts, particularly within analytic philosophy, "ontology" has come to denote a theoretical framework that, often through formal languages, establishes criteria for positing the existence of particular kinds of entities.

As mentioned in previous chapters, "ontology" in this context is not employed in an essentialist, metaphysical, or normative meaning. Rather, it refers to the functional subdivisions of reality under investigation, which are not fixed but evolving along with the processes of knowledge. This approach, far from being abstract, addresses the practical need to differentiate and categorize elements for research purposes, given the heterogeneity of observable phenomena and the varied methods used to describe them. Such a perspective aligns with analyses in the philosophy of physics and biology, which have grouped together elements, systems, or processes relevant to various scientific theories (for example, reference has been made to "fundamental ontologies" to indicate the entities posited by a specific theory).

This level of analysis is instrumental in progressively clarifying what is being investigated, how it is being studied, and in what terms. Ontological considerations are thus closely intertwined with epistemological reflections (cf. Bianchi, 2024). Indeed, it is probably the epistemological aspect that is the actual starting point.

As recalled by Rovelli (2014, p. 221) in his account of the evolution of theories seeking to explain "what the world is made of," or what reality is, it becomes evident that what is commonly identified as a wave, a mountain, or a physical system is less a discrete "object" in the classical sense and more a conventional partitioning of reality, employed to facilitate description and discourse. These are conventional "boundaries" and serve as methods for organizing the informational structures perceived.

In physics, various theories have proposed different accounts of the structure of reality, of which we, too, are a part. According to Newtonian mechanics, the fundamental elements were space, time, and particles. In the theories of Faraday and Maxwell, fields were added to this framework. With Einstein's 1905 theory of special relativity, space and time merged into spacetime, which, together with fields and particles, constituted the building blocks of reality. In Einstein's 1915 general relativity, spacetime itself became a dynamic field interacting with matter. Finally, in standard quantum field theory, the basic entities are quantum fields defined over a fixed spatiotemporal background. However, this formulation assumes the existence of a continuous spacetime and does not account for the possibility that spacetime itself might be quantized or emergent, a possibility that becomes central in more recent approaches to quantum gravity, such as loop quantum gravity (ibid., pp. 73; 114). This evolution marks a shift from a universe described in terms of particles or fields existing within space and evolving over time, to a quantum reality in which space itself is understood as a field, and time emerges from the interactions within this field. In this view, everything that exists can be seen as manifestations of a single type of entity: quantum field interactions. Quantum mechanics presents a world in which, at the smallest scales, everything is in constant vibration, with all variables undergoing perpetual fluctuation (ibid., pp. 116; 166–167). Rovelli also traces the development from the solutions of the Wheeler-DeWitt equation to the formulation of loop quantum gravity, a theory that promotes an interrelational understanding of space grounded in the explanatory power of networks. In loop quantum gravity, the central prediction is that space is not continuous but discrete. A spin network, composed of nodes, links, and graphs, represents a quantum state of the gravitational field, that is, a possible quantum configuration of granular space.[1] Beyond the future developments of this theory (which may eventually be confirmed or falsified), what should be emphasized here is the need to conceive of spin networks not as discrete "entities," but rather in terms of their interactions. The lines within these networks do not exist in isolation; instead, they are defined through their relations, and it is these collective interactions that constitute space itself. Similarly, the quanta of gravity do not evolve within time; rather, time, like space, emerges from their interactions.

Naturally, the theories used to comprehend reality, such as general relativity, quantum mechanics, and information theory, may evolve or be replaced in the future. This parenthesis serves only as a reminder of how conceptions of entities—or more accurately, the processes and methods used to describe reality (through theories and models)—have shifted over time; in other words, how the underlying ontologies have changed.

[1] Specifically, the nodes represent the elementary quanta that constitute space: each node in the graph corresponds to a quantum grain of space, and two nodes connected by a link represent two adjacent quanta. When these grains of space interact, they collectively give rise to the structure of space itself. By moving from grain to grain and completing a circuit that returns to the original node, a ring, or "loop," is formed (ibid., pp. 139–154).

4.1.1 *Habits of Thought: Utility and Simplification*

Returning to the argument, the aim is not to speculatively impose a rigid categorization of reality or to preserve a fixed schema of existing entities. Rather, the intention is to foster a theoretically grounded and empirically informed reflection, one that remains responsive to the often nonlinear and cumulative nature of scientific practice and progress. This approach allows for the distinction and delineation of certain constellations of elements or processes, whose boundaries and interactions can be explored at various levels. Underling this point is relevant because the advancement of theoretical research can itself be seen as a form of practice, playing a significant role in shaping and guiding experimental investigation. Of course, beyond scientific results, the extent and manner of our subdivisions and categorizations are, to some degree, influenced by our own choices. Perspectives are shaped by perceptual, constitutive, and epistemic constraints, as well as by the starting questions posed, the specific methods and analytical procedures employed, and the cultural contexts within which knowledge is intersubjectively shared (cf. Bianchi et al., 2025a).

Especially in recent years, the need to clarify what is being studied—the objects of analysis to be considered and the terms under which they are examined—has become increasingly important within the field of cognitive science. In particular, there is a growing demand to determine which organisms can be said to possess forms of cognition, what organs and processes are involved, and which characteristics and behaviors qualify as "cognitive." The case of plants is especially thought-provoking in this regard. Some argue that plants do not meet the necessary criteria to be included in the so-called "cognitive club," while others maintain that plants may be considered within cognitive frameworks, but only insofar as they exhibit certain relevant features (for an overview of this debate, see Lee, 2023).

The uncertainties and challenges involved in conceptualizing the plant world—what aspects to assess, which capacities and behaviors to examine, and how to clarify the characteristics and modes of plant interaction—can offer valuable insights. Such an analysis may contribute to identifying the minimal elements and processes required for cognition, delineating its boundaries (if any), exploring its intra- and extra-organismic extensions, and assessing its effects on the surrounding environment (cf. Bianchi, 2021). It may also enhance the understanding of the timescales of cognitive behavior within the lifespan of an individual organism (cf. Bianchi et al., 2025a). The task of distinguishing and identifying fundamental processes is crucial because it has practical implications for subsequent research. In the case of plants, if they are not recognized as cognitive or agentive systems, an entire range of processes and behaviors that fall within the study of behavior and cognition cannot be properly investigated; certain processes may remain unidentified or be examined through alternative frameworks.

When considering the reasons behind the tendency to make distinctions, categorize, and classify reality, the most intuitive explanation is that it arises from a drive toward simplification, a form of cognitive efficiency. The aim is to bring order to and make sense of complex system structures and the intricate relationships among phenomena. In essence, this process functions as a means of organizing and managing different types of information (cf. Bianchi, 2021). There are far more ways for things to be "disordered" than "ordered" (cf. Bateson, 1976, pp. 33–38). Anything that does not conform to a particular organizational scheme, or that stands out against a given framework of reasoning, is often interpreted either as a further subdivision of reality—whether structural, material, functional, or conceptual—or as a portion of reality that remains unknown or is perceived as chaotic.

How are segments of reality carved out? What guidelines are followed? What criteria determine one method of subdivision over another? The answer is variable. It depends on the scope and level of analysis, the initial questions and objectives, the subjects and objects of investigation, and other contextual factors (cf. Bianchi, 2024).

4.1.2 Subdivisions Methods

The set of differences between things constitutes reality, and "diversity" is both essential to our existence and something that must be understood and, to some extent, managed. The act of distinguishing and classifying becomes crucial in engaging with such differences. Thus, it appears necessary to identify distinguishing traits that enable the grouping of aspects of reality into categories or classes based on similarities (cf. Bianchi, 2021).

As discussed in Chap. 2, the concepts of similarity and difference are inherently vague and require clarification. Since two non-identical things share certain traits while differing in others, similarities and differences cannot be considered in isolation. Instead, they must always be evaluated within contexts of comparison, where the point of comparison, the specific trait under consideration, the adopted approach, and the initial question are clearly articulated. The selection of characteristics for comparison is therefore a considered outcome, influenced by specific contextual factors and practical aims.

Human culture includes numerous classifications. Common examples are the categorization of living species, languages, literary and film genres, commercial products, and more. As Mayr (1982, pp. 146–147) noted, recognizing the existence of basic classification procedures is neither strange nor particularly difficult, yet determining which criteria are most effective for classification has often proven challenging. The history of macrotaxonomy illustrates how answers to these questions have evolved over time. A paradigmatic example of the necessity, and complexity, of subdividing reality can be found in the development of species systematics and taxonomy, the methodologies by which organisms are grouped into categories. From ancient times, philosophers and naturalists such as Aristotle, Theophrastus, Dioscorides, Linnaeus, Cuvier, Lamarck, Haeckel, and Huxley have sought to classify animals and plants into genera, families, and orders. These efforts continued into the post-Darwinian era, when more systematic groupings began to reflect principles of common descent and degrees of evolutionary divergence. This journey of comparison, inclusion, and distinction is a narrative of experimentation, mistakes, and evolving criteria for making distinctions (ibid., p. 143).

Identification and classification are distinct concepts (cf. Mayr, 1982). Identification is a deductive process that can only take place once a classification system has been established. It involves assigning an individual organism to one of the pre-defined classes based on specific traits observed in the specimen (Mayr, 1969, pp. 4, 66, 112–115). In this sense, individual organisms are identified when they are assigned to a species. Classification is an organizing process that concerns populations rather than individuals. It aims to group populations and taxa into broader sets based on their similarities and relationships, using a wide range of characteristics (cf. Mayr, 1982, pp. 147, 237). Classification is determined by the taxonomic traits of organisms. However, from the early days of taxonomy, there has been no consensus on which traits should be prioritized in taxonomic analysis. A brief look at the historical controversies and transformations sheds light on this issue. For instance, even in pre-Darwinian times, some ancient Greek philosophers and early Renaissance botanists understood that characteristics tied to human utility—such as a plant's medicinal properties or the presence of thorns—were arbitrary and insufficient for gaining a deeper understanding of a plant's nature and its relationships with other plant species. Even essentialists, who adhered to the scholastic method of logical division by distinguishing between essential and accidental traits, recognized that some features held greater taxonomic significance than others. Over time, particularly with the work of Andrea Cesalpino (1519–1603), the importance of classifying plants based on morphological characteristics, rather than traits linked to their usefulness to humans, became increasingly apparent. Since the sixteenth century, debates have emerged about whether to rely on a single key characteristic or consider multiple taxonomic traits simultaneously. If multiple traits are used, questions arise about the relative importance of each trait and the rationale behind these choices (in other words, the quantitative and qualitative aspects of prioritization). Furthermore, there has been ongoing discussion about whether to restrict analysis to morphological or physiological traits or to include behavioral and ecological characteristics as well (ibid., p. 235), even though the levels are ultimately connected.

A recurring question in taxonomy has been, and continues to be: What criteria should be used to determine the relevance of a trait in a classification procedure? In grappling with these issues, certain challenges have emerged. For instance, separating traits that indicate deeper evolutionary divergences from ad hoc adaptations in organisms with highly standardized phenotypes (such as the hundreds of species of frogs, which provide few morphological clues of relatedness) has proven difficult. In these cases, the concern is distinguishing functional changes from other phylogenetic adaptations, a point that Darwin (1859, p. 414) had already emphasized. Even with the advent of genetics and the greater accuracy provided by DNA analysis, in the various methods of classification (in Mayr's time, numerical phenetics, cladistics, and evolutionary taxonomy), it was still essential to clarify the taxonomic criteria for comparison, especially when there was a lack of informative traits sufficient to resolve disagreements among the different classification methods.

Another example that illustrates the difficulty of subdivision, drawn from biological studies, is the complexity involved in defining the concept of a biological species. This challenge led to a redefinition of species, no longer as ideal types but as groups of populations composed of real individuals. It required a move away from essentialist thinking, toward a conception of species based on reproductive isolation and on behavioral, ecological, and, later, genetic relationships with other coexisting species, rather than on intrinsic properties. Additionally, it was fundamental to avoid conflating issues related to the taxon with those concerning the species as a conceptual category. These challenges become even more pronounced in botany, where the identification of species is often more complex than in most animal groups, due to factors such as the formation of ecotypes, polyploidy, hybridization, and self-fertilization (ibid., pp. 271–278).

On the other hand, the examples just mentioned pertain to theoretical-scientific analyses, accompanied by corresponding methodological considerations, in which the very practices of differentiation and classification are subject to scrutiny. In everyday life, however, it is common to rely on crude categorizations and arbitrary subdivisions of reality that lack scientific grounding, remnants of outdated habits of thought. These patterns can sometimes influence even the most rigorous experimental research, giving rise to entrenched disputes over what should be considered and studied across various fields. Such tendencies can result in oversimplifications, trivializations, or generalizations of issues that, upon closer examination, prove to be far more complex (cf. Bianchi, 2021).

For instance, in the study of animal behavior, when faced with ambiguous or poorly understood cases, there is a tendency to use anthropomorphizing labels, terms and concepts that describe typically human behaviors and emotions to explain animal behaviors (e.g., saying an animal is "jealous"). Yet, on closer examination, it becomes apparent that these labels do not necessarily imply that the animal experiences feelings or sensations in the same way humans do. As De Waal (1999) noted, anthropomorphizing expressions, much like metaphors in science, often serve to emphasize perceived similarities. However, the risk lies in a literal interpretation of these expressions, which may lead to a distorted understanding.

Other misleading examples that reveal how many distinctions are more multifaceted than they initially appear include the widespread assumption that plants, unlike heterotrophic animals, are entirely self-sufficient as autotrophs. Even setting aside parasitic species, which obtain nutrients synthesized by other organisms, plants still require inorganic substances from their environment. Using solar energy, they synthesize the organic molecules essential for building their structures and maintaining life. In this sense, a plant's "nutritional needs" are also met through active processes such as "foraging" for minerals in the soil (cf. Affifi, 2013). Acknowledging this, of course, does not eliminate the differences between heterotrophs and autotrophs. A different misconception—this time stemming from superficial similarities related to appearance—involves the confusion between certain aquatic plants and some species of green algae. For example, *Posidonia oceanica* L., a flowering plant that evolved from terrestrial ancestors and returned to the marine environment over 100 million years ago, is often mistaken by non-experts for algae simply because both occupy the same ecological niche, the seabed (cf. Mazzolai, 2021, pp. 152–153).

Beyond these specific examples, this discussion highlights the importance of critically examining and exploring the issues in greater depth. It brings attention to common reasoning errors and reminds us that the process of identifying and distinguishing relevant elements is inherently shaped by the initial questions posed, the context in which the analysis takes place, and the research domain itself. Depending on the processes being considered, the number of shared elements between two categories of objects may vary, and what is regarded as essential, relevant, or irrelevant may shift accordingly. Likewise, the potential for drawing parallels and establishing connections can evolve, extending beyond species-specific distinctions and differences (cf. Bianchi, 2021).

Regarding the origins of the distinction between "cognitive" and "noncognitive," the seemingly intuitive and tacit separation between animals and plants (and other living systems) remains evident in the history of philosophy and science. This division introduced the differentiation between locomotor and sessile species. However, based on current evidence, it appears that the line separating plants from animals cannot also serve as a clear "boundary" between systems that are entirely cognitive or non-cognitive. This would only hold true if cognition were defined solely by the physiological and behavioral traits specific to certain animal species (or uniquely to humans). Such a restrictive approach would imply that only organisms possessing these traits could inform models of cognition, which seems excessively limiting (cf. Linson & Calvo, 2020).

Furthermore, on a broader level, as Baker (2017) pointed out, there are often deep-seated tendencies to generalize the human-animal division, largely due to a lack of in-depth understanding of the systems being considered. This often leads to overlooking the fact that humans are, fundamentally, animals. More critically, the category of "animal" is often employed in an oppositional manner, to define, differentiate, and even elevate traits considered uniquely human. While this tendency to simplify reality is, to some extent, understandable and practically useful, it often comes at the cost of overlooking the distinct characteristics of other species and their specific living environments. Consequently, non-human animals are frequently perceived as lacking the qualities that are valued in humans. Even when attempts are made to overcome these reductive categorizations, new risks arise, such as anthropomorphizing certain animal species by attributing human-like qualities to them, thereby bringing them closer to the human side of the divide while continuing to exclude those species considered more distant or simpler.

When it comes to plants, this tendency has often been even more pronounced (ibid., pp. 138–139). The plant world is frequently regarded as a homogeneous category of undifferentiated organisms, as if, for instance, there were no meaningful differences between *Aesculus hippocastanum* and *Mirabilis jalapa*. Alternatively, plants are sometimes viewed as simplified or reduced versions of animals, such as being likened to "very slow" animals (cf. Godfrey-Smith, 2021, p. 259). It is, therefore, necessary to recognize the biological and ecological (including behavioral) distinctions between different plant species. Nevertheless, it is understandable, and even practical to some extent, to use a broad macro-category for the plant kingdom, especially when time or resources limit the ability to delve into detailed intra- and inter-specific distinctions. This is because plant species generally share greater evolutionary similarities in terms of structure and ecological organization with one another than they do with species from the animal kingdom (cf. Bianchi, 2021, p. 65).

In the field of "plant intelligence" studies, several questions have been raised that prompt reflection on the presence of misleading or inadequate categorizations. Efforts to define what constitutes intelligent behavior, where it should be sought, and at what level it should be explained provide an interesting example of how aspects of living reality are grouped (cf. Bianchi, 2024). To engage in discussions about "intelligence" it is essential to challenge both forms of crypto-essentialism and the linguistic tendency to overgeneralize phenomena (cf. Baker, 2017). In our everyday concepts, it is rare to find a single trait shared by all the referents of a given word. A classic example is the concept of "game." It seems impossible to identify a universal "essence" common to all games or to provide a single, comprehensive definition without exceptions. From a non-essentialist perspective, however, there must still be something that connects all instances of games. In this context, within broader theories of concepts, Ludwig Wittgenstein (1889–1951), in *Philosophical Investigations* (1953), introduced the notion of "family resemblance." He used this idea to explain how, as with members of a family, there may not be a single trait shared by all, but rather a network of overlapping similarities (cf. Baker, 2017, pp. 137–140).

A similar approach could be applied to the concept of biological intelligence across different organisms. Building on Baker's (2017) argument, it is worth asking: What is intelligence? What role does it play in a living system? Which organisms are intelligent, and in what ways do they exhibit intelligence? Is intelligence a higher-order faculty or a capability common to all organisms, expressed in diverse ways? Is it reflected in a single behavior or a spectrum of behaviors? Is intelligence what triggers a behavior, or is it the behavior itself? Is it localized in specific organs, such as the nervous systems of animals, or is it a process distributed throughout an organism's body, emerging from the interaction between an agent and its environment (cf. Bianchi, 2024)?

As discussed in Chap. 3, certain conceptions are more suitable and inclusive than others for addressing the diverse forms of intelligence found in nature, including plant intelligence. Conversely, when considering approaches to avoid, Baker's (2017) argument highlights certain reasoning errors and lingering Cartesian-inspired modes of thought regarding what is considered intelligent, frameworks that prove difficult to apply to plants. As is often the case, it is easier to reach consensus on what is unlikely to be useful, that is, which paths to avoid, than on what would be beneficial to pursue, particularly in the early stages of research. For instance, assuming that plant behaviors cannot be intelligent simply because plants lack brains reflects a neurocentric perspective, one that implies intelligence resides exclusively in brains or central nervous systems. However, from a more biologically and ecologically grounded standpoint, brains and nervous systems, when considered in isolation, are neither "intelligent" nor capable of "communication" (the brain communicates with the rest of the body). Rather, it is the organism as a whole, whether animal or plant, that engages in intelligent behavior and communication through its bodily mechanisms in interaction with its environment (cf. Bianchi, 2024).

One issue often linked to the elements and boundaries of intelligent behavior is the *homunculus fallacy*, the mistaken belief that there must be a "little man" or internal agent within an organism controlling its actions, which leads to an infinite regress. If one assumes the existence of an intelligent homunculus inside the brain, one could then posit another homunculus inside that one, and so on, without end.

Another subtle error of reasoning is related to the "problem of other minds," where behavior is viewed merely as an external manifestation of an inner cause, such as intelligence. This approach does not resolve the issue but merely postpones it. A more compelling perspective is to understand intelligent behavior not as a superficial expression of intelligence, but as intelligence itself in action, emerging from biological processes and interactions. This view can be applied to various forms of biological intelligence, including that observed in plants.

A further persistent issue is the tendency to associate intelligence solely with aspects of inner life. However, it is now recognized that even paradigmatic examples of human intelligence do not always correspond to subjective self-conscious experience. In the case of plants, as discussed in the previous chapter, it can be argued that they display "intelligent behavior," even though their form of intelligence, as currently understood, cannot be explained in terms of metacognitive interiority (Baker, 2017, pp. 149–150).

A critical approach, informed by the most recent evidence—one that is neither naïve nor overly enthusiastic, and capable of making comparisons without reducing the case at hand to an animal or human paradigmatic model—acknowledges that every trait or ability, including intelligent behaviors, is the outcome of unique and contingent evolutionary paths. These paths shape how various communication, reproduction, and survival needs, common across living systems, are expressed through slightly different means and strategies from species to species and even from individual to individual. Considering the specificities of each living organism, the meta-analytic effort to identify traits relevant to the study of particular capacities—including cognitive abilities—makes it possible to incorporate plant intelligent behaviors into the analysis as well.

From this perspective, plants do not lack the capacities present in humans or other animals, nor do humans or animals lack those found in plants; rather, these capacities are simply different. Plant intelligence reflects their unique bio-ecological needs. This inclusive and multifaceted approach will enable a richer analysis of how intelligence is addressed, to which organisms it can be attributed, and in what ways it might manifest, thereby clarifying how it can be studied (cf. Bianchi, 2024).

4.1.3 Cognitive Ontologies: The Hallmarks of Cognition

A similar debate concerning the diverse contributions of scholars from various backgrounds is currently unfolding in the vast and heterogeneous research area of cognitive science. This field endeavors to integrate the latest theoretical, methodological, and technological advancements across a range of disciplines.

Within this broad framework, the analysis of "cognitive ontology" can also have a clarifying and directing impact on the objectives of cognitive science (cf. Keijzer, 2021). Such reflections may not be aimed at determining the existence or the nature of reality per se, rather, they focus on identifying the most effective or comprehensive ways to conceptualize processes and behaviors that can be considered cognitive, while assessing their defining features. This analysis, emphasizing differentiation, could significantly influence future research directions. It could shed light on where to draw the line between cognitive and non-cognitive phenomena, and on whether the distinction between animal and plant life forms serves as a meaningful criterion. Alternatively, one might argue that these distinctions and boundaries are largely arbitrary and lack substantive meaning.

The need for this type of analysis consistently arises in various contexts, particularly in philosophical discussions within the philosophy of biology concerning the different capacities of living species. This need becomes even more complex in light of perspectives suggesting that all living systems are inherently cognitive (cf. Maturana & Varela, 1980). Consequently, the challenge of distinguishing between cognitive and non-cognitive systems may ultimately hinge on the deeper question of how to differentiate between living and non-living systems (cf. Thompson, 2007).

In contemporary philosophical analyses of cognition, a recurring question is what makes a phenomenon distinctly "cognitive." As a preliminary remark, it is important to recognize that the study of cognition is a complex, multifactorial, and multilevel phenomenon, which does not lend itself to reductionist analyses focused on a single aspect (cf. Bianchi, 2021). That said, a prevalent stance in cognitive science holds that only systems capable of "offline behavior" qualify as proper cognitive systems (cf. Clark & Grush, 1999). Within the realm of plant cognition research, this question is often raised: Can plants, with their inherent differences, exhibit behaviors beyond mere online responses? Online behaviors are those triggered directly by the presence of a stimulus, such as phototropism, where plants orient themselves toward a light source. An example of plant behavior that appears to involve offline aspects, previously mentioned in discussions of anticipatory capabilities, is the nocturnal heliotropic reorientation of *Lavatera cretica* leaves. This process enables the plant to anticipate the direction of dawn and optimize light intake (cf. Kreps & Kay, 1997). To incorporate environmental rhythms, such as day-night cycles, into its routine, the plant must generate regular oscillations that align approximately with the 24-h cycle. Remarkably, this nocturnal reorientation can be maintained for a few days even in the absence of light stimuli, suggesting an underlying mechanism beyond simple reactive response.

A central question concerns the most appropriate explanation for this process: whether it should be attributed to some form of internal representation or accounted for in terms of simple causal covariation (cf. Garcia Rodriguez & Calvo Garzon, 2010). A widely accepted view holds that online behaviors establish a direct causal coupling between organisms and their environment through dynamic feedback loops. By contrast, offline processing is thought to involve some form of internal representational capacity (cf. Clark & Toribio, 1994). If this latter view is adopted, it suggests a more sophisticated conception of cognition. Two critical questions emerge: First, is the association between offline processing and representational aspects necessary, and if so, in what sense? Second, must plants be capable of manipulating representations to exhibit offline behaviors? This issue will be revisited later in the chapter. It is worth noting, however, that beyond representation-based models, alternative explanatory frameworks have been proposed, including those grounded in oscillator systems and dynamical approaches (cf. Heras-Escribano & Calvo, 2019). For now, it should be noted that plant life remains profoundly embedded in and responsive to its environmental context (cf. Parise & Marder, 2024).

Beyond the representational issue, reflections on distinctions of this kind—those indicating differences between cognitive and non-cognitive systems and aspects that make a process truly cognitive—have, in recent years, led to more inclusive approaches. These approaches challenge the notion of a rigid division between "cognitive" and "non-cognitive" systems, broadening the scope of what is considered cognitive. They encompass various forms of adaptive behavior and modes of communication, emphasizing the diverse capacities of different organisms to interact with their environments and solve survival-related challenges and problems (cf. Bianchi et al., 2025a). This perspective resists reductionist views and instead conceptualizes cognition as a multi-processual and dynamic phenomenon, including also offline elements.

Studying cognitive behaviors as a set of processes that are partly conserved and partly variable over time, during the organism's life and in relation to what is transmitted to future generations and to the conditioning traces left in environments, capable of influencing interspecific and interkingdom relationships, paves the way for a process-oriented and ecological understanding of cognition (cf. Bianchi, 2022). Naturally, the organism's capacity for flexible behavior remains a central consideration.

At a theoretical level, these considerations highlight that the very concept of cognition varies depending on the adopted framework and the range of factors considered (cf. Bianchi, 2021).

4.1.4 Process Ontologies

Within the context of ontological studies—which are closely linked to the evolution of scientific practices and theories, and therefore to epistemological inquiry—it may be useful to establish a connection with research on process ontologies. These approaches emphasize the fluidity and spatiotemporal transformation of systems as relevant aspects to be considered (cf. Dupré, 2020). In this context, it is also pertinent to engage with the criticisms directed at processual approaches, which are sometimes viewed as ultimately relying on the substance-based principles they aim to replace (cf. Austin, 2020; Koutroufinis, 2017). One might ask, for instance, whether systems biology—which identifies interacting yet distinct "components" or "elements"—and process ontology must necessarily be seen as antagonistic, or whether, as is more likely, they can coexist and evolve together, each contributing to a broader understanding of the complexity of life, depending on the level of analysis and the specific aspect under investigation (cf. Austin, 2016).

By narrowing the focus from sets of behavioral interactions typical of living systems to smaller fragments of reality, as Nelson Goodman (1906–1998) suggested, even seemingly simple "objects," like a stone, can be understood as outcomes of repetitive, ongoing processes. At the level of physical explanation, a rock can be described as a vibrating assemblage of quanta that maintains its structure only for a while, like a wave in the sea (cf. Rovelli, 2014).

In the study of cognition, process-oriented approaches have been integral to epistemological thought since the work of Maturana and Varela (1980). This perspective resonates with the contributions of Conrad H. Waddington (1905–1975), who investigated the dynamics of interdependence and signaling inherent in living systems, within the broader field of complex systems research. A coherent theoretical account of the relevant elements and processes can be developed without resorting to chaotic interpretations, and in alignment with contemporary scientific advances.

A processual approach appears well-suited for addressing forms of plant cognition, and it could, in turn, contribute to the further development of processual ontologies. Embracing such an approach does not prevent us from drawing "porous" boundaries, even if they are transitory, especially when considering the long-term transformation of behaviors and the reciprocal conditioning between organisms/agents and their niches. This approach enables the examination of endogenous and exogenous processes: the physiological signaling mechanisms operating within an organism, and the ecological dynamics associated with perception–action couplings that facilitate agent-environment interactions. It emphasizes the transient maintenance of processual organization across the organism's lifespan, an organization that supports energy acquisition and harm avoidance (cf. Nicholson & Dupré, 2018).

Illustrating the richness of this multilevel and multiprocess approach would mark an advancement in studying the diverse forms of cognition. Consequently, as a provisional conclusion to this discussion, the point I intend to emphasize is the importance of moving away from rigid categorizations of reality. Even in the study of cognitive systems and processes, how reality is described, and what aspects are emphasized, will shape which features or relationships are explored in depth, whether they are individual elements or networks of interactions and processes (cf. Bianchi, 2022). In this way, greater attention is directed toward a specific domain of reality, or toward a particular part or organization of it—sometimes even, in the study of living systems, toward emergent behaviors or aspects previously not considered related—in order to test, or more precisely, to attempt to falsify the best available hypotheses aimed at explaining them.

In the field of plant intelligence and cognition studies, this exercise in thoughtful and contextual comparison can have far-reaching implications. Beyond reshaping our personal perceptions and understanding of plants, this perspective also carries practical and methodological significance for future experimental research (cf. Bianchi et al., 2025a). By deepening our understanding of the diverse aspects of "being a plant," its traits and abilities, the types of processes and abilities considered relevant to specific behaviors will inform distinct investigative approaches. These, in turn, will shape future theoretical questions and guide subsequent lines of research.

4.2 Experiential Issues: Feeling and Acting

Continuing the exploration of the potential implications and avenues for in-depth research prompted by ongoing studies on plant cognitive behaviors, multidisciplinary investigations into plant sensory capacities could greatly expand our understanding of the various perception-action possibilities across different living systems, with or without a nervous system (cf. Bianchi et al., 2025a).

Environmental uncertainty—characterized by fluctuating life conditions, resource scarcity, and competition—combined with unique and contingent eco-evolutionary histories that shape diverse structures and functions, underpins the various perceptual-agentive modalities essential for organisms to respond effectively to environmental challenges and, ultimately, to survive within their respective interaction niches (cf. Bianchi, 2022).

As outlined in the previous chapter, plants possess sensory capacities analogous to those of animals, although these are organized and arranged differently and are supplemented by additional capabilities (cf. Chamovitz, 2012). Engaging with this distinct sensory spectrum and the associated perception-action mechanisms invites imaginative reflection on the presence of modes of interaction that are qualitatively different from our own and from those of organisms more closely related to us phylogenetically. Indeed, given these perceptual-cognitive differences, it seems reasonable to infer an "experiential plurality," encompassing the diverse ways in which organisms "exist in the world," perceive reality, and experience it (cf. Bianchi, 2024).

Research into plant sensory modalities contributes to the ongoing re-evaluation of the perception-action connection. This is especially relevant when examining organisms that lack a nervous system and are incapable of locomotion.

4.2.1 Perception-Action

Before delving into the link between the abilities to perceive and act, it is worth briefly examining how "perception" has been understood. In philosophy, one notable figure who moved away from the Cartesian criterion of evidence and emphasized the perceptual possibilities of acting bodies is, for instance, Maurice Merleau-Ponty (1908–1961), though a detailed exploration of his work is beyond the scope of this discussion. By framing perception as an original mode of experience—one that is primary, pre-discursive, and pre-reflective—the philosopher attributed a fundamental role to sensory-motor activity within the mental domain. This allows to recognize the importance of implicit processes in knowledge formation, situating corporeality as the locus of intentionality. In this and similar frameworks, perception is conceived as a form of unmediated engagement with the world, unencumbered by explicit categorizations or representational structures. Within this perspective, the distinction between subject and object has not yet been established, nor are other conceptual dichotomies operative. This view emphasizes the centrality of the body as a necessary condition for any perceptual or cognitive experience. The body constitutes the original means of interacting with the environment, enabling the differentiation of salient features and guiding the repertoire of potential motor actions. As organisms navigate their surroundings, they not only modify their kinesthetic and perceptual experiences, but also transform the very environment in which they operate in pursuit of their goals, thus opening up new possibilities for action. What emerges is a dynamic circularity in the organism-environment relationship: environmental stimuli and interactions manifest as such in relation to the characteristics and functions of the agent, which, in turn, shape the surrounding environment (cf. Palmiero & Borsellino, 2018, pp. 45–66).

The significance of the link between bodily perceptual characteristics and the potential for action has been extensively explored, particularly in research on vision. Notably, Gibson's work (1966) emphasized the inseparability of perceptual processes from motor actions. This insight contributed to a deeper understanding of the concept of *affordance*, which refers to the action possibilities that the environment offers to an organism, relative to its physical constitution and capabilities. Affordances are relational properties that emerge from the dynamic interaction between an organism and its surroundings. They are shaped not only by the bodily features and intentions of the agent, but also by the specific characteristics of the environment, the particular element with which the organism interacts.

Gibson's approach and its subsequent development within ecological psychology introduced the idea of "direct" perception, which contrasts with the notion of perception as the "processing" of information. As reconstructed by Michaels and Carello (1981), it becomes evident that this latter view of "indirect" perception—often referred to in perceptual studies as the doctrine of impoverished sensory input—has been a defining characteristic of traditional experimental psychology (cf. Neisser, 1967, pp. 7–8). Even earlier, in the structuralist psychology of the late nineteenth century, perception was understood as the cumulative result of individually meaningless sensations, whose overall significance, especially for object identification, could only be reconstructed with the aid of memory.[2] This approach later found implicit expression in Gestalt psychology of the twentieth century and was subsequently adopted more explicitly in classical cognitivism. In general, theories of indirect perception or information processing posit that the senses provide only an impoverished description of the world. According to these theories, sensory inputs alone cannot offer accurate or complete information about objects or events;

[2] For instance, according to this view, memory processes play a fundamental role in reconstructing the visual world, as retinal stimulation alone is insufficient to convey essential aspects of reality, such as its three-dimensional nature (cf. Michaels & Carello, 1981), a point still supported, though now framed within predictive and inferential models of perception.

instead, they provide limited cues or bundles of sensations. Consequently, individual inputs are seen as inadequate or insufficient, necessitating further processing by the nervous system (cf. Michaels & Carello, 1981, pp. 2–3). In contrast, advocates of "direct" perception argue that the richness of experience comes not from cognitive processing but from the stimulation itself, which already contains enough information to specify the nature of objects, locations, and events (cf. Gibson, 1950). Michaels and Carello (1981, p. 184) clarified that "direct" does not mean "unmediated." This does not imply that learning can occur without the involvement of the senses. Instead, the structure of environmental objects, as it relates to the specific behavioral interactions of animals (or organisms more broadly), is preserved in the energy patterns that stimulate perceptual systems. Essentially, the environmental information available to an organism—shaped by its evolutionary history and current perceptual capabilities—is considered inherently sufficient to guide behavior, without the need for internal construction or inferential processing by the nervous system. Organisms actively explore their environments, detecting meaningful classes of structured energy (e.g., sound sequences), which enable them to act effectively.

Another point of contrast between indirect and direct perception involves the active role of the perceiver. While both approaches recognize perceivers as "active," they emphasize different aspects of this activity. In the information processing model, perceivers are considered active because they "construct" and enrich their perceptual experiences through internal processing. In contrast, the direct perception model views perceivers as active in their capacity to explore and perceptually discriminate the environments they inhabit, by looking, touching, listening, smelling, or tasting. From this perspective, perceivers actively gather information that is directly useful for guiding action within the environment (ibid., pp. 46; 15). From this perspective, the perceiver and agent are one and the same (Michaels & Carello, 1981, p. 54). Beyond a simplistic interpretation of direct perception, this approach, even when considering the varying abilities of systems to process initially fragmented and insufficient information for perceptual and agentive effectiveness, is not without criticism. One critique, for instance, concerns the difficulty of empirically testing how perception varies in response to different sources of sensory stimulation (cf. Souza et al., 2018, p. 171). Nevertheless, research continues to investigate the extent to which various processes are involved and how best to describe what occurs within organisms as they adapt to changes in their environmental interactions.

For the moment, one could draw a balanced insight from these different approaches: perception is neither completely determined by the external environment, as though the organism were merely a passive reflector, nor entirely constructed by the perceiver. Instead, perception emerges from the relational dynamics between organisms and their environments, wherein the perceiver-agent plays an active role in distinguishing, selecting, and specifying aspects of reality. However, what the perceivers can detect or what drive them to act is constrained by their bodily constitution and behavioral capabilities, shaped by a unique eco-evolutionary path (cf. Bianchi, 2021).

These different approaches to perceptual processes, which continue to offer valuable insights, contribute to a deeper understanding of the complexity of perception. They are useful both for advancing our general comprehension of how perceptual mechanisms function in nature and for enhancing our study of aneural systems, such as plants. Specifically, these positions can illuminate the behavioral, agentive, and cognitive possibilities of such organisms. Furthermore, a central issue in the theoretical discourse on perception is the variability in sensory reliance both across different species and within the same species, as well as across different developmental stages (or age) of a single organism (cf. Bianchi et al., 2025a). Different forms of life often exhibit a relative "predominance" of certain senses over others, with some sensory modalities being more developed and more frequently used. For instance, consider the stark difference between the olfactory abilities of a dog compared to a human (cf. De Waal, 1999). Beyond species-level comparisons, individual differences also influence perceptual capacities, which can, to some extent, affect modes of action and behavior within certain limits.

The role of the body in spatial perception is another key factor. For example, in humans, the perception of proximal space extends outward with the length of a person's arms, indicating that individual physical differences lead to variations in the perception of space (cf. Longo & Lourenco, 2007; Palmiero & Borsellino, 2018, p. 142). On a broader scale of interspecific comparison, organisms possess sensory-perceptual capacities developed to different extents, shaped by their physical constitution and ecological interactions, which, in turn, affect their possibilities for action and interaction. A striking example from the life of whales, and cetaceans in general, is their ability to "see" through sound, via a sophisticated form of underwater echolocation. Hearing is their most developed sense, allowing them to perceive their surroundings, navigate, locate prey, communicate, and avoid predators, even in the darkness of deep ocean waters (cf. Mazzolai, 2021, pp. 182–184).

Even more distant and counterintuitive is the study of sensory-perceptual abilities in plants. Much is still to be learned about their potential. An interesting area of investigation could be whether plants, when performing various tasks, are capable—given their multisensory perception (i.e., the ability to perceive and utilize multiple senses simultaneously)—of prioritizing a specific sense, or communication mode, in particular contexts. In other words, it would be valuable to determine whether plants exhibit a degree of sensory selection that enables them to focus on different perceptible features in recurring situations, depending on the modality engaged, or whether their responses are more tightly constrained (cf. Bianchi et al., 2025a).

The relationship between cognition, perception, and action constitutes a growing area of interest in contemporary psychological and philosophical research and now warrants closer examination. Since at least the final two decades of the twentieth century, neurobiological research has increasingly challenged the so-called "sandwich" model of the mind (cf. Hurley, 1998). This classical cognitive science model explains the relationship between perception, mental processing, and action (cf.

Morabito, 2020, p. 103). In contrast, the concept of "perception-action," which views perception as inseparably linked to action, has diverged from the traditional "perceive-think-act" cycle (cf. Paternoster, 2011, pp. 175–176). The older model depicted perception, cognition, and action as distinct systems for exchanging information, inputs, and outputs, with less intercommunication than is currently believed. Recent insights emphasize the dynamic interactions between agents and their environments. Consequently, there has been a gradual theoretical shift from viewing perception primarily as a guide to action to embracing the idea of a circular and reciprocal influence among these processes (cf. Bianchi, 2021).

In contemporary cognitive sciences, particularly in discussions surrounding sensation and action, certain strands of enactivism align with this more complex perspective, aiming to explain perception as a form of action itself (for instance, "seeing" is described as a "way of acting") (cf. Godfrey-Smith, 2021, p. 133). From this standpoint, aspects of cognition are understood to emerge from recurrent sensorimotor patterns that guide perception in facilitating action (cf. Varela et al., 1991, p. 173; Morabito, 2020, p. 89). In summary, perceptual systems are thought to have evolved to optimize organisms' interactions with their environment, a process that can be partially explained in terms of potential bodily patterns of action (cf. Glenberg, 1997, p. 3). In this context, action is intrinsically embedded within perception, a concept that has led several scholars to argue that the primary (biological) purpose in life is acting rather than thinking. This perspective has prompted in-depth studies on the nervous system, viewed as a collection of mechanisms and processes designed to regulate sensory-motor activity (cf. Sperry, 1952, p. 310).

In line with this view, Rizzolatti and Sinigaglia (2006) developed the concept of "motor cognition" to describe the role of the motor system in understanding and interacting with the world. Gallese (2009) further explored related forms of "pre-conceptual" or embodied understanding (cf. Morabito, 2020, pp. 109–110). Within this framework, research on the nervous system has provided strong evidence of the interconnections between cognition, movement, and intention, particularly through studies on mirror neurons.[3] These studies support the view that the relationship with

[3] Specifically, in the 1990s, Giacomo Rizzolatti and his colleagues identified a specific type of neurons in the ventral premotor cortex of macaques, known today as mirror neurons. These neurons provided evidence for the hypothesis of shared cortical representations between perception and action (cf. Rizzolatti et al., 1996; Gallese et al., 1996; Rizzolatti & Arbib, 1998). This discovery contributed to what is now referred to as "motor cognition," highlighting the sensory and motor roles of the premotor and parietal areas of the cortex. Within these areas, two distinct types of nerve cells are active: canonical neurons and mirror neurons. Canonical neurons generate sensorimotor responses—such as interactions and potential motor acts—in response to perceived objects. Mirror neurons, on the other hand, activate both when an individual performs a specific action (e.g., gripping a bottle in a certain way) and when that individual observes or hears another performing the same action, simulating the behavior of others internally (cf. Morabito, 2020, pp. 103–104). Moreover, these neurons activate even when the object in question is partially obscured (cf. Umiltà et al., 2001) or when the subject hears the sound associated with the action, regardless of the sensory modality, testifying to the multimodal nature of the process (cf. Kohler et al., 2002).

the environment, including salient objects and other organisms, is fundamentally grounded in motor processes (cf. Rizzolatti et al., 2004; Morabito, 2020, p. 120).[4] Although these discoveries have sparked a range of interpretations and some controversy (cf. Hickok, 2009), proponents of this view argue that mirror neurons represent a fundamental mechanism underlying social cognition (cf. Gallese, 2009; Goldman & de Vignemont, 2009). Moreover, this body of research has highlighted the complexity and multilayered nature of action processes, in contrast to the classical mechanistic sensorimotor model, which posited that actions simply follow sensory inputs.

This expanded understanding of perception, in relation to movement, action, and motor intention, allows for an integrated perspective that acknowledges the diverse types of "action" found in nature, distinguishing among various forms and possibilities.[5]

Recognition of the importance of action in cognition is gaining ground. According to some authors, the purpose of life is fundamentally to *act* rather than to *think*, or, at the very least, it is primarily tied to various capacities for action rather than uniquely human intellectual activity (cf. Gazzaniga, 2019, p. 171). A broader examination of the capacity to act highlights the interconnectedness of organisms, their environment, and their actions. In general, all forms of action involve some degree of transformation. Internally, organisms shift from one state at the onset of an action to another upon its completion. Externally, they bring about changes in the environment, which can, in turn, influence the trajectory of their actions and further modify their internal states (cf. Frazier, 2021).

[4] In this sense, the meaning of others' actions is grasped through direct internal understanding, described as "motor resonance" or "embodied simulation," rather than through deliberate cognitive processes (cf. Palmiero & Borsellino, 2018, pp. 112–123). The mirror neuron system thus creates a shared action space, allowing for the inscription and comprehension of others' intentional movement sequences (cf. Rizzolatti & Sinigaglia, 2006, p. 65). If these interpretations are further validated, they could support the notion of a level of understanding and interaction with others that does not rely on "mentalizing" (cf. Spaulding, 2011; Shapiro, 2021).

[5] Moreover, there is a difference between simple movement and action: an action is finalized, endowed with purpose, caused by a need, and generally results from a configuration of movements, and is also characterized by an anticipatory component of the stimulus (cf. Morabito, 2020, p. 14).

4.2.2 The Concept of Agency

There is a common tendency to view the capacity for intentional behavior and action as exclusively tied to the presence of a nervous system. However, as Godfrey-Smith (2021, pp. 67–70) pointed out, even at the level of single-celled organisms, one can identify behaviors that are functionally equivalent to what we would consider forms of "action" or goal-directed activity in more complex, brain-bearing organisms. For instance, unicellular organisms exhibit behaviors such as swimming, capturing and absorbing prey, and displaying goal-oriented flexibility in order to ensure survival, manifesting an ability to select among different options. Even in analyses of these more elementary "forms of action," there is often a tendency to focus primarily on movement-related aspects, those tied to the unfolding of actions, while overlooking an equally fundamental modality that *could* also be considered a form of behavioral action, particularly in some living beings such as plants. This refers to their ability to regulate the synthesis, emission, and recognition of a wide range of chemical compounds (cf. Bianchi, 2022, 2024). Could this capability be considered a form of plant "action"? It should be noted that the reference here is not to self-conscious action, nor to behavior presupposing a notion of "free will," but rather to forms of activity that are not fully predetermined, characterized instead by a capacity to modulate the surrounding environment and solve survival-related problems. Recognizing, within increasingly inclusive approaches, this additional level at which certain natural actions are expressed does not undermine or diminish the efficiency, complexity, and sophistication of behaviors typically associated with goal-directed movements controlled by nervous systems. Instead, it invites reflection on the idea that "natural action" is not exclusively tied to muscular-motor behavior (even in the case of plants, there is motor activity that does not involve muscles). The distinction lies in how organisms interact with their environment: moving a part of the body or synthesizing chemicals to achieve a survival goal and influence the surroundings represent two different but equally effective ways of acting within one's environment, enabling interaction and the resolution of survival challenges.

A concept worth exploring in future research is that of "plant agency." This concept suggests that, regardless of the mechanism through which it is initiated or the strategy with which it is expressed, plant activity can still bring about changes in the surrounding environment, leaving impactful traces that influence both the context of inter-action and the behavior of other organisms. Thus, the notion of "action," understood in this broader sense, has at times been linked to the idea of agency. In a minimal sense, "agency" refers to an organism's capacity to engage in goal-directed behavior, that is, to act in ways that tend toward specific outcomes (cf. Lyon, 2006).

This concept, particularly when applied to plant activities, remains somewhat vague, as it was only recently introduced to explore the behavior of various non-human organisms. However, its heuristic potential might stem from this very vagueness and broad scope, allowing for exploration in largely uncharted areas. Unlike the notion of "subjectivity," which pertains to first-person experiential states, what happens to an individual, the concept of agency refers to the initiation and control of actions or action sequences whose effects are, at least to some extent, anticipated. It describes what an organism is able to make happen (cf. Godfrey-Smith, 2021, p. 74).

In the study of behavior and cognition, scholars from the ecological school and enactivist approaches—despite focusing on distinct elements—converge on the idea that agency must be understood in terms of the dynamic relationship between an organism and its environment. This orientation has led to the development of more inclusive alternatives to the so-called "standard theory" of agency, frequently discussed in philosophy (cf. Schlosser, 2018, 2019). According to the standard view, an event qualifies as a genuine action only if it is not accidental and involves mental and intentional states that represent the goal to be achieved (cf. Adams, 2018). However, this account may be overly restrictive, as there appear to be organisms capable of performing actions or behaving even in the apparent absence of representational mental states (cf. Segundo-Ortin, 2020).[6]

Even in the philosophy of action, the orthodox view has traditionally defined the capacity to act as the ability of an agent to carry out intentional actions (cf. Anscombe, 1963). This capacity is generally associated with the presence of mental states, beliefs and desires, that generate appropriate behaviors to satisfy initial needs. It is clear how such an approach makes it challenging to classify many organisms, including plants, as (cognitive) agents, given that they do not possess mental states in this traditional sense.

To address these limitations, more inclusive post-cognitivist approaches—such as 4E cognition, ecological psychology, and dynamical systems theory—have expanded the analysis of agency. These frameworks propose criteria with trans-specific validity, emphasizing the embodied, situated, and extended aspects of action (cf. Malafouris, 2013). From this perspective, agency is seen as emerging from the dynamic interaction between a system and its environment, which serves as the basis for the sensory-perceptual and motor possibilities available to different organisms.

According to Barandiaran et al. (2009), a broad conception of agency can be applied to diverse types of acting systems, provided they meet three necessary conditions. First, the agent must exhibit individuality, being distinguishable from its surroundings in order to act upon them. Second, the system must be able to modulate its coupling with the environment, for instance, by initiating processes, withholding behavior, or resisting external influences. Third, the agent must pursue goals through its actions, such that success or failure can be evaluated in relation to its initial aims (cf. Segundo-Ortin, 2020).

[6] For instance, in line with the ecological approach, perceptual information—which enables organisms to access environmental structures and act accordingly—consists of stimulus patterns that specify the properties relevant to the system's sensorimotor possibilities, without requiring inferences based on mental representations (cf. Segundo-Ortin, 2020), even if inference in nature may exist in different ways from the traditional conception of mentalistic inference.

With these expanded and refined conceptions of agency (cf. Buhrmann & Di Paolo, 2017), some researchers have begun to explore its application to the goal-directed activities of plants. These activities, ranging from movement to chemical signaling, are rooted in the plasticity of phenotypic change during development (cf. Gilroy & Trewavas, 2023). In this broader and more inclusive sense, agency—particularly when understood as the capacity to perform successful actions and interactions within a niche to satisfy survival needs—has increasingly been invoked to account for basic forms of cognition.[7]

Further advancements in this direction could contribute to establishing a framework for multispecies analysis of diverse forms of agency, enhancing our understanding of how goal-oriented behaviors, and the capacity to act meaningfully within an environment, are expressed differently depending on the functional, ecological, and relational structures of both neural and non-neural organisms.

4.2.3 *Are the Elements of Behavior Innate, Learned, or Both?*

When discussing "actions," or "agency," the aspect of non-coercion and relative autonomy is particularly important: if an outcome is inevitabile, it cannot be properly considered an action but rather a behavioral reaction. In the study of cognitive behavior, ongoing debates examine the extent to which different organisms exhibit spontaneity and flexibility in responding to environmental challenges.

Within the broader debate between empiricism and innatism, particularly concerning the nature of genuine individual experience, researchers have, since the early days of ethology, sought to distinguish between the elements of behavior that are innate and those acquired through individual development. However, depending on the temporal scale considered, it often proves unrealistic to maintain a strict dichotomy between the two. Based on the insights of eighteenth-century empiricist philosophers, it can be argued that experience, broadly conceived, underlies all aspects of behavior in living systems. Even so-called "innate" behaviors, when examined on evolutionary time scales, can be seen as rooted in experience, not individual, but in the collective experience accumulated at the species level (cf. Monod, 2017, pp. 147–149). Ultimately, these two aspects are interconnected. Nevertheless,

[7] In studies examining the relationship between cognition and action, particularly in animals, various forms of action have been identified. Among these, the actions considered genuinely cognitive are referred to as "epistemic actions." These actions modify the environment in specific ways to improve the efficiency of cognitive processes or to simplify the execution of cognitive tasks, often through the use of tools (cf. Di Francesco et al., 2017, p. 189). Some scholars argue that the cognitive abilities of our species have largely developed thanks to our capacity to create and use symbols, manipulate tools, transform our surroundings, leave external traces, and reduce the memory load associated with complex tasks (cf. Morabito, 2020, p. 95). Increasingly, these kinds of capabilities, though expressed in varying forms and degrees, are being recognized across a wide range of species.

despite the overlap, it remains essential to recognize the differences in types of responses across various forms of reactions and behaviors.

One way to distinguish between the inorganic and biological worlds, for instance, is to contrast mere reactions (occurrence of chemical and electrical processes) with genuine, goal-directed actions or responses of an organism aimed at maintaining, altering, or improving its condition in each situation (cf. Ponzio & Petrilli, 2002, p. 156). Unlike mere reactions, purposeful responses involve a consideration of "the why" behind them (cf. Baker, 2017, p. 142). On the other hand, reactions also occur in the organic world, though they differ from simple chemical-electrical processes. At the behavioral level, rapid and automatic responses are categorized as "reflexes," as they occur without deliberation or conscious control. The concept of the reflex emerged in the mid-seventeenth century with Thomas Willis (1621–1675), who in 1664 used the term to describe the immediate and involuntary link between a stimulus and a motor response. A similar idea appears in Descartes' work, where these reactions are conceived as a three-stage process involving an afferent input, central processing by nervous matter, and an efferent output. Later, John Dewey (1859–1952) critiqued this mechanistic and linear model. In his 1896 paper, he rejected the separation of stimulus and response as distinct psychic entities, proposing instead the notion of an "organic circuit" in which sensory input and motor output are dynamically interdependent (cf. Morabito, 2020, pp. 35; 60–61). Acknowledging this interdependence does not deny the widespread presence of reflexes in biological systems; rather, it situates them within a broader framework of behavioral coordination.[8] Today, the reflex arc is understood as a neural mechanism underlying rapid, automatic, and stereotyped responses to stimuli, typically characterized by a lack of behavioral flexibility. The response tends to be consistent across different occurrences of the same stimulus (cf. Trewavas, 2014, pp. 243–245; Gazzaniga, 2019, pp. 172–173). These stimulus-response patterns usually involve direct neural pathways connecting sensory input to motor output. For instance, in a classic withdrawal reflex, the arm retracts immediately after contact with a noxious heat source, often before the sensation of pain is consciously registered, through a spinal feedback loop. Another well-known example is the patellar reflex, where a sudden extension of the lower leg is elicited by a mechanical tap just below the kneecap, mediated by a spinal circuit without cortical mediation.

In addition to simple reflex responses, researchers such as Lorenz and Tinbergen have described fixed action patterns, more complex, yet still automatic, behavioral sequences triggered by specific stimuli (cf. LeDoux, 2020, p. 56).

At a general level, reflexes are often contrasted with "voluntary" actions in broader explanatory frameworks. Reflex actions occur automatically, cannot be consciously controlled, do not require learning, and are not influenced by prior

[8] According to Trewavas (2014, p. 245) the idea of reflex varies in its meaning according to the type of structural constraints of an organism. For example, in the case of an amoeba (formless or, rather, not anchored to a stable form), despite the limited behavioral range of this organism, it is difficult to apply the standard idea of reflex as an automatic reaction, precisely because of the mutability of the amoeba's body.

experience. In contrast, voluntary actions are typically assumed to be consciously performed and to involve mental processes (cf. Maher, 2017, pp. 100–101). Examples of apparently voluntary actions include walking toward an object, playing chess, drumming a rhythm, or writing a page of a book. By comparison, processes such as salivation, digestion, or aging are considered involuntary and are governed by physiological mechanisms rather than intentional actions. There are also hybrid cases that blur the line between voluntary and involuntary behaviors. Yawning, for instance, is not usually planned or deliberately initiated, but it can be subject to some degree of voluntary control.

Moreover, a term that has often been used in the past to refer to behaviors not arising from individual learning or experience is "instinct." As Bateson (1976, pp. 70–92) noted, the concept of instinct was formulated prior to the theoretical and experimental advancements brought about by the fields of genetics and the study of complex systems, including ecological-relational interpretations of communication and organization. The term 'instinctive' typically describes behaviors where all members of a species act very similarly under similar conditions. However, this term has also been applied to describe behavior whose underlying causes and processes were not fully understood (cf. Trewavas, 2014, p. 248). Due to its inherent ambiguity, the concept of instinct has faced criticism, such as from Lehrman (1953) and Schneirla (1966), and more recently from proponents of developmental systems theory (cf. Oyama, 2000). Although the term is still used to describe capacities that organisms possess automatically—i.e., those they do not have to learn to use—this type of criticism has certainly helped integrate the analysis of bioecological activities with the identification of more specific and circumscribed mechanisms, processes, circuits and behavioral sequences.

Another term commonly used to describe stereotyped responses that are not modified by experience is 'innate.' A classic example of innate behavior is provided by Lorenz (1952), who described the reaction of jackdaws attacking individuals who allowed a black object, resembling the bodies of conspecifics, to dangle. This response, likely an adaptive mechanism for protecting mates from predators, particularly birds of prey, illustrates an innate defensive behavior. However, organisms exhibiting such behaviors do not always respond identically to the same stimuli; rather, they tend to display a general pattern or predisposition. In these behaviors, although there are patterns that remain relatively fixed in the evolutionary history of species, at least since the mid-twentieth century, there has been debate on the types of factors involved and the degree of variability of many behaviors previously considered completely invariable and preprogrammed. As LeDoux (2020, pp. 56–58) noted, contemporary views acknowledge that even behaviors with a strong innate component, rooted in the genetic coding history of certain traits, can be influenced by environmental factors and are inevitably shaped by the organism's perceptual and sensory state at the time of response. This consideration of different types of behavioral responses underscores the idea that behaviors cannot be fully explained as rigid, automatic processes. Instead, they result from a complex interplay of internal states and external influences. Depending on the behavior in question, even within the same species, there can be a significant degree of autonomy and variability. Some responses remain more constrained and less flexible, while others allow for a higher degree of adaptability (cf. Bianchi et al., 2025a).

Regarding plants, these considerations are relevant because, for a long time, it was believed that these organisms, lacking a nervous system, could perform reproductive and survival functions solely by relying on predetermined and inflexible genetic instructions. However, in recent decades, it has increasingly been recognized that plants too can modify and contextually adapt some of their responses to solve survival-related challenges. Nevertheless, this does not suggest that plants exhibit unlimited behavioral flexibility in all situations. Many of their capabilities derive from the species' collective experience, an aspect that should not be viewed as a limitation, but rather as a distinctive mode of life. This way of life entails a lower degree of independence from a specific environment, partly due to their rooted nature, compared to other living species. The more demands an organism faces in its environment, the more complex that environment becomes, and the greater the range of interactions required for the organism to function within its niche (cf. Sherrington, 1906, pp. 7–8; Morabito, 2020, pp. 44–45). Plants' most significant behavioral flexibility likely arises from their capacity to produce, detect, and modulate chemical compounds (cf. Bianchi et al., 2025a).

Based on current evidence, not all plant responses are entirely preprogrammed. Instead, they exhibit a certain degree of contextual variability. This flexibility supports the existence of communication abilities involving various types of signals—and, consequently, information exchange at multiple levels—as well as processes related to bodily memory and learning.

4.2.4 *To Be a Plant*

If one assumes that an organism can act in ways not entirely determined by its internal processes or external environmental conditions, it becomes pertinent to ask what kind of "experience" it is capable of having. The premise is that the living world includes an experiential plurality stemming from the existence of diverse (similar yet not identical) sensory-perceptual modalities. Furthermore, each organism possesses a unique constitution and life history (individual path). Thus, it appears that every living system, whether neural or non-neural, has its own specific ways of accessing reality, enabling it to perceive, act, and consequently have forms of "experience." These processes are not necessarily self-aware, but involve mechanisms of memory, anticipation, and learning, which allow the organism to engage in multiple relationships, whether by preserving, refining, or correcting them (cf. Bianchi, 2024).

Experience at certain levels of complexity enables organisms, throughout their lifetimes, to connect what was once perceived as disconnected and to distinguish what was previously indistinct (cf. Cellucci, 2008). On a basic level, even simple interactions could equate to having perceptions of sensory variations. At a more complex level, an organism may become aware of its own interactions, to the point of being aware of its awareness of having experiences, a form of reflection (cf.

Frazier, 2021). However, genuine "experience" has often been interpreted as the subjective viewpoint of an organism. For instance, Godfrey-Smith (2021, pp. 130; 320–321) suggested that organisms capable of true experience, in the sense of "feeling" their own lives (a form of felt experience), are likely limited to those with a nervous system. Nevertheless, at the current stage of research, it does not seem possible to entirely rule out the hypothesis that aneural organisms may possess some form of "experience." Even Godfrey-Smith himself leaves room for doubt, acknowledging the potential for future investigations that could provide new evidence to support this hypothesis.

Naturally, it is impossible to directly access or feel what another organism experiences—not even that of a member of our own species—since we are not that organism. However, when it comes to another human being, our bio-ecological similarity makes it easier to imagine their experiences. We can never have a direct, internal experience of what others go through, because our subjective consciousness is limited to our own perspective. Nonetheless, acknowledging this does not imply that different organisms are without their own unique ways of perceiving and interacting with their surroundings. At the very least, organisms can sense what is happening both internally and in their environment. This question touches on the unique and individual sensation that an organism has simply by being itself, a concept often closely linked to consciousness in philosophical reflection. At the roots of this philosophical issue is Thomas Nagel's well-known essay "*What is it Like to Be a Bat?*" (1974). Nagel argued that even if we had a detailed understanding of how a bat perceives its environment through echolocation—a sensory modality fundamentally different from our own—this knowledge would still not allow us to grasp what it actually feels like to be a bat. Being a bat "feels like something," but the only way to truly know that feeling would be to *be* the bat (cf. Marraffa & Paternoster, 2013, pp. 49–50). An external observer can only ever approach the behavior of another organism from an outsider's perspective (cf. Maher, 2017, p. 59). This indicates an unbridgeable gap between scientific explanations of observable phenomena and the inherently subjective sensations and experiences of individuals (cf. Chalmers, 1995; Perconti, 2017, pp. 98–99; Di Francesco et al., 2017, pp. 99–100).

This type of consideration has also been explored in emerging work on the notion of "plant consciousness." For instance, in his 2017 paper *"What is it Like to Be a Plant?"*—inspired by Nagel's famous essay—Calvo discussed the inherent impossibility of truly knowing what it feels like to be a plant. Nevertheless, this does not preclude the hypothesis that plants may possess some form of sensitivity related to their behavior and interactions, suggesting that there is indeed "something it is like" to be or to feel as a plant. When plants perceive something, they do so in their own way; if they possess any form of awareness, it would be uniquely plant-like (cf. Maher, 2017, p. 50). If that were the case, they would not be "conscious" like humans, but rather like plants (cf. Warwick, 2000, p. 184).

Making such a claim from an inevitably external perspective may be reasonable when considering the behavioral coherence observed by researchers. This could suggest that various organisms possess some form of awareness of their environment and needs. What researchers are able to observe are solely changes in plant physiology, patterns of interaction, and behavioral responses, nothing beyond that (cf. Trewavas, 2014, p. 259). Even if it were possible to fully access what other organisms feel, which it is not, such access would not necessarily *explain* their abilities. Scientific investigation, therefore, must proceed on a different level of analysis.

4.2.5 Observing Behavior: A Behaviorist Reintroduction?

These considerations—positioned at the intersection of research on perception-action capacities, varying levels of sensory, conscious, and experiential abilities, and the epistemological limits of what an observer can genuinely comprehend about plant behavior—inevitably raise the question of selecting the most suitable approaches for studying the variations in plant behavioral activities. This inquiry diverges from the initial aim of classical behaviorism, because is not to deny the existence of cognitive activities but, rather, to understand how different abilities are concretely expressed. It seeks to identify the factors at play when different types of interactions and environmental challenges arise. Ultimately, this line of research holds the potential to shape a renewed and updated perspective on studying behavior, contributing to a deeper understanding of certain cognitive aspects.

As discussed in Chaps. 2 and 3, the study of behavior has traditionally been rooted in research within ethology and comparative psychology, fields that often overlap thematically. Recent approaches to cognition, aligned with advancements in the empirical sciences, draw heavily from literature grounded in both physiological and ethological analyses, situated within a broader framework of biological, ecological, and evolutionary studies. A psychology or philosophy that fails to engage with empirical findings and contemporary developments, in fact, would not progress very far.

Historically, the best-known and often criticized approach for studying behavior is what is commonly referred to as behaviorism. However, considering the historical evolution of various approaches that have emphasized different aspects, it is more accurate to speak of "behaviorisms," each driven by slightly different theoretical motivations. The body of behaviorist thought thus constitutes a more complex and multifaceted phenomenon than it is often portrayed to be. Nevertheless, on a general level, behaviorist approaches share common elements, such as the principle that behavior should be described without invoking the concept of a mind or any notion of mental control (cf. Bianchi, 2021).

The study of animal behavior has presented significant theoretical and methodological challenges, particularly due to the difficulty of verifying cognitive states or mental phenomena in the absence of verbal language, like human language, which has traditionally been associated with self-consciousness and intelligence. One strategy to address this challenge emerged in early twentieth-century America, leading to the development of approaches aimed at translating psychological or mentalistic claims into empirically verifiable, physicalistic assertions. In this context, an experimenter was required to understand animal behaviors, which reflected their various capacities, based solely on their observable external manifestations (cf. Burkhardt, 2005). These behaviors could also be experimentally conditioned, demonstrating that they could be altered through practice and learning (cf. Mazzolai, 2021, pp. 171–173).

The critical engagement with the concept of the "mind" has been interpreted differently by various researchers. For instance, Ivan P. Pavlov (1849–1936), renowned for discovering the conditioned reflex in dogs, observed behaviors such as salivation in response to an auditory signal that had been repeatedly paired with food. While Pavlov himself focused on the physiological mechanisms of this reflex, some later interpretations suggested that the conditioned stimulus might evoke an image or memory of food, thus triggering an unconditioned response via associative memory (cf. Maher, 2017, pp. 101–102). John B. Watson (1878–1958), in his seminal 1913 paper *Psychology as the Behaviorist Views It*—often considered the manifesto of behaviorism—argued that the only legitimate subject matter of psychology was observable behavior. In his view, the concept of the "mind," if acknowledged at all, was reducible to behavioral dispositions, not internal mental states. Edward Thorndike (1874–1949), a contemporary of Watson, developed the law of effect, presented in *Animal Intelligence: An Experimental Study of the Associative Processes in Animals* (1898). He demonstrated that behaviors followed by satisfying outcomes were more likely to be repeated. In experiments where animals (such as cats) had to manipulate a latch to escape a puzzle box, learning was explained as the strengthening of stimulus-response connections. Thorndike rejected the need for mentalistic explanations, suggesting that learning occurred without invoking internal "ideas" or conscious thought. This mechanistic and anti-mentalistic approach was further developed by B.F. Skinner (1904–1990), a leading figure in radical behaviorism. Skinner proposed an empirical framework that eschewed metaphysical or mentalistic constructs, asserting that all meaningful psychological statements should be reducible to descriptions of observable behavior (cf. Maher, 2017, pp. 13–17; Gazzaniga, 2019, pp. 79–80). In sum, despite their differing emphases, these theorists—building on Pavlov's model of conditioning and Thorndike's law of effect—converged in reducing psychological explanations to associations between environmental stimuli and behavioral responses. They collectively dismissed introspective or cognitive accounts, treating the processes within the "black box" (the mind) as either irrelevant or the domain of physiology (cf. Di Francesco, Marraffa, & Tomasetta, 2017, pp. 69–71).

To further distinguish the theoretical influences on the behaviorist approach, it is possible to identify various intellectual currents that shaped and conditioned its evolution, paving the way for subsequent differentiations. One such current that significantly influenced neobehaviorism—especially during the 1930s to 1950s in the works of Edward C. Tolman, Clark L. Hull, and B.F. Skinner—was logical positivism, particularly associated with figures such as Rudolf Carnap (1891–1970) and Carl G. Hempel (1905–1997). This philosophical movement contributed to a shift away from Watson's original behaviorism, promoting a vision of psychology as a science grounded in physicalist language, operational definitions, and empirical verification. It aligned with the broader ambition to integrate psychology into the framework of the natural sciences. However, as noted by Di Francesco, Marraffa, and Tomasetta (2017, pp. 70–71), the relationship between logical positivism and psychological behaviorism is complex. Neobehaviorists did not fully adopt the stricter tenets of logical empiricism, often finding its formalism and linguistic criteria too rigid or impractical for psychological application. Therefore, the influence was largely methodological rather than doctrinal, promoting a more rigorous empirical approach to behavior and a heightened emphasis on "objectivity," testability, and operational clarity in psychological theories. Taken together, these precautions have contributed to the advancement of experimental research in psychology.

Another way to categorize behaviorisms is by distinguishing between a "psychological" version, rooted in the empirical work of Pavlov, Watson, and Skinner, and a "philosophical" version, most prominently represented by Gilbert Ryle (1900–1976). In *The Concept of Mind* (1949), Ryle proposed what later came to be known as "analytical behaviorism," an approach that, influenced in some respects by Wittgenstein's views on language, argued against explaining behavior through private internal states. Ryle's analysis aimed, on a more general level, to identify the errors associated with mentalistic language and reduce the disputes arising from the mind-body dichotomy. Ryle contended that understanding one's own mental activity occurs in much the same way as understanding the mental activities of others, not through introspective access to private mental contents, but through patterns of behavior, contextual cues, and linguistic expression (cf. Di Francesco et al., 2017, pp. 72–77; 117).

Traditional forms of behaviorism gradually faded from prominence—or more precisely, evolved beyond their original formulations—paving the way for the emergence and consolidation of cognitive science. One of the major limitations of behaviorism was its reliance on models that overlooked individual differences, differences that reflect cognitive capacities and emerge from a complex interplay of internal and external processes. These processes are closely linked to specific bodily constitutions and the variability of environmental dynamics. Although classical behaviorism has been replaced (or at least overshadowed) by advancements in cognitive science, some elements of neobehaviorism seem to persist in some contemporary approaches for studying cognitive behaviors (cf. Bianchi, 2021).

Some contemporary approaches still draw, either directly or indirectly, on behaviorist ideas, or at least exhibit certain methodological affinities. For instance, Daniel Dennett's proposal of heterophenomenology for attributing states of consciousness from a third-person perspective (cf. Dennett, 2007). Similarly, certain assumptions in radical embodied approaches, which challenge traditional representationalist accounts of the mind, echo behaviorist critiques, although within a markedly different conceptual framework (cf. Di Francesco et al., 2017, p. 77).

While it is important to acknowledge the homogenizing and reductive tendencies of some classical behaviorist models, reconsidering certain behaviorist principles in a contemporary, evidence-based framework could be valuable for explaining aspects of plant cognitive behavior. However, in this context, the goal would not be to deny, omit, or sidestep the cognitive dimension. Unlike the original behaviorist intent—rooted in skepticism about the mind's existence and the impossibility of verifying notions related to consciousness or subjective inner experiences—the revised aim would be to highlight cognitive aspects that manifest in behavioral complexity and can be studied at the behavioral level. Thus, I propose that this refined approach could be beneficial for analyzing plant behavior. Although proponents of plant cognition emphasize that they are not "behaviorists," they do, in fact, identify cognition through a multilevel explanation of behavior, understanding it as motivated, goal-directed, and context-dependent variations. This would not entail developing a new form of behaviorism but rather engaging in a critical dialogue with theoretical and methodological approaches that have historically addressed the study of behavior, in comparison with more recent perspectives such as behavioral ecology.

4.3 Epistemological Issues: Situated and Natural Practices of Knowing

In this section, dedicated to the theme of knowledge, my aim is not to present or align with any particular theory of knowledge but rather to reflect on some conceptions of "knowledge" with a view to developing a more inclusive perspective. A distinction to emphasize is between "knowledge" as something one possesses and "knowing" as an active, constantly evolving practice. This second understanding is more inclusive and less tied to intellectualist approaches.

The goal is to try to develop a perspective that can explain how phylogenetically distant organisms, such as plants, can behave consistently within their environment (cf. Bianchi, 2024). This potential outcome is by no means a certainty; indeed, it may even be inappropriate to talk of "knowledge" in the context of plants. However, the value of this exploration lies in the analysis itself, which seeks to be more integrated, detailed, and critically engaged with different forms of knowing. A philosophical analysis of plant cognition could, indeed, broaden or productively challenge our understanding of the differences, possibilities, and constraints inherent in the diverse cognitive strategies found in nature. These adaptive capacities are employed by various living systems to navigate survival challenges in dynamic environments. As mentioned above, throughout these pages, I have chosen to focus on the terms 'behavior' and 'cognition' rather than 'intelligence' or 'knowledge,' as the latter are more commonly associated with explanations describing typically human capabilities. Nevertheless, as with the concept of intelligence, the meanings and applications of terms like knowledge—and related concepts such as "logic," which will also be discussed in this part—can be extended or refined to suit specific contexts.

I wonder, then, whether this focus on the relationship between plants and forms of knowledge should be regarded merely as metaphorical, or whether it can genuinely contribute to advancing research in the epistemological domain. Such an approach may illuminate cognitive processes, methods, and strategies that remain underexplored, overlooked, or marginalized, particularly when contrasted with the more intellectualist and propositional dimensions typically emphasized in studies of human cognition.

4.3.1 Epistemological Analyses

At a general level, when we think of "knowledge," what often comes to mind is its most reliable form: scientific knowledge. Traditionally, epistemology has been the branch of philosophy that critically examines the conditions, methods, and limits of acquiring this type of knowledge. More recently, however, the term has broadened to encompass the general study of knowledge, not just scientific one, including the scope of gnoseology. Gnoseology, or the general theory of knowledge, examines the nature of knowledge, with a focus on the possibility of verifying its forms and constraints, particularly with regard to human, situated, and contextual understanding (cf. Giolito, 2011, p. 7).

Contemporary epistemology has complicated the investigation of what scientific knowledge entails, evolving into a "question about the genesis of questions." Over time, the notion that the meanings of scientific concepts change has become widely accepted, highlighting the empirical and transformative nature of scientific practices (cf. Frezza & Gagliasso, 2014, p. 29).

As is increasingly emphasized in contemporary relational, situated, and post-positivist epistemologies, one of the main reasons scientific knowledge has moved away from the certainties of the past is the growing recognition and critical reflection on its own investigative and measurement methods. Scientific inquiry, which operates by isolating, explaining, and organizing phenomena, has come to acknowledge that its methodological interventions influence the very objects under investigation. In this sense, the process of knowing is not separate from its object, but co-constitutive. In a similar vein—though formulated within the framework of quantum physics—Heisenberg noted: "*we have to remember that what we observe is not nature herself, but nature exposed to our method of questioning*" (Heisenberg, 1958).

Additionally, the level at which phenomena are observed is very important. While indeterminacy is primarily a feature of microscopic, quantum-level events, unpredictability can also emerge in a very different context: that of nonlinear dynamics. Concepts such as "sensitive dependence on initial conditions," commonly known as the "butterfly effect" (cf. Hilborn, 2004), demonstrate how small variations in initial parameters can sometimes produce large-scale consequences over time. As revealed by the science of complex systems, many macroscopic processes are inherently difficult to predict due to their multifactorial, multilevel, and non-linear nature. This unpredictability helps explain why deterministic models of behavior often fail when applied to complex biological and ecological systems, a reality well recognized by researchers working in these fields (cf. Heylighen, 2023).

Another factor that has challenged the simplistic, non-relational view of scientific knowledge—as a direct and objective means of accessing reality—is the acknowledgment of theory-laden data. There is no "*View from Nowhere*," not situated, or decontextualized, from which to answer a certain question. This concept suggests that there are no purely "raw" data, free from theoretical interpretation. This becomes evident when we consider the assumptions underlying specific theories that guide researchers, particularly in the design and functionality of technologies used to conduct experiments, which then yield specific results (cf. Giolito, 2011, p. 15).

In sum, scientific evidence—the product of data analysis carried out through scientific practice—remains the most reliable source from which research progresses. Nevertheless, this process inherently involves careful and context-sensitive selection of information. The complexity and diversity of phenomena cannot be fully captured from a single perspective, and any method of organizing knowledge is shaped both by the observational process itself (cf. Boncinelli, 2020) and by the tools employed by researchers, including models, theories, and specific technologies (cf. Bateson, 1976, p. 23).

Recognizing these limitations does not diminish the effectiveness of the scientific method. It remains our best, though inherently fallible and continuously revisable, means of understanding the world. The current inability to identify a unified and all-encompassing explanatory framework in science is, in fact, a defining feature of scientific practices themselves (cf. Giolito, 2011, p. 63). Moreover, this variability is essential; it is a constitutive requirement of the scientific enterprise.[9]

It is precisely the interplay of these aspects that makes epistemological reflection continually relevant and necessary. There will always be a need to analyze the limits and possibilities of knowledge, the conditions that enable its emergence, the research contexts, and the kinds of questions being asked. Equally important is the recognition of various sources of bias, partiality, and error—whether conceptual, methodological, or related to measurement—given their potential, and perhaps inevitable, recurrence over time in similar or slightly modified forms.

4.3.2 Conceptions and Types of Knowledge

Beyond the analysis of scientific forms of knowledge, various philosophical approaches have distinguished between different types of knowledge, often emphasizing either the rational-propositional or the practical-experiential dimension. However, within an eco-evolutionary framework, such distinctions are situated within a broader *continuum* of specialized means and capacities that organisms employ to navigate and interact with their environments, whether these are more or less (socially) complex.

[9] Scientific thinking does not involve the acceptance of single, indisputable, truths but is grounded in the continuous testing of the elements upon which the best available theories are progressively built. From a scientific perspective, acknowledging that there is still much in the universe we do not understand—and that many of our doubts will persist—does not prevent us from offering the most reliable answers available at any given time, based on the evidence at hand. However, in doing so, there is an inherent recognition of our limitations and the gaps in our understanding across many areas.

As Tommaso Piazza (2017, pp. 7–12; 111) explained, knowledge can be categorized into at least three forms: propositional knowledge, which is truth-oriented (e.g., knowing that a particular city is the capital of a certain nation); direct knowledge (e.g., knowing what a specific food tastes like, or more generally, knowledge of things); and procedural knowledge, or "knowing how" (e.g., knowing how to practically perform a certain task). This categorization (cf. Russell, 1912) is not universally accepted, as some argue that direct knowledge and "knowing how" are ultimately reducible to propositional knowledge. Direct knowledge, or "knowledge by acquaintance," is often better understood as a non-propositional and non-conceptual cognitive relation to an object. From this initial relation, propositional knowledge about the object may subsequently be developed by the subject. Nonetheless, some maintain that only propositional knowledge qualifies as genuine or authentic "knowledge" (cf. Piazza, 2017, pp. 15–16).

When human knowledge is the focus, it is typically regarded as propositional or conceptual in nature, expressed through statements about the state of affairs in the world. According to the foundationalist view, knowledge is seen as consisting of justified true beliefs. By restricting the analysis in this way, there is a risk of excluding forms of perceptual knowledge (if it is appropriate to call them that) which cannot be considered true forms of justified beliefs. Moreover, a particularly significant consequence of this view is the potential inability to attribute any form of environmental knowledge to non-human animals, or even more so, to aneural organisms, given that their behaviors do not convey propositional content (cf. Cellucci, 2008, pp. 62; 72–73). Yet, in the absence of some form of environmental awareness, it seems unlikely that any organism could successfully survive. Furthermore, cognitive abilities vary not only across species, due to differences in their perceptual and cognitive systems, but also throughout an individual organism's lifetime, as its forms of "knowledge" may change over time, for instance, between juvenile stages and those of maturity or senescence.

While it is entirely appropriate in the context of human cognition to focus on the rational-propositional dimension of knowledge (which some argue is the only true form of knowledge), such a focus would be overly reductive for the purposes of exploring modes of knowing in non-human organisms, and even more so in plants (even if these cases do not constitute a real "knowledge" in the strictest sense). In this context, the non-propositional, practical, and action-oriented aspects of knowledge become more relevant. Even within the scope of human knowledge, the need to broaden the framework of analysis to include forms of practical knowledge, distinct from propositional ones, has been acknowledged. For example, some anti-intellectualist perspectives—such as Gilbert Ryle's concept of "knowing how" (e.g., knowing how to swim or ride a bicycle)—challenge the centrality of propositional attitudes in the understanding of knowledge (cf. Di Francesco et al., 2017, p. 157; Perconti, 2017, p. 72). According to this view, procedural knowledge is not reducible to propositional knowledge about the ability in question. In contrast, Stanley and Williamson (2001) defended an intellectualist position, arguing that "knowing how" entails propositional knowledge. To illustrate the anti-intellectualist stance, Carl Ginet (1975) noted that performing a simple action, such as leaving a room, does not require one to consciously consider propositions like 'there is a door' or 'I must turn the handle.' One typically acts without explicitly entertaining these propositions. In essence, this suggests that acting on a known proposition does not necessarily require conscious reflection on that proposition (cf. Piazza, 2017, pp. 18–20).

Another critical issue in the contemporary debate on anti-intellectualism concerns thought experiments that challenge the idea that knowing how to perform a task is equivalent to having the ability to perform it. One such example involves a professional pianist who, after a severe accident, loses both arms. While some might claim that the pianist no longer has the ability to play, he arguably retains the knowledge of how to play the instrument. This suggests that even if the skill is no longer exercisable, the knowledge persists, indicating a conceptual distinction between the two. Addressing this type of objection requires distinguishing between the possession of a skill and the capacity to exercise it. Further support for the non-identity of propositional knowledge and practical know-how arises from cases in which individuals can perform a task but cannot articulate how they do it, or conversely, can explain how to perform a task but are unable to carry it out themselves. This distinction underscores two separate dimensions of knowledge: one descriptive and the other practical (ibid., pp. 21–25).

Despite the various objections and counterarguments between intellectualist and anti-intellectualist approaches (which cannot be fully explored here), examining the nuances of these debates can deepen our understanding of different types and conceptions of knowledge. This exploration could help determine whether, and in what ways, cognitive abilities can be studied in non-human organisms, and perhaps even in aneural organisms such as plants. Such an inquiry has the potential to broaden the scope of philosophical research.

Future studies could explore whether, in the case of plants, we are dealing with a form of non-propositional, inferential, or representational knowledge. Alternatively, it might involve direct knowledge, procedural knowledge, knowledge by acquaintance, a form of know-how, or perhaps none of these. What is certain is that plants, like other living systems, possess means and strategies for interacting with their environments and solving survival-related problems. In other words, they exhibit species-specific abilities to acquire, organize, and transmit the information necessary for their activities. Clearly, something is taking place, the challenge lies in how to describe it, and whether it can meaningfully be understood as a form of "knowing" (cf. Bianchi, 2024).

The pragmatic side of knowledge, or "performative knowledge," certainly deserves deeper exploration. Regarding human knowledge, certain approaches have demonstrated that knowledge is not solely dependent on epistemic-conceptual or theoretical-propositional factors. Two cases of these approaches are pragmatic encroachment and epistemological contextualism (cf. Piazza, 2017, pp. 103–109).

Pragmatic encroachment refers to the idea that practical considerations can influence the constitution of knowledge. In other words, whether a subject knows something may depend on their practical interests (cf. Fantl & McGrath, 2012). For instance, a person who loves sweets is more likely to know where to find a good pastry shop than someone who dislikes sweets or avoids them for health reasons. However, this example does not necessarily prove a direct link between practical interests and knowledge; rather, it suggests a greater likelihood that the sweet lover possesses relevant information, without excluding the possibility that someone uninterested in sweets might have the same knowledge. A less trivial interpretation, however, considers cases in which two individuals possess the same information but differ in what they can be said to know due to their differing practical stakes. In such scenarios, practical concerns shape what counts as knowledge, even when informational content is the same.

In the case of epistemological contextualism—particularly in relation to the semantics of expressions—significant emphasis has been placed on the context in which utterances occur. Considering the role of context and the salient alternatives that shape cognitive and semantic processes could also provide valuable insights into the forms of communication and behavior exhibited by other organisms, emphasizing how these elements are deeply integrated into broader cognitive and communicative practices (cf. Bianchi et al., 2025b).

To extend this reasoning to the cognitive abilities of other organisms, as already mentioned, it may be helpful to distinguish between "knowledge" and "knowing." Although I often use these terms interchangeably throughout this text—given that what is at stake is a bio-ecological, naturalized conception of knowledge—the intended meaning aligns more closely with "knowing" understood as a set of capacities actively enacted by organisms. For example, Michaels and Carello (1981) emphasized the importance of this distinction, noting that in epistemological studies, theories of knowledge, and psychology, "knowledge" is often viewed as something internal to an animal, something that can be possessed or that serves to represent reality (knowledge as a representation of the world). In contrast, ecological psychology prefers to refer to "knowing" as an active process, something organisms *do* rather than *have* (ibid., p. 62). Even if expressed in different terms, other post-cognitivist approaches might also agree with this distinction.

4.3.3 Knowledge and Action: Relational Dimensions and Practical Forms of Knowing

In accordance with this perspective, there have been efforts to move beyond rigid dualistic distinctions between the knowing subject and the known object, or between the organism and its environment, by focusing on processes of codetermination, in which, nevertheless, the systems are still distinguishable. In this regard, the work of Maturana and Varela (1987) is particularly significant. They proposed an understanding of cognition and knowledge as rooted in biological processes, as the active enactment of a world in constant flux, rather than as representational knowledge that passively reflects an independently existing reality (cf. Nucara, 2014, pp. 102–110). Viewed in this light, the right question to ask is not what is inside the head, but rather what the head is inside of (cf. Turvey et al., 1981), or more precisely, what the entire living system is embedded in, what surrounds it and influences it, and what it, in turn, affects. In the same vein, one might argue that the world is not simply "in front of us" but "around us." This is important because if the organism is the knower and the environment is what is known, then a full understanding of knowledge cannot be achieved by analyzing only one of these elements; each exists in dynamic relation to the other (cf. Tyree & Sperry, 1989). This way of understanding the organism-environment relationship, and framing knowledge as system-based, contributes to a deeper understanding of the agentive and processual dimensions of knowing. One could argue that knowing is an action, an operational reality within the domain of existence of a living system (cf. Maturana & Varela, 1987). From this perspective, knowledge does not merely register outcomes; while it does keep track of what the organism learns, it continuously produces visible results through the variation of ongoing actions (cf. Rispoli, 2012, p. 103).

Framing knowledge in terms of practical activity and the development of procedural processes helps to orient the analysis toward action-based conceptions. Knowledge can be understood not as a primarily contemplative, conceptual, or reflective phenomenon, but rather as a set of fundamentally practical processes and activities, an understanding that is both scientifically grounded and widely acknowledged. In the "world of life," cognitive activity is not primarily aimed at abstract theorization or the pursuit of "truth;" instead, it is oriented toward determining how to act, remember, and prioritize in ways that produce effective results in specific situations (cf. Palmiero & Borsellino, 2018, p. 44). In this sense, different cognitive modalities—necessary for distinguishing, acting, and adapting in dynamic contexts—can be interpreted as actions that are appropriate to the diverse circumstances (cf. Nucara, 2014, p. 102).

From an experimental standpoint, particularly in the realm of nonconceptual or procedural knowledge, considerable attention has been focused on sensorimotor aspects and the "tacit knowledge of the body" (cf. Marraffa & Paternoster, 2011), an idea that traces back at least to the observations of Jean Piaget (1896–1980). Piaget, for instance, argued that cognition is not defined by abstract mental faculties but is deeply rooted in the organism's overall activity, specifically in the sensorimotor coupling ("assimilation" and "accommodation") between the body-mind and its environment. While Piaget's work did not explicitly focus on the subcortical or subsymbolic levels of motor knowledge, his ideas laid the foundation for later developments in this area. At the dawn of the twenty-first century, the concept of motor cognition of the body, which had previously been implicit, was brought to the forefront. This knowledge has since been primarily investigated at the subcortical and subsymbolic levels, which underpin more complex behavioral expressions (cf. Morabito, 2020, p. 76; 115).[10]

One might then ask: what does a more integrated and comprehensive view of knowledge processes offer?

[10] A concrete example of how in our species certain forms of knowledge are inextricably linked to practical and motor action is provided by Thelen and Smith (1994), who have shown that when children are placed in front of an inclined plane, they progressively learn to crawl safely only as their motor skills develop, allowing them to avoid the danger of falling. This finding illustrates that in this type of cognitive interaction with the environment, motor action is not a neutral element but a crucial component (cf. Di Francesco & Piredda, 2012, pp. 69–70).

4.3.4 Eco-Evolutionary Framework and the Biological Role of Knowledge

Progressively incorporating this multiplicity of factors allows for the relocation of cognitive processes within the framework of eco-evolutionary naturalism. Experience, action, and context become integral components of a comparative investigation into the possibilities of knowing found in nature. In the case of our species, this includes intellectual capacities and propositional aspects, which have evolved within our unique eco-cultural niche. Even theoretical actions can themselves be interpreted as a form of practice. However, this perspective does not exclude other modes of cognitive interaction characteristic of different forms of life, each adapted to their particular "way of being in the world."

To underlie this aspect, it is helpful to consider the ecological psychology approach, which aligns in some respects with empiricism and posits that the potential for knowledge arises from the senses, experience, and the effectiveness of action. Extending the analysis beyond human cognitive modalities to encompass the diverse cognitive possibilities of living organisms from an evolutionary perspective reveals that pragmatic knowledge of the environment (or forms of practical knowing) corresponds to an organism's actions within its environment (cf. Shaw & McIntyre, 1974). Consequently, asserting that knowledge consists of "pragmatically true actions" implies that an organism's ability to act is compatible with and suited to environmental challenges. It does not mean that actions or behaviors can be labeled "true" or "false," "correct" or "incorrect," as propositions can be. The hypothesis, therefore, is that for actions to be appropriate and effective in preserving the organism's life and achieving its goals, the "knowledge" of its environment must be at least pragmatically "adequate" for the task. Furthermore, within an evolutionary framework for analyzing behavior, practical knowledge, tied to the concrete ability to act effectively within an organism's niche, is rooted not only in an individual's interactions with the environment but also in the practical challenges their ancestors overcame (cf. Michaels & Carello, 1981, pp. 108–109).

In this way, within the epistemological context and broader knowledge studies, the focus shifts from questioning the nature of knowledge to examining its reliability and manifestation across living systems. Following Cellucci's (2008) proposal, instead of asking what knowledge is, it may be more useful to inquire about its role in nature. For our species, this approach implies a rethinking and integration of the justificationist view alone, which typically reduces the problem of knowledge to justifying existing knowledge. Instead, it emphasizes a heuristic conception of knowledge, focusing on discovering new procedures, organizing information differently, and creating new connections across fields to generate new knowledge (ibid., p. 74).

Aware that the inquiry will always be "human," one can nonetheless expand the scope to include the cognitive strategies of other life forms and to explore the evolving and processual nature of the "living practice" of knowing. From this perspective, cognitive practices appear not as isolated or merely reflective phenomena.

Rather, they serve a biological and vital role,[11] as essential as the ability to acquire resources for sustaining metabolic processes (cf. Cellucci, 2008, p. 193). In this sense, the role of knowledge is problem-solving, with survival as the primary aim. In some way, organisms need to "know" their environment to survive, obtain energy for their activities, and minimize dangers that could potentially harm or kill them. This is more achievable if they establish a reasonably adequate relationship with their surroundings; in a sense, they seem to "know" what to do, *how* to move, *when* to act, and *what* to avoid. When considered in this way, the concepts of knowledge and life (of knowing and living) appear to be closely interconnected (ibid., p. 194; p. 213; cf. Palmiero & Borsellino, 2018, p. 94).

From an evolutionary perspective, if the primary (not the only one) role of knowledge is bio-ecological, the hypothesis suggests that some form of it must exist in all organisms, even those without a nervous system (cf. Nucara, 2014, pp. 113–114), such as plants or unicellular organisms. These organisms, in different ways, though lacking nervous systems, function as cognitive systems through the entirety of their organizational structure, forming a core relationship with their environment. Indeed, this relationship is not only based on the activity of intermediate systems, such as the nervous system in animals or the genome in prokaryotes. This relationship includes the internal and external processes that drive them to behave in specific ways rather than others (cf. Cellucci, 2008, pp. 194–198). From this perspective, life itself, in its capacity for continuation, can be interpreted as a differentiated cognitive process (cf. Maturana & Varela, 1980).

Asserting that knowledge plays a bio-ecological role does not exclude it from also having a "cultural role," where "culture" refers to the set of knowledge transmitted across generations not through genetics but through actions and communication. This is particularly evident in our species, where knowledge has specialized into highly abstract forms, also becoming a field of research. However, approaching knowledge in the way just suggested means taking a naturalistic and more inclusive perspective that encompasses various species. The cultural role of knowledge is not in opposition to the biological role; rather, it serves as an extension or significant enhancement of it (cf. Cellucci, 2008, p. 200).

Adopting such a broad and empirically informed perspective, while also exploring different conceptions and types of knowledge, highlights another important aspect: the various bodily, structural, functional, and ecological constraints of different living systems, each with unique perceptual-cognitive mechanisms that shape their relation with reality. This implies that while the capacity for understanding one's environment is widespread across the tree of life, it exists to varying degrees, as the methods, strategies, and scope of different cognitive systems differ. In brief, what an organism can "know" is specific to its bodily capabilities and related ecological needs.

[11] Or "metabiological" role, since knowledge serves to optimize the biological functions of organisms for survival. However, a metabiological role is still fundamentally a biological one, tied to the organizational and interactive capabilities of living systems (cf. Cellucci, 2008, p. 198).

4.3.5 Knowing: Constraints and Possibilities

Even when limited to the human case alone, cognitive processes must account for the various constraints, bodily constitutive factors, ecological possibilities, and the essential operations of distinction performed by knowing agents. As Maturana and Varela (1980) noted, within the cognitive acts of distinction and selection, a unit (or a set of elements and processes) is specified as an entity distinct from its background (everything that the unit is not), thereby defining a phenomenal domain that is uniquely tied to the agent's interests and needs.

Of interest here is Ortega y Gasset's (1994, pp. 132–134) claim that every human life may have "a point of view on the universe." While Ortega's focus was on human individuality, the notion can be extended, by analogy, to other living organisms, each with its own situated mode of perceiving and engaging with reality. This suggests that all knowledge is inherently situated and framed from a particular perspective. Each form of knowledge represents a specific way in which an agent filters and selects elements to focus on, while disregarding or overlooking others (cf. Nucara, 2014, pp. 167–170). Thus, there is no unmediated or entirely neutral access to reality; reality is always interpreted from the perspective of a given system.

Moreover, another aspect related to the possibility of knowledge is the constant presence of some form of communication (whether verbal language, mathematical, gestural, or otherwise). Communication constitutes the "domain of existence," serving as the starting point and origin of possibilities. It is something that cannot be done without, not only for human beings but also for other life forms, as it is indispensable. The potential for knowledge appears to be contingent on the mastery (recognition and use) of a form of communication, necessary for learning and transmitting information. From this perspective, communication is more than just a cognitive tool; it embodies the very possibility of knowledge (ibid., p. 188).

In any case, even our most reliable form of knowledge—namely scientific knowledge—remains a human endeavor, with inherent possibilities and limitations, subject to species-specific constraints. As Cellucci (2008, p. 94) observed, "all our science of the world is human science," meaning it is influenced not only by the constitution of the world but also, inextricably, by our own physical constitution and ecological interactions. In this context, knowledge emerges from the relationship between us and reality. Although scientific knowledge is our most effective tool for understanding the world—enabling us to transcend certain limitations (such as computational complexity, levels of analysis, or spatial-temporal distance) using various epistemic tools like mathematical models, theories, and technologies—it is deeply rooted in human experience. It remains intertwined with our perceptions, intuitions, questions, thoughts, and representations. This does not imply that our best efforts and paths of knowledge are arbitrary. Instead, they are intersubjectively valid, plausible, and compatible with the findings of current research. In other words, the justification for supporting a hypothesis must outweigh the reasons against it. Scientific knowledge advances through the formulation of fruitful and often risky hypotheses, characterized not by certainty but by uncertainty, and subject to revision or refutation with the discovery of new evidence (ibid., pp. 99; 177–178).

Beyond scientific knowledge, the fact remains that all forms of human understanding are constrained by our physical constitution and modes of communication. Our cognitive processes, encompassing both thought and action, are intrinsically tied to the sensory and motor capabilities of our physical system. The way we conceptualize reality is partly shaped by our anatomical structure and physical organization. This reflection on the relationship between knowledge and our bodily, behavioral, and ecological constraints allows us to consider, through thought experiments, how even everyday concepts might differ if, for instance, we could perceive electromagnetic radiation at different wavelengths (ibid., p. 163), or if we relied more heavily on hearing or smell rather than sight (cf. Bianchi, 2021, 2024).

Reasoning about the perceptual, cognitive, and agentive limits and possibilities in relation to knowledge practices encourages an exploration of the various means, strategies, and potentialities of cognitive practices across anatomically and functionally diverse systems, particularly in the case of simple, aneural organisms such as bacteria and plants.

When analyzing the behavioral and cognitive capacities of other species, a key question arises: do these capacities truly constitute forms of "knowledge"? This issue has sparked considerable debate. For instance, some argued that the behaviors exhibited by bacteria should not be considered genuine knowledge of the environment (cf. Adams, 2018). Instead, these behaviors are mechanisms for storing information about past interactions with the environment. According to this view, the terms "know" or "knowledge" should be reserved for more complex activities that require mental representations and involve sophisticated discriminatory and propositional abilities. In the absence of such features—often associated with self-awareness, reflective-conceptual thinking, computational and abstraction skills, metacognitive capabilities—these behaviors cannot be regarded as true knowledge. Alternatively, some have suggested that what appears to be knowledge in these organisms may be better described as special forms of adaptation, with instructions encoded in their genome. For example, a shark's ability to navigate skillfully through the ocean does not imply that the animal possesses explicit or propositional information about the principles of hydrodynamics (cf. Bateson, 1976, p. 173). Instead, it is an evolved capability that enables effective interaction with its environment without involving sophisticated cognitive processes.

According to other authors (cf. McClintock, 1984), a goal for the future is to better understand the nature and extent of the "knowledge" a cell possesses about itself and how it utilizes this information when faced with various challenges, through a sophisticated system of self-organization and self-recognition (Trewavas, 2014, p. V). In these and similar discussions, the primary difference lies in how knowledge, or knowing, is conceptualized. Depending on the rigor and specificity of the definition, a greater or lesser number of organisms will be included in the analysis, and their distinct capabilities will be explored.

It is important to note, however, that many analytical difficulties likely stem from the use of a conceptually loaded term such as 'knowledge,' which is generally employed to explain abilities that are typically human, rather than, for instance, 'cognition' or 'behavior.'

Another critical factor in this analysis, which shapes the cognitive potential of different organisms, is the complexity of their environments. The type and degree of knowledge a system requires are heavily influenced by whether it lives in a simple or complex environment. In simple environments, little knowledge (few actions and interactions) is needed to survive; in complex environments, many aspects must be (re)known or re-cognized to ensure survival (cf. Cellucci, 2017). Of course, the nature of an environment and the interactional richness of a given niche are closely tied to the bodily and behavioral capacities of the organism. For example, our eco-cultural environment is particularly complex because our needs and interaction possibilities are extensive, leading to a wide range of cognitive opportunities. In contrast, in the case of distant life forms without nervous systems, such as prokaryotes, much of their capacity to respond effectively to environmental stimuli is specified in their genome rather than acquired through individual experience, due to their relatively short lifespans (although the lifespan is relative to the specific bio-ecological characteristics and requirements. What matters in this case is the rapid intergenerational change). Their responses, though highly efficient and effective, are less flexible compared to those of more complex species (cf. Cellucci, 2008, p. 309). The criteria for assessment will vary depending on the species under consideration (the basis for comparison) and the specific characteristics being analysed (cf. Bianchi, 2021).

Regarding plants, their living environment is quite complex and ever-changing. Due to their sessile nature, plants are required to engage in unavoidable interactions with a wide array of species, both within and outside the plant kingdom. Being rooted organisms, plants must coexist with other organisms, including those that pose antagonistic threats. Additionally, plant bodies are divided into aerial and underground parts, each exposed to different local environments with distinct interactions and chemical variations. For instance, in the above ground environment (aerial environment), various physical factors such as wind, rain, light, and biotic signals, including herbivory and volatile chemicals released by other species, can elicit diverse responses (cf. Trewavas, 2014, p. 176). The interplay of these factors, through the interactions between the organism and its surroundings, drives metabolic and growth changes in plants, which, in turn, influence and help shape their environment, contributing to the construction of ecological niches (cf. Laland et al., 2001). All living things, in some way, modify their environments (cf. Bianchi, 2022).

It has thus been proposed that plants and their environments should be viewed as an integrated system of processes, with fluid boundaries of analysis (cf. Trewavas, 2014, pp. 53–55). To develop a biologically and ecologically informed understanding of how plants' capacities to perceive and respond to their environment evolve and are modulated, it is essential to consider this intricate network of relationships and specify which elements or processes take precedence for a given behavior.

4.3.6 *The Logic of Knowledge Processes*

I would now like to steer the analysis toward a potentially fruitful argumentative direction, leaving room for future refinement, with this attempt serving merely as an initial starting point.

Different organisms exhibit various ways of engaging with their surroundings, interacting, and addressing survival challenges in their respective environments. Based on the information available to an organism, its constitution, and its ecological relationships, it acts accordingly, employing specific strategies. The fluidity of these dynamics, at both individual and interspecific levels, should be recognized. This recognition is why, despite varying means and degrees, many forms of life are thought to exhibit some capacity for "attuning to their environment," a kind of "practical knowledge" or "knowing how to act" that involves tracking and adjusting to relevant conditions in order to modulate their responses (cf. Bianchi, 2022).

One might wonder if there is another, sufficiently inclusive and valid way to describe the relational activity underlying the behavioral strategies of different species. An approach to describing these dynamics, which may reveal interesting connections, could involve exploring and expanding on the concept of "logic." This inquiry could then lead to further developing a particular version of it, specifically a form of "natural logic," as proposed by Cellucci (2008).

The importance of logic and reason, understood here as the ability to engage in behaviors deemed "rational" or "sensible," does not solely refer to human forms of rational thought. Rather, it encompasses the capacity to communicate effectively through various strategies and to behave successfully in interactions to achieve specific goals. This is especially relevant because the living environments of organisms, which include interactions with other species, are constantly changing in not entirely predictable ways. Thus, organisms must be able to select and adapt strategies to meet new challenges. In this context, the use of the term 'logic' is intended to guide this multilevel investigation in a specific direction, one that, as we will see, diverges from the notion of "formal logic."

Before delving into the possibilities and modalities of interspecific interactions among various organisms, particularly plants, it is helpful to briefly clarify what is generally understood by "logic" in research on human cognitive abilities. According to Giolito (2011), logic has traditionally focused on the study of human reasoning, with the primary goal of specifying the way in which one should reason. At a normative level, logic has aimed to identify and articulate models that ensure the correctness of arguments. However, the need to investigate how humans actually reason in everyday life has gradually emerged. As a result of this shift in focus, the objectives of logic have expanded to include the delineation of models that are not only formally correct but also realistic, accounting for the peculiarities of human cognition (ibid., p. 34). A similar approach could also be explored in other forms of life. In formal logic, as the name suggests, the emphasis is placed on the structural correctness of reasoning—its form—rather than on the specific content of the statements involved (ibid., p. 36). However, incorporating considerations of content and context has made it possible to analyze how everyday human reasoning actually occurs, reasoning that often does not rely on deduction, where conclusions logically follow from premises. Indeed, much of everyday communication is not based on formally rigorous reasoning and is instead filled with logical fallacies, such as affirming the consequent, denying the antecedent, false dilemmas, *argumentum ad hominem*, *tu quoque*, appeals to authority, and others. These logical "errors," which cannot be examined in detail here, are committed with notable regularity. They do not appear to be random but rather seem to be typical of our communicative practices. Elements that hold little value from a strictly formal logical perspective are often nonetheless persuasive. For instance, a statement or argument may be accepted simply because it is endorsed by someone held in high regard or trusted for personal reasons (ibid., pp. 47–52).

Therefore, based on what is empirically observed in various forms of communication, it appears that the principles of classical logic may not be entirely adequate for capturing the full scope of everyday reasoning and communication. Thin observation has led to the exploration of alternative approaches to the study of logic. Additionally, a reconsideration of classical logic has also extended to the field of more rigorous scientific reasoning, contributing to the development of so-called "nonclassical logics" (or alternative logics) aimed at addressing a range of different cases and issues. Even within the structured context of scientific inquiry, it is evident that scientific reasoning cannot be reduced solely to deduction. Instead, the inductive method plays a central role, prioritizing probability rather than certainty. Moreover, in probabilistic reasoning, a spectrum of possibilities exists, with "true" and "false" representing the two boundary cases, extremes on a continuum of varying degrees of probability. In probabilistic logic, this gradation of values between 0 and 1 reflects the inherent incompleteness of our knowledge regarding the occurrence of a given event. Another type of logic that addresses the vagueness characteristic of much human reasoning is "fuzzy logic." In this case, the gradation of truth values pertains to the intrinsic vagueness of the event or concept being analyzed. The term 'fuzzy' refers to the idea that a predicate can assume a value that is neither strictly true nor false, but rather "nuanced," existing somewhere in between. For example, the expression "to be young" can apply to an individual to varying and imprecise degrees, lacking a clear-cut boundary (ibid., pp. 56–61).

From what has been discussed thus far, it becomes evident that human reasoning is not fully logical in the sense of adhering strictly to models of classical or formal logic, whether in everyday life or in scientific contexts, both of which are forms of *human praxis*. These observations carry a twofold significance. The first relates to the role of reflection within the field of logic, distinguishing between a formal and an informal version (cf. Ippoliti & Cellucci, 2016). The latter, conceived from a broader perspective, serves as an attempt to analyze and minimize the errors that typically characterize human reasoning, while also aiming to understand the other dimensions of human communication and interaction (cf. Giolito, 2011, pp. 66–69). When logic is conceived in broader and more inclusive terms, it opens the door to a second, potentially valuable outcome: the extension of logical analysis to the behavioral strategies of various living systems. This could lead to the identification of shared elements among multiple species or even across all forms of life. Reasoning in these more inclusive terms shifts the focus from exploring the types of logic inherent to our species to examining the role that logic plays in nature across kingdoms. In this case, the reference is to a form of "natural" or "pragmatic" logic, approached without the pretension of demonstrative proof but rather as an orienting element. Of course, this serves as a means of guiding the analysis, which, especially in its initial stages, is always, to some extent, a gamble.

4.3.7 A Natural Logic?

Emphasizing the practical aspect of interactions, with its inherent limitations and possibilities for error, can help us focus on forms of nonrandom, goal-directed behavior present in nature, identifying where the best strategies have led to success (what has proven useful and continues to be maintained) and where it is certainly not possible to go (the paths that have proved to be unsuccessful).

Therefore, even in the case of human language, what emerges is primarily a form of logic that is "naturally" acquired and deeply pragmatic (cf. Cellucci, 2013). This does not imply, however, that human thought processes are "illogical." Rather, they are grounded in pragmatic and often tacitly intersubjective rules that ensure successful communication and effective interactions. For instance, in everyday conversations, unlike formal contexts that rely solely on the internal information provided or that needs to be proven, participants frequently and implicitly draw upon external or situational knowledge specific to the individuals involved. Indeed, natural conversations operate on cooperative principles shared by interlocutors. These principles include: providing information that is sufficiently relevant to the speakers' intentions (the quantitative aspect); ensuring that what is said is true or plausible and backed by adequate evidence (the qualitative aspect); maintaining coherence in the discourse (the relational aspect); and avoiding ambiguity (the aspect of modes of expression). The overarching objective is to facilitate understanding of the intended meaning, though, of course, there may be other communicative goals.[12] Thus, there would be a certain practical and contextual rationality underlying forms of communication and, in general, human behavior, as well as in those of other species. Each species exhibits its own "rationality" in line with its goals and capabilities. However, this is not the kind of rationality associated with Aristotelian formal rules or other forms of formal logic (cf. Grice, 1975; Walton, 2005; Smokrović, 2018).

[12] A case of pragmatic inference in a conversation occurs when one person asks another if a mutual friend has returned home and the other responds by saying that he saw the friend's car parked outside the house. Although the answer may initially seem incongruous with the question, it immediately prompts a series of deductions, leading to the likely conclusion that the friend has returned home since their car is there. In cases like this, the conversation relies on implicit rules, learned through practical experience, that make the response appropriate and meaningful to both speakers (cf. Catastini [*slides online*], 2015).

"Rational" behavior, as understood within the framework of natural logic, refers to the capacity to follow rules of cooperation in interactions with others, as appropriate to each context. Thus, behaving "rationally" may simply mean being successful in communication: making oneself understood, achieving a desired outcome, or reaching a goal through the knowledge of how to act and what to say. This conception can also be extended to the behavioral strategies of other organisms. In this regard, Cellucci (2008) argued that hypotheses for problem-solving are generated through inferences. It would be fascinating to explore whether or not, and in what way, plants and other aneural organisms utilize practical "inferences" in their behavior. Questioning the type of logic to which these inferences might belong leads to the suggestion that they could be rooted in a form of "natural" logic, conceivably an evolutionary outcome associated with the problem-solving strategies developed by various forms of life, primarily in relation to survival. Clearly, this natural logic cannot be equated with "artificial" logic, the latter being a human cultural construct and a discipline focused on the study of reasoning, the rules of argumentation, the causal links and the identification of types of relationships between different elements and categories. Logic, therefore, can be understood in two ways: as the academic study of reasoning and argumentation or as the set of adaptive processes common to various species, enabling them to select effective means to achieve their goals. This natural capacity helps organisms overcome obstacles and respond to environmental challenges, enhancing their overall adaptivity, a trait seemingly shared by all living beings, even those without a nervous system, albeit in diverse ways (ibid., pp. 261; 278–279).

From this perspective, the potential of plants and other organisms should be examined in relation to their species-specific capabilities and the distinct sets of information they possess in different contexts. For example, when threatened by herbivores or under pathogenic attack, plants can activate complex, multilevel defense strategies, producing substances that are repellent, toxic, or difficult for herbivores to digest. Alternatively, they may employ rescue strategies by attracting insects that are "natural enemies" of their predators, using specific volatile organic compounds as signals. These processes of chemical production and emission, as discussed in the section on plant signaling and communication, can also influence the behavior and bolster the defenses of nearby plants that detect these risk-related signals.

4.3.8 Distinction of Quantity and Difference

What is the relationship between the environment, forms of knowledge, and logical capacities?

From an ecological-relational perspective, "the environment" is also "the others," an aspect that requires various organisms to exhibit behavioral flexibility. This flexibility allows them to adapt, modify their strategies, and solve the different kind of problems. To perform these essential activities, organisms appear to require the capacity to discern quantities and differences. For instance, they need to distinguish between what is beneficial or harmful, and to what extent, as well as what is to be acquired versus what may be safely disregarded. Consequently, it has been hypothesized that these basic cognitive capacities are distributed, in various forms and degrees, across different systems throughout the tree of life (cf. Lyon et al., 2021).

Research in animal cognition has revealed that "mathematical" abilities, such as numerical and geometric understanding,[13] are not limited to humans (including both adults and preverbal infants) (cf. Gallistel & Gelman, 1992). These abilities are also found in many other animal species, which demonstrate the capacity to recognize space, size, shape, numerical order, and the recurrence of specific events (cf. Anobile et al., 2021). Many animals can classify relevant objects within their environment and understand the relationships between them based on abstract categories, such as spatial-geometric characteristics, independent of other perceptible features like color or size. This is evidenced by the abilities of creatures like rats, octopuses, and bees (cf. Monod, 2017, p. 147).

Specifically, numerical skills—at their simplest level, linked to the ability to distinguish differences in size and quantity between two or more sets of elements (e.g., determining if something is equal to, more than, or less than something else) (cf. Davis & Pérusse, 1988)—did not first emerge in humans with verbal language abilities. Instead, according to Rugani et al. (2015), these skills developed from an original nonverbal precursor system and are shared across a wide variety of phyla. Currently, most evidence on quantity distinction, numerical abilities, and numerical-spatial mapping has been documented primarily in mammals and birds (cf. Rugani et al., 2015) and, to a lesser extent, in fish, amphibians,[14] and invertebrates (exceptions include cephalopods, as well as more in-depth studies on the abilities of insects, such as honey bees) (cf. Howard et al., 2019; Rapp et al., 2020).

In general, these analyses of numerical and proto-numerical discrimination are often used to study foraging dynamics in the animal kingdom, one of the main ecological scenarios in which organisms often choose between food sources of the same type but differing in quantity, or vice versa. According to the theory of optimal foraging (cf. Krebs et al., 1974), organisms, when able, select the nutritional option that provides the highest energy gain, depending on the initial conditions and context (cf. Stancher et al., 2015).

Research on these variables should also be extended further into the plant world, particularly with regard to foraging dynamics (cf. Novoplansky, 2009), which

[13] As with knowledge and logic, the reference here is to a form of "natural mathematics." Indeed, mathematics, in a broad sense, can be interpreted as the study of quantity, relationships, and differences.

[14] A study conducted on amphibians, specifically on frogs (*Bombina orientalis*), investigated their ability to discriminate between different quantities of prey. The goal was to determine whether this ability reflects a sensitivity to number or physical variables. The findings revealed that frogs can distinguish between small quantities (e.g., 1 vs. 2, 2 vs. 3, but not 3 vs. 4) as well as larger numbers (e.g., 3 vs. 6, 4 vs. 8, but not 4 vs. 6). The experiments suggest that different cognitive systems are likely engaged for processing small versus large numerical quantities, with varying sensitivity to quantitative stimuli (cf. Stancher et al., 2015).

appear to involve quantity discrimination. This is observed in root behavior and in the ability of carnivorous plants to trap prey, an alternative foraging strategy. For instance, the well-known *Dionaea muscipula* (Venus flytrap) responds selectively to the number of times its "trigger" hairs—located within its snap-trap leaves—are stimulated within a defined temporal window, typically requiring two or more stimulations to initiate closure (cf. Darwin, 1875a, 1875b; Forterre et al., 2005).

This fascinating line of research is still in its early stages, and further studies are required to confirm whether plants possess numerical abilities related to the discrimination of size and quantity. The aim is to better understand the variability in the mechanisms, processes, and strategies plants use and to assess how sophisticated these abilities are compared to those found in animals (cf. Segundo-Ortin & Calvo, 2022). At present, it can only be hypothesized that plant sensitivity to both quantity and the quality of environmental elements may play a crucial role in enhancing their decision-making and adaptive capacities in ecologically significant contexts. Even if rudimentary, this sensitivity could help plants perform tasks necessary for survival (cf. Nieder, 2020).

4.3.9 Social Life and Game Theories

When distinguishing quantities, differences, and other relevant elements in the environment, one cannot avoid considering—within a relational context—the behavior of others to respond appropriately. For this reason, in recent years, connections have been made between the behavioral strategies of various organisms, including plants, and concepts developed within the research area of game theory.

Broadly speaking, game theory is a field of applied mathematics that examines models of strategic interactions between two or more agents, which make decisions based on the actions of others to maximize their own benefits or, from an operational perspective, achieve mutual advantages (cf. Nash, 1950). For "optimal behavior," the best individual strategy depends also on the strategies adopted by others in each situation; thus, it must take those strategies into account.

For instance, in social interactions where multiple agents often have conflicting interests, the most effective strategy is frequently based on the heuristic known as "tit for tat" (cf. Rapoport, 1980). This expression—often rendered as "an eye for an eye" or "blow for blow"—refers to a principle of reciprocal behavior: mirroring the actions of the other agent, whether cooperative or hostile, typically starting with a cooperative move (cf. Axelrod & Hamilton, 1981; Axelrod, 1984). Such interactions rely on the outcomes of previous encounters and are shaped by expectations of future ones (cf. Di Francesco et al., 2017, p. 158). Accordingly, behavior is regulated relationally, determined by the actions of others, rather than introspectively, in response to internal mental states.

In biological studies, comparisons, sometimes metaphorical, have been drawn between these and similar dynamics of mutual altruism and cooperation found in nature (cf. Dudley, 2015). For instance, this refers to scenarios in which an organism sacrifices part of its resources—and thus the immediate energy gain it could obtain—or incurs a certain risk in order to benefit another organism with which it shares no close genetic relationship, in the expectation that the favor will be reciprocated in the future (cf. Castiello, 2019, p. 142).

Given the potential of these studies, the dynamics of resource exchange and varying levels of cooperation/competition between plants have recently been analyzed through the lens of game theory (cf. Trewavas, 2017). For example, root interactions have been studied to predict the behavior of *Capsicum annuum* competing for soil resources (cf. Cabal et al., 2020). It has generally been observed that root behavior exhibits a strategy similar to "tit for tat." A comparable phenomenon occurs at the shoot level, where plants are sensitive to shading dynamics caused by neighboring plants (cf. Trewavas, 2014, p. 169). As explored by McNickle and Dybzinski (2013), the most adaptable traits of plants in interaction-rich environments can be modeled on the basis of the behavior of surrounding plants and other organisms. According to these authors, game theory may offer informative tools to characterize such interactions, depending on species density and the type and frequency of various strategies, leading to a more realistic understanding of plant ecology in future research (cf. Bianchi, 2022).

Applying these theoretical models to resource allocation dynamics in shoots and roots can predict growth investments or phenotypic changes that extend beyond what would be optimal in the absence of competition. Additionally, models that investigate defense and competition dynamics suggest that energy investment in these processes may not only help mitigate damage but could also play a role in redirecting potential threats—such as predators or antagonistic species—toward competing organisms. Furthermore, theoretical models examining interactions with mutualistic partners propose that optimal resource acquisition may, under certain conditions, give rise to competitive dynamics among plants competing for access to the same mutualistic partners.

In plant ecology, the application of game theory models is characterized by four main elements: the identification of the most effective strategy, the influence of one or more biotic factors that limit fitness, population density as a factor altering plant development conditions, and the impact of neighbors' behavioral strategies in shaping the development and environmental conditions of the plant (cf. McNickle & Dybzinski, 2013).

The primary resources that plants need to thrive are light, water, and soil nutrients (cf. Casper & Jackson, 1997). Securing the appropriate amount of these resources enables different organs and tissues to grow according to the prevailing conditions. For instance, in competition for light, the emphasis is on increasing stem height (cf. Falster & Westoby, 2003). This strategy is not aimed at achieving absolute height—since, under non-shaded and favorable conditions, even herbaceous plants can access light similarly to tall trees—but rather at attaining relative height. Even a slight height advantage over neighboring plants can confer a competitive edge while minimizing the energy costs associated with further growth (cf. Murphy & Dudley, 2007).

In the context of evolutionary studies, the relative effectiveness of strategies used by cooperative/competing agents plays a significant role in game theory. Similarly to how height development is prioritized, resource allocation can also lead to modifications in other above-ground organs or tissues, affecting leaf size and quantity, self-shading dynamics, and interactions with environmental factors. With regard to underground dynamics, theoretical models of active root allocation have provided insights into the development of root biomass (cf. Gersani et al., 2001; Dybzinski et al., 2011). These models account for scenarios both with and without competitors. When root density is minimal, roots still perform essential tasks such as anchoring and nutrient foraging while preventing the leaching of scarce resources (e.g. of minerals). However, when competition is anticipated, strategies often involve a greater allocation of resources to root growth to meet future needs. A relatively high root density appears to be advantageous in competitive environments (cf. McNickle & Dybzinski, 2013).

Additionally, mutualistic associations between plants and other taxa—such as between plants and nitrogen-fixing bacteria, with mycorrhizal fungi (cf. Hoeksema et al., 2010), or various pollinators (cf. Fenster et al., 2004)—can significantly influence plant development and well-being. These associations impact population growth, community structure, and ecosystem functioning (cf. Archetti et al., 2011). Given the heterogeneity of the species involved and the temporal and spatial variability of these interactions, it is impossible to rely on a single or fixed explanatory model (cf. Bianchi et al., 2025b). This highlights the need for continued ecological research alongside the development of models designed to analyze different types of dynamic interactions across a variety of contexts (cf. McNickle & Dybzinski, 2013).

Other examples of plant behaviors that have been explained through interactive dynamics include the responses of roots under drought conditions (cf. Falik et al., 2012). Information exchange between neighboring plants enables those not yet affected by water stress to prepare in advance. The communication of impending drought triggers an “energy-saving mode,” helping to conserve resources and prevent depletion. This collaboration could be interpreted as a strategic move by the plant experiencing stress: by slowing its own growth, it ensures that neighboring plants do not gain an excessive advantage from its weakened state (cf. Castiello, 2019, p. 143).

Another aspect to emphasize is that, given the multifactorial and multilevel complexity of relationships within an ecosystem, while some organisms may benefit more at the expense of others, it is often difficult to identify true “free riders.” This is because, in some way, the activities performed by an organism, although not beneficial in all circumstances or for all agents involved, usually provide some advantage in certain respects. For instance, the investment in and exchange of resources within a common mycorrhizal network form the basis of cooperative behaviors that enhance resource availability. Generally, the costs are outweighed by the long-term benefits (cf. Anten & Chen, 2021). Even if mycorrhizae do not provide phosphate or other supplementary elements, they still promote plant survival in various ways, such as by reducing disease susceptibility or increasing stress tolerance (cf. Trewavas, 2014, p. 174).

Plants are thus engaged in a dense web of interactions that also involve organisms from other kingdoms. Besides mechanisms of "reward," which offer mutual benefits to one or more of the species involved, there may be also "sanctioning" mechanisms (cf. Raguso & Kessler, 2017, p. 32; Castiello, 2019, p. 145). Naturally, just as species-specific traits have emerged from coevolutionary histories, there are limitations to the type, duration, intensity, and flexibility of plant interactions. These interactions are dependent on specific conditions and function to reinforce and memorize useful patterns for future encounters and challenges (cf. Lehman & Keller, 2006; Dudley, 2015).

From what has been discussed in this section, it appears that studying plant behavior while considering insights from research on practical, action-oriented forms of knowledge—in the context of natural logic and variable behavioral strategies in both intra- and interspecific interactions—can enrich and advance these approaches. Simultaneously, it can deepen our understanding of the dynamic and context-dependent variability of plant activities. Furthermore, research focusing on the behavioral and communication strategies of plants, both among themselves and with other organisms, holds significant potential for practical applications in agriculture, such as reducing the reliance on chemical fertilizers (cf. Castiello, 2019, p. 146).

4.4 Plants in Cognitive Sciences

Within the broader framework of cognitive science, the discussion now turns to a more detailed examination of the research programs most pertinent to understanding plant behavior. I investigate how ongoing studies of plant capacities can illuminate aspects of behavior and cognition that are often overlooked or insufficiently understood, revealing the diverse ways in which cognition may manifest across the natural world.

I first discuss some of the most comprehensive and relevant theoretical approaches to cognition, which have the potential to be extended to brainless organisms, such as plants. I then focus on the concept of biological cognition as it emerges from a interdisciplinary and pluralistic analysis. The goal is to integrate plant organisms into the cognitive science debate and to develop a systemic-relational, process-based, and extended conception of cognition that is particularly attuned to the bodily characteristics of organisms and their environmental interactions.

Specifically, I delve into the descriptive and explanatory power of the concepts of network and ramification to illustrate a range of relational dynamics, peculiar to non-neural and sessile organisms, such as plants. These capabilities rely on the ability to exchange and modulate information, varying the activity of modular bodies to solve survival problems. From an empirically informed and comparative perspective, this reflection may clarify why it is meaningful to refer to plant "cognition." Importantly, this does not dilute or trivialize the concept of cognition but instead highlights its broader applicability across the living world.

4.4.1 Historical Evolution of Approaches to Cognition

This section begins with a brief retracing of the salient stages in the historical evolution of the approaches adopted within the diverse field of cognitive sciences to identify the most promising perspectives for studying plant capacities.

Based on Varela's account, cognitive science can be framed as having developed through four main turning point: its origins in cybernetic studies, the establishment of classical computational cognitive science, the emergence of modularism and connectionism, and finally the expansion of embodied and enactive approaches to cognition (cf. Palmiero & Borsellino, 2018, pp. 71–72). This last phase is now integrated into the broader, multidisciplinary research program known as 4E cognition (embodied, embedded, enacted, and extended cognition), whose aim is to understand cognition in terms of its bodily, relational, and contextual dimensions. These stages, however, should be viewed as indicative rather than definitive. The development of cognitive science is not straightforward or linear; rather, it is shaped by multiple conceptual hybridizations and a spectrum of both radical and intermediate positions, often involving middle-ground solutions drawn from diverse theoretical perspectives (cf. Bianchi, 2021). One aspect, however, stands out: theoretical, scientific, and technological advancements have led to the gradual incorporation of several factors and processes that were once unknown or overlooked (cf. Bianchi et al., 2025a). This broader perspective has rendered the study of cognition more empirically grounded, especially because research has shifted from focusing predominantly on human cognitive capacities to investigating the behavioral dynamics, modes of interaction, communication, learning, and problem-solving abilities of other life forms.

Observing the limitations and shortcomings of previous or alternative approaches is entirely normal; after all, each perspective inevitably emphasizes certain aspects of cognition while neglecting others. Different methods yield distinct results at varying levels of analysis, often driven by differing experimental settings and technologies. Therefore, for a more comprehensive understanding of cognition, it seems appropriate to critically consider even those theoretical frameworks less suited to studying plants. Such an approach facilitates meaningful comparison, as many concepts may resurface, evolve, or be reintegrated over time, often under new names but retaining a core resemblance. Consequently, before exploring the most promising and inclusive frameworks for studying plant activities, it is helpful to outline the aims and content of these relevant previous approaches (cf. Bianchi, 2021).

The field of cybernetics has contributed significantly to the understanding of the concept of information and its organization in different types of systems. This was also made possible by Claude Shannon (1916–2001) and Warren Weaver (1894–1978) of Bell Telephone Laboratories. Their formulation of information theory laid a more rigorous foundation for addressing the challenges associated with information. Cognitive science, on the other hand, emerged during the second half of the twentieth century, between the 1950s and 1980s, in response to the decline of the behaviorist paradigm within the broader context of experimental psychology.

The conventional birth of this new approach to studying mental processes is often associated with two pivotal events in the 1950s: the 1956 Symposium on Information Theory held at the Massachusetts Institute of Technology, attended by key figures such as George A. Miller (1920–2012), Allen Newell (1927–1992), Herbert A. Simon (1916–2001), and Noam Chomsky, all of whom would become central to the "cognitive revolution;" and Chomsky's 1959 critical review of B.F. Skinner's book *Verbal Behavior*, which challenged the behaviorist perspective (cf. Perconti, 2017, p. 61). This burgeoning field was largely motivated by philosophical inquiries that emerged alongside the development of the first electronic computers in the 1940s, prompting questions about the parallels between these machines and human intelligence. This line of inquiry led to the development of computer programs capable of solving complex problems and, notably, to the creation by Newell and Simon of early problem-solving systems, designed to replicate aspects of human cognitive functioning. These efforts were instrumental in laying the groundwork for artificial intelligence (cf. Di Francesco et al., 2017, p. 124). The pioneering work of Alan M. Turing (1912–1954) further explored this parallel, proposing that simple physical devices could compute highly complex algorithms and, under certain conditions, even mimic human responses in verbal communication (cf. Di Francesco & Piredda, 2012, pp. 44–45). Within this context, philosophers such as Hilary Putnam (1926–2016), Ned Block, Daniel C. Dennett, and Jerry A. Fodor (1935–2017) played a crucial role in establishing philosophy as a key component of the so-called "cognitive hexagon." Their contributions helped articulate and critically examine the central thesis that there exists a distinct level of information processing, intermediate between the phenomenological domain of conscious experience and the neurophysiological substrate of the brain (cf. Dennett, 1993, p. 195; Di Francesco et al., 2017, p. 194).

An often-overlooked aspect of cognitive science is its inherently multidisciplinary nature. From its inception, it integrated various fields, though it later experienced a computationalist narrowing that made it less attuned to the biological characteristics of different organisms. Initially, however, cognitive science thrived on the collaboration between disciplines like psychology, early neuroscience, and emerging artificial intelligence. This cooperative spirit, evident in the works of figures like Norbert Wiener and Warren S. McCulloch (1898–1969), established the foundation for a comprehensive research agenda. Yet, by the 1980s, research became dominated by symbolic aspects of computational functionalism, often sidelining neurobiological considerations, even though the relevance of brain knowledge for understanding cognitive processes was not entirely dismissed.

The gap between artificial intelligence, psychology, and neuroscience began to narrow in the 1970s, driven by groundbreaking research such as the discovery of long-term potentiation mechanisms by Bliss and Lømo (1973), and Eric Kandel's (1976) studies on the cellular basis of memory and associative learning. This research, which elucidated presynaptic mechanisms involved in neurotransmitter release, contributed to explaining associative learning in more anatomically complex systems (cf. Hawkins & Kandel, 1984), thereby fostering greater integration among disciplines (cf. Di Francesco et al., 2017, pp. 128–130). Today, cognitive science continues to evolve as an interdisciplinary field, increasingly informed by insights from the biological, ecological, and social sciences, as will be further discussed.

In terms of content, even after considering the various contributions and transformations that have occurred within classical cognitive science, it remains possible to identify some of the core shared principles that underpinned the initial endeavor and helped shape subsequent approaches. One of the fundamental insights of early cognitive science was that the cognitive processes of both natural and artificial systems could be studied as operations performed on symbolic representations. The introduction of concepts such as computation, information processing, and representation aimed to offer a naturalized, scientifically tractable account of mental processes, processes that had previously seemed resistant to quantification and formal analysis (cf. Keijzer, 2021). According to the computational-representational theory of mind, cognition is understood as information processing, specifically, the execution of syntactic operations that encode information into mental representations with semantic content (cf. Fodor, 1987). Broadly speaking, this theory equates thinking with calculating, where internal representations serve as mechanisms for guiding behavior (cf. Di Francesco & Piredda, 2012, pp. 43–50). From this perspective, cognition is fundamentally a problem-solving activity (cf. Nucara, 2014, p. 44). Additionally, the modularist hypothesis posits that the mind is organized into distinct functional modules, each highly specialized for specific cognitive abilities. Functionalism, on the other hand, holds that cognitive processes can be realized on different types of "hardware," with the physical characteristics of the underlying substrate playing a secondary role in the study of those processes (cf. Perconti, 2017, p. 63).

Beginning in the 1980s, a gradual shift occurred away from the notion of a strict functional equivalence between computers and the human mind, though research continued within the framework of computational analysis. During this period, connectionism gained traction, offering models more consistent with the biological structure and functioning of nervous systems. One of the central aims of connectionism—whose explanatory potential may also be relevant to the study of signaling networks within plant systems—was to develop models that replicate key aspects of neural activity. These models proposed that computational processes occur in distributed, parallel, and subsymbolic ways. Rather than relying on the manipulation of discrete symbolic units (as in the "language of thought" hypothesis), cognition is understood as emerging from the synchronous interactions among multiple interconnected units within a network (cf. Bruni & Fugali, 2012).

By the early 1990s, a growing emphasis on biological realism in the analysis of cognitive processes marked the beginning of a new phase, one that underscored the importance of organisms' bodily characteristics in relation to their ecological and social contexts. This shift highlighted that cognition must be examined as an embodied phenomenon, shaped by the environment in which an organism behaves. In brief, the physical and contextual conditions of an organism are essential for understanding its cognitive abilities, not just from an evolutionary standpoint concerning the origin, selection, and transmission of behavioral traits, but also from an ecological-relational perspective. This perspective emphasizes the organism's capacity to maintain and adjust its behavioral capabilities in response to its surroundings (cf. Bianchi, 2022).

Traditionally, the work that laid the groundwork for addressing these needs and expansions in the field of cognition can be found in *The Embodied Mind: Cognitive Science and Human Experience* (1991) by philosophers Francisco Varela and Evan Thompson, along with psychologist Eleanor Rosch. In this text, one of the primary objectives was to move beyond the Cartesian framework, which has traditionally placed particular emphasis on the mind as the central source and expression of cognitive processes. Instead, the authors argued that cognitive capacities are fundamentally grounded in the organism's bodily organization, particularly in its sensorimotor interactions with the environment. They also emphasized the importance of self-organization in cognitive systems, the active role of perception-action dynamics, and the influence of environmental factors in shaping cognition. From this viewpoint, cognition is not a mere reflection of a pre-existing world where cognitive states align with the representations of an isolated system reacting to external stimuli. Rather, cognition is a co-emergent phenomenon, shaped by the history of interactions between the organism and its environment, arising from the dynamic relationship between an agent and its surroundings (cf. Varela, Thompson, & Rosch., 1991, p. 35; Bianchi, 2021, pp. 144–145).

This transformation in the understanding of cognitive systems can also be described through the so-called "vertical" and "horizontal" expansions of cognitive science. Vertical expansion refers to a deeper exploration of the functioning of nervous systems, which elevated neuroscience to a central role in the study of the mind. This gradually led to an increased recognition of the rest of the body's importance in cognitive functions. Horizontal expansion, on the other hand, involves considering the role of physical (and, in some cases, social) environments in which organisms operate. This shift made it clear that cognitive processes cannot be fully understood without taking environmental dynamics into account (cf. Caruana & Borghi, 2016; Zipoli Caiani, 2016; Di Francesco et al., 2017, p. 130; 183).

It is also worth noting that even as some embodied, enactive, and extended approaches attempt to integrate or reject key theoretical commitments of classical cognitive science, they vary significantly in their degree of radicalism. For instance, certain forms of connectionism offer an alternative to classical computationalism by rejecting rule-based models of inference, while still maintaining a commitment to representational states, albeit in a distributed and subsymbolic form. At the more radical end of the spectrum, some enactivist approaches and perspectives inspired by ecological psychology seek to replace or eliminate both representational and computational frameworks in explaining cognition (cf. Shapiro, 2021). As a result, depending on the framework adopted, different aspects of cognition are foregrounded or downplayed to varying degrees.

4.4.2 Bodily, Enactive, and Extended Turn of Cognition

At this point, it is helpful to briefly revisit some of the core assumptions of the approaches that either facilitated the transition toward, or directly laid the groundwork for the embodied, enactive, and extended turn in cognitive science. Collectively, these approaches have provided fertile ground for studying plant capacities. I then examine how certain theoretical frameworks can be adapted and further refined in the context of analyzing plant abilities.

Among the approaches that have challenged the representational and symbolic-computational view of classical cognitive science, paving the way for the embodied turn in cognition, are connectionism and ecological psychology (cf. Shapiro, 2021). These approaches have contributed to a deeper understanding of how nervous systems, and more broadly, whole organisms, function in dynamic relation to their environments.

As for connectionism, among the relevant contributions are Hopfield's (1982) work on neural networks and physical systems with emergent collective computational abilities, and the parallel distributed processing framework developed by Rumelhart and McClelland (1986). In contemporary cognitive science, neoconnectionist approaches seek to explain mental processes through models of artificial neural networks.[15] Achieving this requires an in-depth understanding of the brain's concrete functioning, particularly the activities of neurons. Although connectionist theories have primarily focused on neural mechanisms rather than other forms of bodily interaction with the environment, they have been instrumental in moving beyond the mind-as-computer analogy dominant in classical cognitive science. Unlike classical computational approaches, which emphasize the rule-based manipulation of symbolic representations, connectionism proposes that cognitive functions emerge from the collective activity of interconnected processing units. These systems rely on distributed, quantitative interactions that approximate certain features of real neural dynamics (cf. Calabretta, 2007).

To replicate cognitive dynamics, connectionist models employ networks of nodes that can be either excited or inhibited depending on the signals they receive. Each unit in the network—linked through connections loosely analogous to axons, dendrites, and synapses—simulates basic neuronal activity. A node becomes activated when the cumulative input exceeds its activation threshold. Connection strengths, represented by numerical values, or "weights," determine how strongly signals are transmitted between nodes, mimicking synaptic efficacy. Different inputs can modulate the network's nodes, resulting in distinct activation patterns across deeper layers (cf. Shapiro, 2021). Learning in artificial neural networks occurs through adjustments of these weights, guided by variations in input and feedback. The most commonly used learning method is backpropagation that computes

[15] The neo-connectionist research program shares important conceptual affinities with the cybernetic models developed in the 1940s, particularly in their emphasis on feedback, distributed control, and adaptive system behavior (cf. Di Francesco et al., 2017).

the error between the network's actual and desired output and modifies the weights to reduce this error (cf. Di Francesco et al., 2017, pp. 165–166). A particularly compelling aspect of neural network activity—and of biological studies based on signaling dynamics in networks—is that information processing occurs in a distributed, parallel, and subsymbolic manner. Rather than relying on the manipulation of symbolic representations, this activity emerges as a global property that is both functionally structured and flexible. It results from the simultaneous interactions of the various elements within the system, without the need for a central processing unit (cf. Palmiero & Borsellino, 2018, p. 76). The potential of this distributed processing model has begun to be explored beyond neural networks, extending to biological networks in diverse organisms, including those without nervous systems.

Reasoning from a different theoretical foundation and level of analysis, ecological psychology (cf. Gibson, 1966)—which began to emerge around the same time computationalism was gaining prominence in cognitive science—also played a pivotal role in articulating a relational, non-representational framework of cognition. This approach is grounded in the perceptual and agentive specificities of organisms in relation to their environments. This approach, which was developed primarily in the context of perception studies, introduced a new way of understanding perception and action, fundamentally rejecting most principles of representational models that mediate between perception and cognition, as characterized by traditional computational cognitive science (cf. Morabito, 2020, p. 81). In visual perception research, for instance, Gibson's ecological approach stands in contrast to the computational theory of David C. Marr (1945–1980), which considers "seeing" as the product of processing stimuli through multiple computational subsystems. Ecological perception, instead, posits that intermediate processing by systems between the body and environment is unnecessary. The information in the environment is complete and does not require further processing. Perception is thus direct and does not need inferential mediation; it is intimately connected to the various actions an organism can perform in its environment. Specifically, information in the form of invariants and structured energy patterns is directly available for detection by the perceptual systems of the organism. In *The Ecological Approach to Visual Perception* (1979), Gibson distinguishes between the physical world—which includes everything from atoms to galaxies—and the environment, defined as the subset of the world that is relevant to, and detectable through, an organism's perceptual interactions. For instance, electromagnetic fields and certain physical properties may provide ecologically meaningful information for sharks but remain imperceptible to other species. Ecological information thus has relational significance, defined by the interaction possibilities offered to different organisms. It specifies opportunities for action, resolving ambiguities not at a mentalistic or representational level but through action itself. Within this framework, cognition is tied to perception and action, enabled by an agent's active exploration of its surroundings (cf. Segundo-Ortin & Calvo, 2019; Segundo-Ortin, 2020). The concept of affordance is used to describe the opportunities for action that organisms can perceive directly within their environments. Depending on an organism's evolutionary history, ecological relationships, and physical perceptual-agentive capabilities, the elements of its environment present distinct possibilities for interaction. For example, a human with hands can "grab," "drink from," or "throw" a bottle, while a tree stump for a person may provide a place to sit but not to nest, as it might for a particular bird species (cf. Michaels & Carello, 1981, pp. 42–45). Thus, different species perceive different affordances. Affordance is not a substitute for representation but a relational concept, a construct used to indicate the constitutive relationality of the possibilities of perception-action. To perceive affordances, an agent needs to actively engage with its environment and selectively attend to specific informational patterns. The proper unit of analysis for perceptual-agentive-cognitive processes, therefore, is the organism-environment system, highlighting the structural and functional coupling between the two (cf. Turvey, 2018; Segundo-Ortin, 2020).

Overall, the ecological approach provides valuable tools for considering the bodily, agentive, and relational specificities of different organisms from an eco-evolutionary perspective. It emphasizes the active role of agents in perceptual-cognitive processes and lends itself to meaningful dialogue with other perspectives developed within the broader field of embodied cognitive science.

More recently, in the cognitive sciences, advancements in the biological and ecological sciences have facilitated the expansion of the broad and heterogeneous research program known as 4E Cognition (embodied, embedded, enacted, and extended). In this multidisciplinary context, while it is possible to identify a shared theoretical core, the use of various labels for cognition reflects a range of distinct and nonoverlapping perspectives (cf. Caruana & Borghi, 2013; Bianchi, 2021). Moreover, beyond those already mentioned, there are also other positions, as grounded cognition (cf. Barsalou, 2008). Depending on the specific theoretical stance, the focus in the study of cognition varies. It could emphasize the structural and functional bodily constitution of cognitive systems; the embeddedness of cognition in environmental contexts; the role of perception and sensorimotor characteristics in exploratory and cognitive interactions with the environment; the rejection of amodal symbolic cognition in favor of simulations; the contexts in which specific behaviors occur; the causal or constitutive significance of external elements beyond the physical boundaries of organisms in shaping and modulating cognitive processes; the potential for reducing cognitive load through the interaction of the agent with certain features of the physical and social environment (cf. Palmiero & Borsellino, 2018, pp. 82–92; 183–184).

In more detail, it is possible to outline the main points of some of these perspectives. For instance, the approaches associated with embodied cognition emphasize theoretical and methodological frameworks that highlight the dependence of cognitive processes on bodily characteristics, particularly sensorimotor functions, to better understand the constraints that bind cognitive systems to their environments (cf. Varela et al., 1991). This approach to cognition does not focus on an external world as a pre-given reality to be discovered, nor on the projection of a fully formed internal world, but rather on the reciprocal relationship and co-specification between the organism and its environment (cf. Bianchi & Castiello, 2023a).

Schematically, the core principles of embodiment can be summarized as follows: a more or less radical departure from the foundations of classical cognitive science and traditional dualisms, such as the mind-body separation, in favor of a conception of cognition that emerges from bodily structures and is embedded in the environment; the recognition of a more realistic link between perception and action, central to the sensorimotor paradigm; the development of a more distributed and multi-processual conception of cognition, which does not refer exclusively to a single centralized system for information processing (e.g., solely brain activity) (cf. Morabito, 2020, pp. 87–88).

However, this approach to cognition has not been without its critics. A common criticism of embodied cognition is that it represents a limited view of cognition, failing to provide a comprehensive alternative to computational cognitive science. As with other approaches within the embodied cognition framework, several complex philosophical debates have emerged. These include questions about the existence, nature, and necessity of mental representations; the meaning and explanation of the concept of "mind;" whether the body plays a causal or constitutive role in cognition; and whether human concepts, regardless of their type, possess an embodied/modal nature (cf. Barsalou & Wiemer-Hastings, 2005; Barsalou, 2008).[16]

In such cases, and in similar ones, explanations grounded in an embodied approach to cognition may be effective at accounting for certain levels of cognitive activity, but not necessarily for all aspects. Furthermore, the appropriateness of these explanations depends on the type of concept under consideration and the specific cognitive system being studied (e.g., animal, plant, or another form of life). This does not undermine the plausibility of the idea that cognition has a bodily, systemic, and relational basis. Regardless of the future trajectory of embodied cognition—whether it becomes more widely adopted, refined, or scaled back—a critical examination of its core assumptions, methods, and objectives can enrich philosophical analysis within the cognitive sciences (cf. Shapiro, 2021).

A closely related approach to embodied cognition that has gained significant attention in recent years is enactive cognition, or enactivism. The core principles of enactivism emphasize the importance of sensorimotor and interactional activity in the development and expression of cognitive capacities. Cognition, from this perspective, arises from the relational dynamics between organisms and their

[16] Regarding this last point, one of the proofs in favor of embodied cognition, according to some authors, would derive precisely from the detection of the embodiment of concepts in a neurological context, linked to the activation of sensorimotor areas associated with specific actions. For instance, when a person reads or hears a word like "kick," the motor areas related to kicking are activated. Additionally, electrical stimulation of these areas can influence speech comprehension (cf. Pulvermüller, 2005). However, other scholars argued that motor activity accompanying a word is merely associated with the concept and does not contribute to its meaning. According to Mahon and Caramazza (2008), the concept of "football" is not inherently modal; rather, its linguistic processing can trigger a series of processes that engage brain areas associated with the word's meaning. Essentially, thinking about a kick or hearing the word "kick" may lead to the mental imagery of leg movement, activating the motor cortex. Nevertheless, the word's meaning remains independent of this motor activation. Furthermore, even if some concepts, like "football," are indeed modal, this does not imply that all concepts, including more abstract ones like "democracy" and "morality", are embodied. The debate surrounding these issues remains ongoing (cf. Barsalou, 1999; Shapiro, 2021).

environments (cf. Di Paolo et al., 2017; Hutto & Myin, 2017). According to enactive theory, a cognitive system is not merely reactive to external stimuli but is instead an open, self-organizing, and adaptive system with a degree of autonomy, which enables the system to maintain its viability while flexibly responding to environmental disturbances, thereby defining the goal of its cognitive activities, exploring its surroundings, and adjusting its sensorimotor coupling as needed (cf. Thompson, 2007). Thus, cognitive structures emerge and self-organize through continuous interactions between the organism and its environment (cf. Palmiero & Borsellino, 2018, pp. 100–102; Castiello, 2021). As Shapiro (2021) noted, enactivism encompasses the general idea that cognition emerges from, or is primarily constituted by, sensorimotor activity. Within enactivism, it is possible to distinguish three main versions: autopoietic, sensorimotor, and radical (cf. Ward et al., 2017). The autopoietic version, originally developed by Varela et al. (1991), views cognition as rooted in the biological dynamics of living systems, underlying the recursive structural and functional coupling between organisms and their environments. Just as the life of organisms (even the simplest ones such as bacteria) is based on processes of energy flows and matter transformation that traverse bodies and environments, similarly, as a biological phenomenon, cognition is formed and maintained by sensorimotor processes at the intersection of brain-body activities and the world. From this perspective, cognitive processes represent an integrated version of biological processes (cf. Di Paolo et al., 2017). The sensorimotor version (cf. O'Regan & Noe, 2001; Noë, 2004), which has been influential in studies of intentionality and visual perception, reconceives perception as an active, exploratory engagement with the environment. It emphasizes the agentive-cognitive capacities of organisms, framing perception as dependent on patterns that link sensory states, movement possibilities, and contextual interactions. This perspective reduces or even eliminates the need for internal models framed in classical representational terms. Lastly, the radical version of enactivism (cf. Chemero, 2016; Hutto & Myin, 2013) focuses on eliminating representational explanations of cognition. It challenges the necessity of reflective cognitive states or mental content, advocating for embodied explanations and proposing naturalized accounts of intentionality. It also critiques certain concepts used in other enactive approaches, such as the notion of sense-making central to the autopoietic version (cf. Chemero, 2016). Despite different theoretical goals and methodologies, these three versions share the unifying idea that cognitive abilities emerge from, and are realized through, the sensorimotor activities of organisms (cf. Shapiro, 2021).

Continuing this summary analysis, one of the most initially controversial positions to emerge in recent cognitive science debates is the concept of extended cognition, which was formally introduced by Clark and Chalmers (1998). This hypothesis proposes that cognition can extend beyond the physical boundaries of the body into the surrounding environment, relying on external elements (which may cause, support, or even constitute, cognitive processes) to condition, manipulate, and enhance cognitive abilities. A central point of contention in this theory is the "parity principle," which holds that if a part of the environment performs a function equivalent to a cognitive process that would typically occur in the brain, then that part of the environment should be considered part of the cognitive system (cf. Clark, 2008).

Clark and Chalmers illustrated this idea with the example of Inga and Otto, a healthy woman and a man living with Alzheimer's disease. Inga recalls the location of the Museum of Modern Art (MoMA) in New York using her biological memory, whereas Otto relies on a notebook to access the same information. According to the parity principle, both are engaging in functionally equivalent cognitive processes, whether the information is stored internally or externally. The argument is that Otto's notebook effectively performs the same role in his cognitive process as Inga's biological memory (cf. Shapiro, 2021). Critics who argue that cognition begins and ends within the brain contend that it is difficult to empirically demonstrate that an external element is an indispensable part of a cognitive system. They argue that external aids, like Otto's notebook, should be considered supportive tools rather than integral components of cognition.

Some scholars have raised concerns about the extended mind thesis, arguing that extended systems may lack what they refer to as the "marks of cognition" (cf. Adams & Aizawa, 2010). In their view, cognitive states should involve intrinsic (rather than derived) content,[17] and should operate according to uniform processes within a clearly defined scientific domain. To illustrate their concerns, critics have offered analogies such as that of a person using an axe to chop wood: just as the axe does not become part of the person's body, Otto's notebook, though it aids his memory, should not be considered part of his cognitive system. According to this view, tools or external aids support cognition but remain outside the cognitive system itself. In contrast, proponents of the extended mind thesis argue that, while the physical composition of cognitive systems may differ, emphasizing their functional roles reveals a meaningful continuum between biological and extended systems (cf. Shapiro, 2021). Some have proposed a more moderate stance, advocating for a "reasonable conservatism" in which cognitive systems are understood as embedded within environments that provide tools to reduce cognitive load and facilitate cognitive processes, without treating these tools as components of cognition itself (cf. Rupert, 2004). Finally, according to Shapiro (2019), the defences of the parity principle, expanded computationalism and the constitution thesis, somewhat related to functionalism, do not fit well, in general, with the key assumptions of embodied cognition, suggesting that the involvement of bodies and their characteristics are not indispensable for the occurrence of cognition and that the type of "hardware" through which it can be expressed is indifferent (cf. Shapiro, 2021). The debate over the extended cognition approach, with all its complexities and implications, has not yet been resolved. Nonetheless, it has undeniably highlighted the limitations of viewing cognitive systems in isolation, detached from their context. This ongoing

[17] In debates within cognitive science, a distinction is often drawn between derived and non-derived content. Non-derived content is intrinsic, it has meaning independently, such as the thoughts within one's mind. Derived content, by contrast, relies on an external system of interpretation, as in the case of words written in a notebook, which only carry meaning for someone who understands the language. A common critique of extended cognition holds that external tools like Otto's notebook cannot be part of the cognitive system because they contain derived content, whereas genuine cognition, according to this view, requires non-derived content.

discussion invites us to reexamine what constitutes cognition and to consider, depending on the research question at hand, how (far) the boundaries of cognitive systems should be extended.

With regard to the systemic, relational, and process-oriented dimensions of cognition, particular emphasis has been placed on them in the development of dynamical systems approaches to cognition (cf. Thelen & Smith, 1994; Van Gelder & Port, 1995; Kelso, 1995; Beer, 2000). Contemporary dynamic cognitive science has modeled cognitive processes using differential equations derived from the broader framework of dynamical systems theory (cf. Shapiro, 2021). A well-known metaphor used to illustrate dynamical cognition is the Watt centrifugal governor,[18] which some authors have later employed to argue against including certain organisms, like aneural systems such as plants, within the scope of cognitive analysis. This exclusion is based on a comparison to more traditional cognitive approaches that require the presence of mental representations and have stricter criteria for cognition. However, when applying the mathematical framework of dynamical systems theory to cognitive processes, this approach may not be sufficient to capture all levels of cognition, especially more complex cognitive phenomena. Additionally, it has been noted that this method tends to be more descriptive than explanatory. For instance, simple behaviors such as finger oscillation and stepping in newborns are considered examples of basic, non-complex forms of behavior that do not involve higher-level cognition (cf. Shapiro, 2013). Studying more intricate phenomena may require integrating epistemic tools from other cognitive theories. On the other hand, authors like Chemero (2001) argued that the differential equations used in dynamical systems cognitive science can serve as genuine explanations of cognitive phenomena. These equations can predict organism behavior and account for behavioral counterfactuals. Conversely, Dietrich and Markman (2001) suggested that these equations merely describe phenomena without offering true explanations. The mathematical methods of dynamical systems cognitive science have also been linked to neo-connectionist models to elucidate the evolution of neural connection activations. This integration has the potential to extend further, incorporating body-environment interactions and aligning with the dynamic processes observed in connectionism (cf. Shapiro, 2021).

In these and other cases, the hybridization of different approaches reveals promising opportunities for studying complex phenomena across several levels of analysis, from microscopic physiological processes to macroscopic behavioral variations. These diverse theoretical positions, each emphasizing different factors, provide valuable tools for analyzing the cognitive capacities of various living systems, including plants. This allows for the possibility of investigating forms of cognition

[18] A centrifugal regulator is a device that controls the speed of a steam engine by automatically adjusting the opening of relief valves over time, without relying on computational or representational processes. However, there is no consensus on this interpretation (cf. Bechtel, 1998; Prinz & Barsalou, 2014). Some argued that there may be "representational" aspects involved in a broader sense, though not in the way typically described by computational theories. Whether this is the case depends largely on the specific conception of representation being used (cf. Shapiro, 2021).

in organisms without nervous systems, particularly concerning abilities such as memory, learning, communication, and general problem-solving for survival. Thus, the exploration of biological (bodily-functional) and ecological (relational-contextual) aspects of cognitive systems appears to support, with appropriate considerations and specifications, the inclusion of plants within this empirically informed, multidisciplinary, and multifactorial framework for the study of cognition (cf. Bianchi, 2024).

4.4.3 4E Cognition and Systemic-Relational Approach to Plant Cognition

What follows is an attempt to outline, within a flexible and interconnected framework, potential research directions for future studies concerning some of the previously discussed approaches to cognition, particularly as they relate to the characteristics and activities of plants.

To begin with, the extensive and diverse research program known as 4E Cognition has been instrumental (and continues to be) in moving beyond a strictly neurocentric analysis of cognitive phenomena. It has broadened the scope of inquiry to include a multiprocessual and multilevel context, expanding beyond the traditional intracranial boundaries that focus solely on brain activity. More broadly, considering the differences in bodily structure and organization contributes to transcending traditional zoocentric models, fostering the development of epistemic tools for studying the cognitive and behavioral activities of various living systems, both neural and non-neural.

In this regard, the foundational ideas of this approach can be traced back to the work of Maturana and colleagues, who posited that cognition can be understood as the self-regulated adaptive capacity of organisms, allowing them to respond to the variability of their living and interactive environments. Simultaneously, the organisms maintain and modify their own organization, the network of relations that define them as living entities (cf. Nucara, 2014, pp. 63–64). Living systems, from unicellular to complex multicellular organisms, can explore their surroundings and extract critical information for communication, reproduction, and survival. For instance, they can identify what is beneficial or harmful, nourishing or dangerous, distinguishing recurring environmental stimuli and engaging in behaviors such as searching, fleeing, or defending. The capacity to distinguish and actively gather information is intrinsic to life and, simultaneously, is governed by life's demands (cf. Morin, 2007, p. 50). In essence, from this standpoint, living systems are cognitive entities, and to live is to "know" (to possess some form of practical knowledge or cognition). Thus, cognitive processes can be understood based on the specific interactions that various organisms have within their ecological niches (cf. Nucara, 2014, pp. 113–117). This comprehensive, systemic-functional perspective on cognition also appears to be readily applicable to the behavioral patterns observed in plants, provided that the biological and ecological aspects of these organisms are adequately considered.

An approach that has more radically advanced the exploration of the connection between action possibilities in the environment and cognitive processes is ecological psychology. Heras-Escribano and Calvo (2019) raised the question of whether some ecological principles can be applied to the study of plant cognition. It is important to note that in Gibson's original approach (1979) this reference was not considered; in fact, it was explicitly dismissed. Gibson argued that plants, as a category encompassing inanimate, insensitive organisms lacking a nervous system and incapable of movement or behavior, should be treated similarly to inorganic materials studied in physics, chemistry, and geology. According to Gibson, organisms without sensory or muscular systems, such as plants, do not engage in a functional cycle like animals do, and thus their environment was not regarded as relevant for studying perception and behavior. However, these views are no longer tenable given the advancements in understanding the characteristics, capabilities, and living environments of plants. Consequently, scholars aligned with the ecological approach have proposed that certain principles of this framework could be extended to the plant world (cf. Carello et al., 2012; Heras-Escribano, 2016). In this context, just as elements in human and animal environments are considered ecologically informative and linked to their respective abilities and action possibilities, there is potential to explore the ecological significance of the interactions between plant organisms and their surroundings.

An example of applying the principles of ecological perception theory to the plant world is found in the work of Calvo, Raja, and Lee (2017), based on time-lapse observations of the nutation dynamics of climbing beans (*Phaseolus vulgaris L.*) as they move toward a support. The researchers examined these movements through the lens of ecological psychology, specifically General Tau Theory. Originally developed by Lee (1976), this theory of sensorimotor control offers a mathematical account of prospectively guided actions, identifying informational variables, such as tau (time to contact), that enable agents to anticipate future states (cf. Lee, 1998; Lee, 2009). In this context, the analysis explains how directly available ecological information guides target-directed movements, helping to disambiguate environmental properties and coordinate behavior. Unlike classical representational theories of the mind, which view perception as internal information processing, this approach emphasizes the dynamic, ecologically embedded relationship between agent and environment. Perception and action are thus grounded in biologically and ecologically meaningful interactions (cf. Carello et al., 2012). Even for plants, behavioral interactions can be explained through the concept of affordances. For instance, a climbing bean may be said to perceive the climbability of a support through its capacity to engage with it. Developed through an interdisciplinary dialogue among ecological psychology, biology, and philosophy, this research aims to contribute to a more unified understanding of diverse species as "agents," without relying solely on models developed for human or animal cognition and behavior (cf. Heras-Escribano & Calvo, 2019).

Alongside these developments, neoconnectionist theories, particularly contemporary developments in network neuroscience, emphasize the dynamic-processual and systemic-relational aspects of biological activity, including cognition, and offer explanatory frameworks that account for behavioral variability. The broader application of network theory to biological systems, especially in analyzing signaling networks, is particularly relevant for investigating the physiological and behavioral modulation capacities of non-neural organisms such as plants (cf. Bianchi, 2021). Plants, whose bodily systems can be analyzed as network structures, present an opportunity to study information exchange both in terms of their modular architecture and in the context of ecological interaction dynamics. Even in the absence of a brain—where the conventional focus is on neural networks and their integration with bodily functions—it is instead possible to interpret the signaling processes within the biological networks of plant bodies. This approach allows for understanding plants as complex systems of information networks. In this framework, the concept extends beyond viewing individual organs as independent networks. Instead, it conceives the entire plant as a multilevel, functionally organized, and relatively flexible system of interconnected signal networks (cf. Bianchi, 2022).

The idea of leveraging the explanatory power of the network concept to describe aspects of biological functioning and cognitive processes—where causal chains of information and conditioning result in diverse behavioral responses—belongs to a rich and varied tradition, which employs network-based models to explain multiple aspects of reality (cf. Trewavas, 2014). The concept of network offers a valuable framework for understanding the dynamic relationships among elements, enabling the identification of different types of connections and modes of information transmission (cf. Bianchi, 2022). By adopting a systemic-relational perspective, reality can be seen as a highly intricate web of processes rather than a linear chain of discrete entities (cf. Bateson, 1976, p. 315). This view allows for flexibility in defining the scope of study, spanning from the analysis of biological systems and ecological interactions to sociological phenomena, economic structures, political systems, human community organization, and web networks (cf. Trewavas, 2014, p. 233; Gazzaniga, 2019, p. 122). Human life can thus be interpreted as the product of a dense network of individual and social interactions shaped by different context. Every living system can be conceived as a "node of a very rich network of mutual information" (Rovelli, 2014, p. 223).

Narrowing the focus of the analysis to the realm of biological and ecological research, it is important to note that the brain, a key subject of neuroscientific research, is composed of and emerges from the complex chemical and electrical dynamics of countless nerve cells. Thus, even the brains themselves are networks. Consequently, intelligent behaviors must be understood as systemic emergent properties. After all, "a single neuron alone can do little," and certainly cannot generate complex activities like human thought; such abilities arise from the intricate relationships between brains (networks of cells and signals), bodies (the set of other cells and related signals), and the external world (all the factors and processes relevant to an organism at a specific moment) (cf. Longo & Montévil, 2014, p. 269).

Moreover, at the functional level of the nervous system, the same behavior or mental event can be generated by different neural activation patterns. This phenomenon, often associated with the concept of multiple realizability, suggests that if one neural pathway is disrupted, alternative pathways may assume its function (cf. Gazzaniga, 2019, p. 168). Like other complex systems, brain networks exhibit degeneracy, the capacity to perform the same function through different pathways. This feature contributes to the wide variability observed in neural connections across individuals (cf. Morabito, 2020, pp. 91–99). As development, experience, and individual learning unfold, neural connections and the overall architecture of the brain undergo reorganization. Specific groups of neurons become more densely interconnected while decoupling from others, giving rise to functional neural circuits capable of flexibly responding to environmental changes (Gazzaniga, 2019, pp. 121–124). Although these dense interconnections often support localized functional specializations, recent findings highlighting the dynamic interactions between cortical and subcortical structures have prompted a rethinking of Fodor's classical modularity theory. Rather than discrete, encapsulated modules, current models emphasize neural circuits involved in broader, more context-sensitive cognitive processes (cf. Morabito, 2020, pp. 118–119).

According to some authors (e.g., Godfrey-Smith, 2021, pp. 224–225), while focusing on network signaling is a crucial starting point for understanding the functioning of the nervous system, it alone is insufficient to fully analyse cognition, especially when exploring forms of human experience or self-awareness. The goal here is not merely to state, albeit truthfully, the simplistic idea that we are the result of a network of processes or that varying behaviors are possible because of dynamic signaling networks within our bodies. Instead, beyond acknowledging the appropriate level of analysis (e.g., molecular, physiological, or macroscopic-behavioral) and clearly defining the questions to be address, a systemic-processual approach to studying biological-cognitive systems, relying on the descriptive and explanatory potential of network models, can offer a more comprehensive understanding. This perspective enables the identification of elements and processes—whether common across species or species-specific—that operate at various levels, from micro to macro, within biological organizations. It can contribute to the observation of how a given capacity or an entire system develops through its interactions, offering a more comprehensive and realistic perspective (cf. Bianchi, 2022).

It has been observed that the study of networks does not concern only intracranial neuronal dynamics. For instance, intra- and intercellular signaling are also analyzed within the framework of signaling networks. Cells are highly complex systems where multiple types of proteins interact to form intricate organizations and perform a variety of functions (cf. Kauffman, 1995). Groups of interconnected proteins, referred to as "modules," often exhibit a hierarchical organization, reflecting nested functional structures within protein networks. These modules also display emergent properties, both transiently and at specific stages of cellular development (cf. Ravasz et al., 2002). Various types and intensities of interactions among cellular proteins have enabled researchers to characterize cellular networks by identifying densely interconnected components, featuring numerous connections and key connectors (hubs), as well as regions with fewer elements (cf. Trewavas, 2014, pp. 23–24). Such characterization contributes to the broader field of biological network analysis, which is still evolving and working toward the development of comprehensive models. Future research is expected to further investigate the functional interactions of genes and proteins, with the aim of elucidating their intra- and intercellular dynamics, especially through multi-omics and network biology approaches, increasingly supported by AI-based technologies.

One might wonder why network explanations are particularly well-suited for describing the dynamics and processes characteristic of life. To answer this question, one may consider their unique characteristics (cf. Bianchi, 2022). Network structures possess several properties, including redundancy, degeneracy, robustness, and a certain degree of elasticity or plasticity (cf. Civello, 2013). Redundancy enables similar elements to perform the same function, thereby enhancing the system's overall resilience. Degeneracy, by contrast, allows different elements to fulfill the same role, enabling the system to generate novel outputs under varying conditions. Robustness arises from a biological system's need to withstand environmental uncertainties, preserving the stability of specific components or functions across diverse contexts (cf. Waddington, 1977; Trewavas, 2005; Shinar & Feinberg, 2010). It is important to note, however, that no feature is universally robust; each new adaptation may introduce potential vulnerabilities into the system (cf. Gazzaniga, 2019, p. 148). Elasticity and plasticity, on the other hand, allow a system either to restore its prior state following disturbances or to flexibly adjust to new or changing environmental conditions. These attributes clearly illustrate why the structural and functional organization of network systems is particularly well-suited for capturing the relational processes and dynamic behaviors inherent to living systems. Indeed, living networks are functionally structured to preserve their core constitution, remaining relatively stable despite contextual perturbations. Yet, because they are also dynamic and flexible, they can adapt partially in response to environmental changes. Excessive rigidity would hinder the system's ability to respond quickly to new challenges, delaying adaptation and jeopardizing survival. Conversely, too much variability from internal fluctuations could disrupt the system's organization, preventing it from reliably performing necessary functions. Thus, networks are generally described as relatively stable yet not rigid systems, making them appropriate models for explaining many life processes, including certain cognitive dynamics (cf. Bianchi, 2021, pp. 17; 135–139; Bianchi, 2022).

In this context, some authors (cf. Trewavas, 2014, pp. 3; 196–202) have proposed that the processes underlying intelligence should be analyzed in terms of the functional organization of network systems. They argued that the essential requirement for intelligent behavior arises from the functioning of a complex, self-organizing networks capable of regulating their own behavior, actively acquiring resources, overcoming environmental challenges, and solving survival problems. In sum, they are systems that can manage and modulate the flow of information available to them, using their metabolism and repertoire of behaviors to maintain or improve their state. In line with this view, it seems plausible to discuss the concept of "plant intelligence." This refers to a form of distributed intelligence or adaptive behavior exhibited by organisms without nervous systems. At the cellular level, it emerges from interactions among various biological networks of chemical and electrical signaling. At the organismal level, it manifests as adaptive morphofunctional behavior that changes throughout the plant's modular life (cf. Bianchi, 2022).

In recent years, there have been increasingly frequent efforts to describe plant behavior using descriptive-explanatory models grounded in functional molecular networks and, more broadly, a systemic perspective on body organization and ecological interactions. Plant organisms serve as prime examples of self-organized systems, despite being highly dependent on numerous external factors. Within these complex networks—where no centralized control dictates growth or final morphology, as structure emerges from interactions among cells, tissues, and the local environment—plants are capable of actively synthesizing the organic compounds they require and responding to environmental challenges, even in the absence of a brain (cf. Trewavas, 2014, pp. 93, 105). Moreover, the analysis of these systems' functioning can be conducted at various levels of network organization: from cellular networks to the individual modules[19] that form different organs, the whole plant, or even larger systems involving the plant's interactions with mycorrhizal fungi. This analysis can extend to encompass entire forest ecosystems, where trees collectively exhibit emergent properties not observed in isolated plants (cf. Bianchi, 2021, 2022). These emergent characteristics arise from continuous metabolic and energetic interactions that constitute a circular exchange system involving the transfer of both resources and information. Within this integrated network, the system as a whole is capable of self-regulation and adaptation through multiple pathways (cf. Bateson, 1976, p. 526).

More specifically, interpretations have been proposed, such as those by Beata Oborny (2019), viewing plants as networks of interacting semiautonomous modules, in line with network theory, which is applied across various levels of biological organization, from internal structures, such as molecules, cells, and organs, to supraindividual systems like ecological networks that include many species.

To prevent terminological ambiguity, the notion of "module" as used here requires clarification. The term has a long-standing tradition, particularly in classical cognitive science, where it has often been interpreted in different ways. Here, modules are defined as nodes in a highly interconnected network. These nodes are specialized but not rigidly fixed, capable of performing one or more functions during plant development, such as reproduction or energy capture and storage. From this perspective, plants can be viewed as organisms composed of multiple modules, originating from totipotent cells located in meristems. Network theory provides a valuable framework for studying the interactions among these developmental modules, depicted as nodes whose connections form the network's links. These semiautonomous modules execute various functions based on resource acquisition and information exchange. The network evolves dynamically over time and space,

[19] See Lüttge (2012) for an analysis of the concepts of modularity and emergence in the study of biological systems, examined across different spatio-temporal levels.

shaped by the birth and death of ramets. The state of each node may be influenced by factors such as age or nutritional status, while the links can vary in their transmission capacity or become fragmented due to environmental stress. Overall, the network can integrate information across multiple scales, adapting its behavior based on environmental structure and characteristics, and interacting with other networks of living organisms (cf. Oborny, 2019; Bianchi, 2022).

What benefits come from interpreting a plant as a network of interacting modules? There are several practical, non-metaphorical advantages to this perspective. For instance, methodologies from network theory could facilitate the study of plant systems by enabling a comprehensive analysis of their overall structure, composed of multiple modules whose functions, such as sexual reproduction or resource acquisition, may vary across time and space (cf. Bianchi et al., 2025b). This approach could also contribute to understanding how signal transmission and resource flow between modules enable events in one module to influence others, either directly or indirectly. Additionally, representing plant bodies as networks provides a useful framework for comparing different species, genotypes within the same species, or various developmental stages of a single plant, taking into account factors such as age, nutrient availability, or exposure to environmental stressors. Finally, in the context of ecological studies, this approach can help differentiate plant species based on their functionally relevant behaviors within a community, providing insights into interaction dynamics in larger ecosystems over extended periods. This applies to everything from the study of modular organisms to the examination of multispecies ecological networks.

Regarding the question of whether networks have a more "solid" or "liquid" nature (in the sense of being more or less flexible), Oborny (2019) suggested that plants present an intermediate case between these extremes. They exhibit characteristics of solid systems, marked by a certain rigidity, as they are anchored in the ground, with functional modules and links generally fixed in space, and relatively rigid bodies. The loss of a connection between modules, typically irreversible, may consequently introduce a certain rigidity into the system's behavior. However, despite these limitations, plant systems remain quite flexible due to the functional variability of their semiautonomous modules and connections, which allow for local specializations. In brief, while much remains to be explored, especially in connection with ecological studies, plants, despite lacking brains, are capable of relatively flexible behaviors, both chemically and in terms of growth and function, due to their bodies interpretable as network systems (cf. Bianchi, 2022).

Additionally, in exploring biological networks, some authors (e.g., Trewavas, 2014, pp. 237–240) have proposed drawing comparisons between the functional characteristics of neural networks and other cellular interaction networks. This approach, using contemporary terminology, highlights parallels between the properties of "connectomes" and "interactomes" to identify shared processes that underpin the adaptability of these networked systems (cf. Bianchi, 2021, pp. 106–107; Bianchi, 2022). The connectome is the comprehensive map of neural connections in the nervous system. Even simple neural systems, composed of only a few neurons, can exchange and store information, correct errors, and, to some extent, solve problems. In the nervous system, information from the environment is processed by neural networks and changes through interactions among nerve cells. When a connectome learns, the flow of information is modified, and the number and strength of synaptic connections change, creating new pathways for directing information. Thus, animal brains exhibit slight structural changes during learning and experience, with these changes persisting as long as the synapses remain intact (cf. Trewavas, 2014, p. 225). As for the interactome, it refers to the complete set of molecular interactions within a cell, primarily involving protein interactions that vary across different developmental stages (cf. Tarassov et al., 2008; Yu et al., 2008). Proteins serve as the principal agents of cellular activity, translating information flows into concrete responses (cf. Bray, 1995). While traditionally focused on direct physical interactions, some authors have proposed a broader conception of the interactome, one that also includes indirect regulatory influences among genes and macromolecular components involved in the coordination of metabolic functions (cf. Trewavas, 2014, pp. 221–222; 237). Understanding the adaptability of a living system thus requires analyzing changes within these dynamic molecular networks, as such alterations can trigger cascading effects that ultimately manifest at the behavioral level (cf. Calvo et al., 2020). Several methods have been developed for analyzing interactomes, including measuring the number and length of paths between nodes, assessing connection strength (cf. Zhu et al., 2007), evaluating node influence, and identifying bottlenecks, nodes through which multiple paths converge. A defining feature of network organization is its capacity for modulation via positive and negative feedback loops (cf. Souza & Lüttge, 2015). In general, highly connected networks allow for more precise regulatory control, while less connected networks may provide greater adaptive flexibility (cf. Calvo et al., 2020). However, the overall plasticity of such systems is influenced by cell and tissue types and varies across species (cf. Souza et al., 2009).

Given these characteristics, cellular network systems based on the circulation of electrical and chemical signals have been the subject of comparative analyses. Both animal nervous systems—essentially cellular networks specialized in signal transmission—and other types of cellular systems exhibit adaptive behavior by altering the number and strength of interactions among components, thereby modulating information flow. Despite differences in specialization and complexity, connectomes and interactomes share core features: the ability to self-organize (within contextual constraints), to exchange information, to modulate signal transduction in response to stimuli, and to adapt their morpho-functional properties following perturbations (cf. Trewavas, 2014, pp. 226; 232).

Concerning plants, although they lack neurons, they possess an information-processing system made up of cells that engage in chemical and electrical signaling and have the capacity to organize information. These features underpin both the robustness and the plasticity of their behavior. A recent analogy, moving beyond the concept of the "root brain," compares the animal signaling system to the entire plant body, which is capable of organizing and transmitting environmental information at multiple levels, and of exhibiting adaptive and intelligent behavior suited to their communication and survival needs (ibid., p. 233; Bianchi, 2022, 2024).

The idea of a form implicitly also contains the history of that form.
(Hallé et al., 1978)

Another connection I propose for further investigation is between enactive approaches and the branching capacities of organisms with different sensorimotor organizations.

In the aerial environment, branching allows plants to expand spatially and increase access to light, while below ground it enables nutrient uptake, anchorage, obstacle avoidance, and communication with a range of other organisms.

Branching processes are not limited to plant roots. For example, mycorrhizal fungi display highly plastic and dynamic branching patterns, unconstrained by the rigid cell walls found in plants. Their hyphal networks extend rapidly through the soil, facilitating nutrient acquisition, environmental exploration, and the establishment of symbiotic interactions.

Branching processes, understood as a form of multidirectional and synchronous growth, occur in aneural, sessile organisms that lack locomotion but not growth or movement. This strategy can serve as a functional equivalent to the survival solutions employed by mobile organisms with nervous systems (cf. Bianchi, 2021, pp. 139–140; Bianchi, 2022).

What can the overall growth structure of a plant reveal about its behavior? Plants are modular organisms composed of repeated elements that possess a degree of autonomy and can specialize in different survival functions. Beyond the repetition of these modules, another aspect that might be considered in understanding plant adaptive behavior is the branching process, which is closely associated with patterns of meristematic differentiation occurring at various stages of development (cf. Nicolini, Chanson, & Bonne, 2001), the basis of the architectural constitution of plants (cf. Barthélémy & Caraglio, 2007).

By considering the branching capabilities of both shoots[20] and roots,[21] it is possible to gain a better understanding of the characteristics and behavioral strategies associated with a sessile mode of life, such as that of plants. For these organisms, the most effective—and perhaps the only available—strategy for overcoming obstacles or avoiding harmful conditions, in conjunction with their capacities for synthe-

[20] Various models of branching have been described, such as terminal and lateral branching, which depend on the position of the active meristem, whether apical or axillary. Branching may also be classified as monopodial or sympodial, based on whether the apical meristem exhibits indeterminate or determinate growth. Additionally, branching can be immediate or delayed, depending on whether axillary meristems are activated right away or remain dormant before initiating growth. Finally, branching may be continuous or rhythmic, depending on whether all axillary meristems along the stem develop into lateral axes, or whether lateral branches are grouped at intervals, resulting in an alternation of branched and unbranched nodes along the main axis (cf. Qu & Zhu, 2009).

[21] In root branching, the root apical meristem plays a central role in regulating the process. It inhibits the initiation of lateral root primordia in the subapical region, influences the spacing between successive branching points, and contributes to the orientation and eventual function of lateral roots. Theoretical branching patterns tend to become more regular in the absence of strong environmental stimuli, which can otherwise lead to the abortion of young lateral root primordia.

sis and chemical modulation, is the initiation of multidirectional and parallel growth in both aerial and root systems. The concept of ramification, even when examined independently, has found application across various fields (e.g., in graphic analysis) to describe growth geometries based on multiple directional changes, achieved through the development of new components that integrate into the existing structure. Thus, branching embodies the idea, both metaphorical and physical, of composite, parallel, and multidirectional growth (cf. Bianchi, 2022).

In recent years, biological processes involving branching dynamics and directed growth at the suborganismic level have been the subject of increasing study. This includes research on cells involved in blood vessel formation, which exhibit growth dynamics within their tissue environments, initiating the development of new vessels under appropriate conditions (cf. Bentley & Chakravartula, 2017). The various branching processes, therefore, warrant in-depth investigation not only in non-neural tissues but also in non-neural organisms (cf. Lyon et al., 2021). This research may also extend to the growth and behavioral processes of plants, which have a capacity to plastically modify their phenotype, particularly when examining their decision-making abilities for solving survival challenges (cf. Trewavas, 2005).

Plants explore both above and belowground environments through the directed growth of branches (primarily for light acquisition) and roots (to access water and minerals) (cf. Hodge et al., 2009). This growth occurs in a constant yet dynamic chemical-tactile interaction with the surrounding biotic and abiotic matter, enabling cooperative or competitive interactions with members of the same or different species. Naturally, not all branching is aimed at an immediate, specific goal; plants can grow into both aerial and underground spaces in a way that anticipates future interactions, occupying available areas (cf. Novoplansky, 2016). Additionally, branched structures play a critical role in providing stability and anchoring the plant to the ground. "Branching events" could thus serve as a key to understanding certain behavioral abilities in plants that emerge from and are refined by their multispecies ecological interactions. At the intersection of plant biology, behavior, and cognition research, a multilevel study of the dynamics that give rise to the characteristic architecture of various species, while acknowledging genetic constraints, could be instrumental in enhancing our understanding of plant behavioral flexibility across different relational contexts (cf. Bianchi, 2022).

To gain a more detailed understanding of the conditions, mechanisms, and processes involved in branching, it is essential to recognize that a plant's branching architecture results from a combination of multiple endogenous factors—such as genomic and ontogenetic development characteristics, meristem activity, and hormonal signaling (cf. Bessonov et al., 2008)—and exogenous factors, including the local microclimate (e.g., light, humidity, and temperature) and specific site conditions (such as resource availability, competition or cooperation for resources, and the presence of competing species, herbivores, and pathogens),[22] that impact individual

---

[22] E.g., rhizomes growing in resource-poor environments are typically thinner and exhibit longer internodes, enabling more rapid growth when favorable conditions arise (cf. Trewavas, 2003, p. 15).

starting conditions and needs (cf. Caraglio & Edelin, 1990). The branching patterns that emerge, with variations both between and within species, can be distinguished by the arrangement of different growth units (also known as metameres or modules). These patterns are the visible outcome of an evolving balance between growth processes, predominantly controlled by meristems, and a range of external stimuli related to both abiotic and biotic climate and environmental factors (cf. Qu & Zhu, 2009). Consequently, branching dynamics are neither rigidly predetermined nor entirely random; rather, they result from interactions among various signals that drive successive phenotypic adaptations. The structural form that a plant develops—largely determined by its branching pattern—has a significant impact on its ability to adapt to environmental conditions and ensure its survival (cf. Leduc et al., 2014).

A review of the scientific literature on plant architecture and branching models shows that, while early approaches were primarily descriptive, often relying on graphical reconstructions of overall structure, subsequent research has increasingly emphasized the relationship between form and function, understood as the result of complex and interacting developmental processes. In recent years, numerous modeling efforts have aimed to better understand branching processes and to visualize, and even predict, the development of plant systems in specific areas or directions. These approaches have combined direct descriptions, graphical models of branching systems, and 3D reconstructions of plant structures, often integrating analyses of the physiological states of meristems. Such elaborations enhance the understanding of branching patterns across different plant species, including individual intraspecific variation. They also shed light on how external stimuli and environmental conditions can influence a plant's structural and developmental plan—a form that is genetically constrained but not strictly determined—by mediating the plant's interactions with its surroundings (cf. Qu & Zhu, 2009).

To provide a concrete example of plant behavior—or, more specifically, of the activity of branched network structures, as a plant can be understood as a branching network—the dynamics of root foraging may be considered, with a focus on the processes of growth and structural modification of plant bodies over time (cf. Bianchi, 2022). Unlike animals, which are generally capable of locomotion from one place to another (from point "A" to "B") within their habitat, plants forage by absorbing resources from different modules positioned across the soil. This often involves the simultaneous initiation of new growth links to develop additional absorption units. As Oborny (2019) pointed out, the most interesting aspect is that plant foraging dynamics (strategies for acquiring water and nutrients) differ from those of animals, which can either stay and exploit a specific area or move to search for resources elsewhere. In contrast, plants engage in a dual strategy: they remain stationary to absorb and allocate resources at the site where a given module has developed, while simultaneously expanding by growing new units to explore and collect resources at additional locations. From this "stay-and-move" behavior, it is possible to infer not only the current behavior of a plant but also the directions of its past growth activities. The complexity of plant behavior arises from the simultaneous occurrence of numerous events. Moreover, in branching processes, epigenetic inheritance can enable developing structures to retain and transmit information about prior physiological or environmental states, thereby influencing the modulation of future growth responses. The shape of a plant thus reflects its "branching history," with many branches and roots continuing to grow and extend toward areas where conditions are more favorable or where resource availability is higher, often preventing other species from occupying that space. Conversely, roots that encountered unfavorable conditions may have ceased growing or died due to senescence. Therefore, the overall branched structure of a plant is not merely a "mirror image" of external reality but rather the result of a multitude of interactions, occurring for different reasons and at various sites throughout the plant's body, in continuous exchange with its environment (ibid.; Bianchi, 2022).

In summary, branching ability can be regarded as a valuable framework for explaining certain peculiarities of the intelligent behavior exhibited by rooted organisms, such as plants. Branching represents a process of phenotypic adaptability, shaped by genetic constraints, epigenetic modifications, hormonal signaling, and environmental influences, and this process is neither random nor entirely predetermined but occurs within variable ecological interactions in which the organism plays an active role. This is made possible by the plant's capacity to detect and flexibly respond to diverse signals. "Branching events," which involve spatially directed extensions, expand the plant's preexisting structure by forming new elements. These elements help the plant better align with its goals (e.g., accessing light and nutrients), establish chemical communication (e.g., interactions between the roots of the same or different species and with other organisms), and navigate obstacles (e.g., rocks in the soil) or avoid potential threats (e.g., pathogens or harmful substances). All these processes seem to suggest goal-directed behaviors rooted in the selective ability to distinguish between elements that are beneficial or detrimental to the plant's development. A more in-depth consideration of these aspects could contribute to a better understanding of plant behaviors, such as competition for light or foraging dynamics, which in turn involve the ability to retain useful information and make strategic growth decisions in certain directions over others. The ability to distinguish between beneficial and harmful elements may be considered a basic trait of cognition, not necessarily dependent on metacognitive or reflective capacities. Instead, it can reflect a kind of "know-how" rather than a "knowing of knowing" (cf. Marraffa, 2011, pp. 200–202). This distinction deserves further exploration, particularly in relation to the study of practical knowledge and action-oriented sensorimotor cognition (cf. Bianchi, 2022).

Lastly, and no less importantly, future research on plant cognitive behavior warrants further exploration, particularly regarding insights offered by analyses developed within the framework of extended cognition (cf. Parise & Marder, 2024). This type of reflection, similarly to the previous cases, offers a means of testing, integrating, and potentially enriching the ongoing debate on the nature and extension of cognition. Plants, indeed, appear capable of extending their cognition through bodily growth. As they develop new tissues or organs, they also expand their abilities. This process includes both active interaction with the environment through the production of biochemical substances and the behavior of avoiding obstacles or unfavorable conditions (cf. Marder & Geremia Parise, 2024). More specifically, a key question in the study of plants' dynamic relational behaviors concerns how organisms sharing the same niche (the biotic environment surrounding the plant) influence plants' behavioral abilities. These influences can at times facilitate, enhance, or even entirely support various interactions that help achieve survival goals.

An interesting case study comes from the hypothesis proposed by Parise et al. (2020), which suggests that plants may extend their cognitive abilities beyond their physical structures through root-level associations with mycorrhizal fungi and certain genera of soil bacteria (e.g., nitrogen-fixing bacteria). These associations, which span not only across different species but also across different biological kingdoms, can improve plants' perceptual, memory, learning, communication, and stress-response abilities (cf. Bianchi, 2022). For instance, through the release of root exudates that accumulate between exploring roots and soil obstacles, plants can inhibit root growth in the direction of these exudates, effectively perceiving and bypassing obstacles (cf. Falik et al., 2005). By emitting root exudates, plants actively modify soil chemistry and influence microbiota activity in the rhizosphere. This interaction creates a form of "soil memory," where the microbial community composition affects the plant's fitness, impacting chemical activity and plant growth through a complex web of interactions.

Relationships of this kind, involving active forms of environmental modification, have previously been studied within the framework of Niche Construction Theory (cf. Odling-Smee et al., 1996, 2003). An example is the connection between spiders and their webs, that are actively manipulated to extend some of their behavioral abilities beyond the physical boundaries of their bodies. Specifically, Japyassu and Laland (2017) observed that artificially altering the tension of the web threads can, in turn, modify the spider's foraging behavior, causing it to pay attention to areas of the web that were previously ignored. Similarly, Parise et al. (2020) suggested that plants may extend certain capabilities into their environment by modifying soil chemistry and forming associations with mycorrhizal fungi and soil microorganisms. These transformations enable plants to improve their adaptive performance in critical survival functions. Through mycorrhizal networks, plants can perceive environmental elements (such as water, nutrients, and stressors) beyond their immediate growth site, making more accurate and timely growth decisions before their roots reach these resources. Plants, therefore, appear to significantly extend their range of perception, resource acquisition, and signal reception by connecting to the dense mycorrhizal networks in the soil. Furthermore, research by Huang et al. (2019) demonstrated that *A. thaliana* utilizes specific pathways, as do other plant species, to synthesize substances that shape and cultivate particular microbial communities within its rhizosphere. These communities, in turn, establish "chemical soil memories" outside the plant body, influencing future plant development based on their composition (see the concept of "soil legacy," cf. Hannula et al., 2021).

In general, these cases are relevant because they involve processes where information commonly associated with cognitive activities is transmitted. These activities are based on mechanisms such as environmental information storage processes, learning, communication, self–non-self recognition, cooperation and competition.

To avoid the risk of anthropomorphizing these plant behaviors, on the other hand, it does not seem appropriate to describe them merely as mechanical or stereotyped reactions. Instead, they should be understood as modulated responses that adapt to specific needs within changing interaction contexts (cf. Bianchi, 2022). Therefore, to more accurately and realistically explain certain plant behaviors—such as trophic dynamics, the perception of neighboring species, and resilience to environmental stressors—it is necessary to consider the role played by mycorrhizal fungi, soil microbiota, and other organisms (cf. Simard, 2018). Indeed, their presence, through symbiotic relationships, is capable of influencing plant activities, enhancing flexibility and expanding the potential for other ecological interactions(cf. Bianchi, 2022).

4.4.4 A Trivialization of the Cognitive?

The processes and abilities described—alongside those of other aneural organisms such as fungi, unicellular eukaryotes, and prokaryotes—may not, according to some perspectives, represent genuine instances of "cognition" (even in its most basic form). Instead, they might be better characterized as mere biological information-processing activities, like the mechanisms that govern the operation of various organs and physiological systems. For instance, it has been suggested that equating cognition with the capacity of a single-celled organism to generate responses to environmental stimuli risks diluting the concept of "cognition" to the point of rendering it meaningless. According to this perspective, cognition is conceived as emerging from biological processes enabled by a nervous system, and therefore as a trait that evolved only in certain animals, specifically those capable of offline behavior, such as forming, storing, and using internal mental representations (cf. LeDoux, 2020, pp. 261–262). Similarly, a distinction has been proposed between genuinely cognitive or mental processes and more basic biological functions, such as digestion, which also involves information processing and problem-solving activities. It has been noted that even digestion, despite the involvement of millions of neurons in the gut, is not typically classified as a cognitive process (cf. Di Francesco & Piredda, 2012, p. 37; pp. 247–249). The authors posed a critical question: How integrated and interconnected must processes be to qualify as cognitive? They observe that contemporary approaches to cognition have led to "vertical" expansions, delving deeper into bodily structures and functions, and "horizontal" expansions, exploring environmental influences. However, these approaches often lack clear criteria for distinguishing cognition from mere biological activity, risking a trivialization of the concepts of mind and cognition. The authors' concern lay in defining the boundary between the mind and the external world, trying to avoid an unrestrained diffusion of the concept of the mind that would render it unrecognizable or devoid of meaning. This is why they refer to the concept of the "personal mind" (cf. Di Francesco & Piredda, 2012). Such inquiries are well-founded, aiming to prevent the notions of mind and cognition from becoming overly broad or ambiguous. These reflections, and those in line with them, have made valuable and critical contributions to discussions of cognition and conceptions of mind. On the other hand, in the context of this analysis, these approaches serve as a starting point for the analysis of notions of cognition and mind. These definitions are often developed to address a narrow range of cognitive or mental capacities, frequently using human cognition or the abilities of organisms with nervous systems as a term of comparison. The central insight to be drawn from these reflections is the importance of specifying, in each case, what is meant by "cognition" and in what sense it is being considered (cf. Schlicht, 2018).

As emphasized in Chap. 2, this proposal fits within the broader framework of the biogenic approach to cognition (cf. Lyon, 2006; Lyon et al., 2021). This perspective asserts that the study of cognitive behavior should begin with identifying and understanding fundamental bio-ecological principles, striving to develop the most inclusive approach within the "cognitive sciences of life" (cf. Keijzer, 2021). An empirically grounded approach, both biologically and ecologically informed, aligns well with a post-cognitivist perspective. It operates within a relational framework for analyzing natural cognition, extending beyond the exclusive focus on the possibilities and capacities associated with animal nervous systems. In this view, various cognitive specializations occur in relation to the real-world contexts in which organisms, including aneural ones, live. These specializations and solutions emerge from species-specific and individual-specific abilities, enabling communication and survival. This is achieved through the integration of multiple levels of information and the coordination of diverse behavioral responses to address the most urgent needs in specific situations. This conception, however, does not preclude the possibility of distinguishing between different cases and levels of analysis and complexity. Consequently, it is possible to rely on the approach that appears most appropriate for examining a specific aspect.

4.4.5 Bio-Ecological Cognition: A Processual Perspective

From this perspective, even studying the cognitive behaviors of plants, given their different biological and ecological organization, can contribute to expanding or deepening our understanding of biological cognition, examined as a multifactorial, processual, and contextual phenomenon throughout the evolutionary tree.

Efforts to integrate cognitive studies into the broader life sciences have led to proposals for new terms that aim to capture a more unified concept of cognition, grounded in the characteristics of living systems. One such proposal is Fred Keijzer's (2021) development of the concept of "cobolism." By cobolism, Keijzer refers to the ways in which each living system encompasses structures, processes, and external events that maintain the fundamental metabolic processes constituting the core of the system. Metabolism, broadly defined, encompasses a series of complex and cyclic biochemical processes forming the core of all living systems. These metabolic processes must remain active, as living systems operate far from thermodynamic equilibrium. As dissipative structures, they require a continuous energy supply to sustain chemical reactions, enable the uptake of useful substances, and facilitate the expulsion of waste products, all while maintaining the physical conditions that preserve the system's integrity. Therefore, Keijzer emphasizes that the continuity of an organism's metabolic organization depends on contextual features that provide and regulate the conditions making these same cyclic metabolic processes possible: that is, its cobolism, a set of structures and processes that maintain the conditions for the continuity of metabolic organization. This integrated approach, which transcends the boundaries of individual systems and focuses on the conditions enabling the maintenance of metabolic functions, aligns with biologically informed positions on cognition (cf. Godfrey-Smith, 2002; Lyon, 2015; Keijzer, 2017). It also resonates with theoretical frameworks that explore the mind-life continuity within autopoietic systems and enactive approaches to cognition (cf. Maturana & Varela, 1980; Thompson, 2007; Di Paolo et al., 2017). However, this approach seeks to sidestep some conceptual challenges of these original frameworks, such as the difficulty in establishing a fundamental link between mind (or cognition) and life. By emphasizing the empirical side, this proposal suggests that a deeper understanding of biological processes and the nature of life can illuminate certain cognitive phenomena. According to Keijzer (2021), the aim is not to define or describe cognition itself but to specify the material constraints that shape it. This approach does not advocate for biological reductionism or dismiss the need for a theoretical foundation. Instead, it emphasizes that even when studying complex cognitive processes, one must first have a clear understanding of the internal and external conditions, structures, and processes that enable such capacities across different living systems. The concept of cobolism is broad, applicable to any living cognitive organization, from humans to bacteria, including intermediate cases such as plants (cf. Keijzer, 2021). This framework is not intended as a prescriptive model, nor as an exhaustive or conclusive solution, but rather as an invitation to consider the possibility of orienting research along these lines.

Another recent proposal for developing a pluralistic biological approach to cognition—one aimed at identifying which processes to observe and how to investigate them in practice—has been advanced by Huebner and Schulkin (2022). They emphasized the abilities of different organisms to adapt to new challenges and overcome survival problems, underscoring the importance of allostatic regulation. This refers to the capacity of living systems to maintain relative physiological stability through change, often employing anticipatory regulatory mechanisms.

A biological and embodied approach to cognition should embrace a form of pluralism that acknowledges how various challenges and opportunities shape the experiences, cognition, and behaviors of living beings from a relational and processual perspective. Organisms exist in dynamic environments and need to regulate their behavior to manage a variety of challenges, employing a wide range of strategies that involve information processing, including neural and chemical signaling systems. One of the primary functions of cognitive systems is to minimize environmental perturbations, addressing a wide range of biological and social needs. To preserve their vital integrity, living organisms must locate and remember where to access essential resources, avoid predators and other dangers, engage effectively with other organisms, and keep track of relationships, sources of stress, and damage based on past experiences. Thus, a comprehensive analysis of cognition should explain how organisms adapt their behaviors to current and future needs, manage bodily changes, and navigate ecological and social relationships. Managing environmental challenges is inherently complex and involves various systems of behavioral control and anticipation. These systems must be responsive to fluctuations in both internal processes and ecological factors, as well as to the interactions between these constraints (cf. Bechtel, 2009a, 2009b).

Although Huebner and Schulkin (2022) focused on the capacities and processes of organisms with nervous systems, their analysis could also be useful for studying non-neural systems. Their approach emphasizes examining neural systems through a systemic, multiprocessual perspective, which considers the regularities modulated by chemical and hormonal systems. This perspective goes beyond simplistic forms of functionalism that attempt to establish direct, exclusive causal links between neuronal activity and cognitive states. Instead, it aims to account for the full range of physiological and ecological constraints that, through chains of effects, influence and differentiate the behavioral responses of various embodied and situated biological systems. Obviously, an idea of process in this context, does not imply an idea of constant change without cyclicality, repetition, organization and maintenance of systemic organization.

Neural activity, which plays a critical role in regulating behavior in response to environmental uncertainty, is influenced by a network of chemical signaling systems, including neurotransmitters, endocrine and neuroendocrine systems, that convey physiological demands from distributed body systems (ibid., pp. 1–6). These widely distributed neural and bodily networks integrate to function efficiently in response to ecological demands. Although chemical signals are often produced locally within specific neural networks, they can also originate from distant regions in the brain or remote areas of the body, such as the gastrointestinal tract, heart, and kidneys, without necessitating centralized control from subcortical systems (cf. Schulkin & Sterling, 2019). To sustain the organism, these networks engage in dynamic exchanges linking the brain, body, and external environment. The brain functions as a heterarchical system, relying on networks of interacting processes that operate in parallel across multiple timescales. These networks draw on diverse sources of information to support the organism's survival in changing environments. As a result, information processing in the brain should not be understood as rigidly modular, with discrete components corresponding to psychological states or functions such as attention, language, or social cognition. Instead, it should be understood as the outcome of complex, distributed signaling systems operating across multiple levels, in which information flow is dynamically regulated according to context. This perspective implies that neural, behavioral, and environmental resources are reused to support a range of cognitive abilities that emerge as needed across different situations. In sum, biological cognition leverages heterarchic networks of processes that constantly adapt to maintain predictive consistency across various cognitive domains of life. In this adaptive, predictive framework, chemical and neural networks respond to changes in the variability of bio-ecological needs, such as the demand for vital substances like oxygen and glucose or, at the behavioral level, the necessity for environmental comfort, such as shelter and protection. Discrepancies between predicted and actual states trigger reactions, prompting animals to engage in behaviors that safeguard their integrity (cf. Huebner & Schulkin, 2022, pp. 8–11).

In their approach to biological cognition, the authors acknowledged that providing a precise and comprehensive account of the biological processes underlying adaptive behavior remains challenging. They concentrated specifically on the systemic and integrated study of body signaling networks and their variability in relation to different ecological contexts. Even though their effort aimed to offer a systemic perspective, incorporating networks of information and processes beyond the nervous system and highlighting environmental interactions that guide behavior, their analysis remained centered on animals, organisms equipped with nervous systems. The question that arises is whether a similar approach can be developed for studying the capacities of aneural systems, such as plants. This would involve considering the distinctive characteristics of plant body structures and organizations, particularly in relation to biological signaling networks within non-neural tissues (cf. Baluška & Levin, 2016), and considering the behavioral possibilities shaped by their interactions with other forms of life. Attention should be given to how plants' specific needs, bodily constraints, and behavioral strategies emerge from different levels and timescales of information organization and integration (cf. Bianchi et al., 2025b).

Future research will shed light on the most promising directions and determine which of the theoretical frameworks discussed here—or, more likely, a combination of them—may best facilitate the inclusion of plants in cognitive analysis from a relational and processual perspective.

4.5 Plants in the Philosophy of Mind

In this final section, within the broader context of the philosophy of mind, I address conceptions of "mind" for two main reasons: first, to identify, through a brief historical overview of the concept, the most inclusive approaches for studying the mental capacities of different living systems, particularly by moving away from a disembodied notion of mind; and second, to explore whether any conception of the mind can be enriched through a focused analysis of the capacities of plants. Throughout this work, as outlined in previous sections and chapters, I have chosen to use terms such as behavior, cognition, or cognitive abilities rather than "mind" or "mental capacities." This choice reflects the typical association of the concept of mind with specific characteristics and abilities attributed to humans and certain animals, as the concept was originally developed to address human faculties. Nevertheless, both informal conversations and academic-scientific discussions increasingly include references not only to the human mind but also to the animal mind, and, occasionally, to the mental capacities of unicellular organisms like prokaryotes. With appropriate caution and a critical perspective—similarly to the approach taken in examining other behaviors and cognitive traits—it is worth considering whether it is useful to discuss plant abilities in terms of the presence of a mind or a proto-mind. This would involve a radically reformed conception, diverging significantly from traditional views and avoiding unproductive or misguided directions.

Another reason to engage in this type of reflection is that the concept of the mind remains widely used and is intricately linked to how we broadly understand opportunities for knowledge, orientation, interaction, communication, and the resolution of various survival challenges (cf. Cellucci, 2017). More importantly, in an empirically informed context, reconsidering the concept of the mind could provide insights into what the mind is, its role in nature, and the degrees and interspecific variations it may exhibit. This could lead to an "ecological understanding of the mind," situating the study of mental processes within the complex network of bodily processes from which they arise, or, more accurately, that constitute them, and within the broader set of processes with which they interact (cf. Bianchi, 2022, 2024).

Therefore, even though it may not be necessary to frame the analysis of plant abilities within the context of mind studies, the theoretical and methodological reconsideration of mental capacities—and the potential evolution of the concept of mind—is not aimed at proving that plants have a mind, but rather at understanding how the study of the mind, especially from an evolutionary-comparative perspective, could benefit from this broadening of horizons. The true value lies not in reaching a definitive conclusion but in the ongoing exploration and analysis of these questions: determining what to add, modify, integrate, or exclude (cf. Bianchi, 2021).

The time is ripe for a serious discussion of plant behavior and cognition. Even if it is not entirely accurate to refer to a plant "mind," reflecting on this topic could illuminate how we conceptualize the mind more broadly. It could guide the development of a bodily, ecological-relational, and process-oriented understanding of mental capacities. Mental capacities, like all other traits of organisms, are products of eco-evolution. Thus, viewing the mind as a bodily capacity—or more accurately, as a set of bodily and ecological capacities (cf. Cellucci, 2008) for integrating information and coordinating behavior toward survival goals—could, in this basic sense, be extended also to plants (cf. Bianchi, 2024).

Another topic addressed in this concluding section is the complex question, raised at various points throughout the preceding analysis, of whether representations are involved in the cognitive behaviors of plants and, if so, what form they might take. I revisit how the concept of representation has been interpreted and the diverse meanings attributed to the term. The aim is to determine whether understanding plant behavior requires invoking some notion of representation, or if it would be more accurate to describe their behaviors using other types of frameworks. Some processes, which include abilities to organize, store, and transmit information in biological signaling networks, may better explain the flexible and adaptive behaviors observed throughout a plant's life. The core question concerns whether, in studying plant capabilities, it is useful and justified to employ the concept of representation, or whether it is preferable to adopt non-representational approaches. Alternatively, it may be necessary to develop new concepts, terms, and methodologies specifically tailored to the study of some aspects of plant cognition.

4.5.1 Contemporary Philosophy of Mind and Historical Reconstruction of the Concept of Mind

The philosophy of mind is now integrated into the broader field of cognitive science, often referred to as the new science of mind. Moving away from a foundational and metaphysical approach, contemporary philosophy of mind has evolved into a less speculative and more evidence-based discipline, aligning closely with scientific advancements (cf. Perconti, 2017). Since at least the 1960s, debates over the ontological status of the mental—although still a significant part of contemporary philosophy of mind, particularly following the development of the cognitive sciences and, more specifically, neuroscience—have gradually given way to the emergence of different models of mental functioning, each arising from distinct philosophical perspectives (cf. Di Francesco et al., 2017, p. 113).

Today, the philosophy of mind encompasses diverse goals and research directions. On one side, analyses of the scope and limitations of postclassical cognitive science have driven the field toward a more radicalized form of naturalism, positioning it as a branch of the philosophy of applied science. Conversely, there has been a parallel trend to maintain some degree of autonomy in researching the nature of the mind, seeking to move beyond a strict, rigid adherence to scientific methods, in order to avoid physiological reductionism. For example, some approaches have employed tools from analytic philosophy, such as logical-linguistic analysis and thought experiments, and, in certain cases, have drawn on subjective accounts of mental life (ibid., pp. 15–16), especially when examining the human mind. Perconti (2017, pp. 8–14) noted that contemporary philosophy of mind fulfills several flexible roles, notably acting as a critical overseer of cognitive science epistemology and serving the function of keeping pace with the evolving conceptions, approaches, and empirical findings in the study of mental capacities. Thus, the philosophy of mind is intertwined with the broader scientific endeavor that, in recent decades, has advanced our understanding of the human mind, the minds of other animals, and the information-processing mechanisms of artificial systems. However, one notable absence immediately stands out: the widespread lack of reference to the abilities of plants. This omission, common even in recent cognitive science and philosophy of mind textbooks and manuals, highlights the need to address this neglected area.

Traditionally, the philosophy of mind is a branch of philosophical analysis that emerged to address the relationship between the mind, once considered immaterial, and the body, the material substance. At the core of this inquiry lies the so-called mind-body problem (cf. Galadí, 2024).

By briefly tracing some key historical developments and transformations in the concept of the mind, it becomes apparent that the “mind” is both a product of evolution and a constructed idea. Indeed, the concept of the mind has not always existed in its current form within human culture; it was introduced and consolidated in modern times in a "disembodied" version (cf. Cellucci, 2008, pp. 154–160), largely influenced by the reflections of Descartes. Descartes' perspective marked a break from the medieval Aristotelian view by proposing a substantial distinction between mind and body. He identified consciousness, rather than rationality (understood as the intellectual exercise focused on investigating the external world), as the defining feature of the mental. Because sensory experiences can be deceptive, Descartes argued that the soul, rather than the body, is what we can know best (cf. Di Francesco et al., 2017, pp. 28–38). This perspective, known as Cartesian dualism, posited the existence of two fundamentally different substances: the mind or psychic reality (*res cogitans*) and the body (*res extensa*). According to this view, humans, unlike other animals, are not identical to their bodies. Specifically, human thought and verbal language cannot be explained as mere operations of a mechanical automaton but must be understood as the product of an immaterial, "non-extended" substance, a concept that has since been cemented in the idea of the mind (cf. Maher, 2017, pp. 7–8).[23]

Beyond the traditional form of dualism, which is difficult to uphold today, other forms of dualism have emerged over time, such as the less ontologically demanding property dualism. Rather than positing the existence of two separate substances, property dualism suggests that mental and physical properties are distinct for us. While the world consists only of physical entities and processes (meaning that all substances are material), the attributes of the mind are considered different from those of physical bodies and are not reducible to physical properties (cf. Perconti, 2017, p. 19). This moderate materialist view seeks to reconcile the scientific understanding of the world, as presented by the natural sciences, with the partial autonomy and irreducibility of mental phenomena (cf. Di Francesco et al., 2017, pp. 30–31; 94), especially for how these are felt firsthand. Another non-reductionist form of physicalism employs the concept of "supervenience" (cf. Davidson, 1980, pp. 293–294) to express the idea that mental characteristics depend on, or are grounded in, physical ones. It holds that no two events can be identical in all physical aspects while differing in mental aspects, and that an object cannot undergo changes in its mental characteristics without corresponding changes in its underlying physical processes. This thesis is compatible with the theory of token identity and the concept of multiple realizability, thereby allowing for the simultaneous acknowledgment of both "mentalistic" and neurobiological levels. This approach attempts to unify a scientific worldview with the partial autonomy of mental

[23] On the other hand, there has never been unanimous agreement on this view, even in past centuries. For instance, Julian Offray de La Mettrie (1709–1751), in *L'Homme Machine* (1747), rejected both dualism and Descartes' assertion of a qualitative difference between humans and other animals. He argued that humans are machines, not only in terms of their bodies but also their minds. Moreover, in *L'Homme-Plant* (1748), La Mettrie emphasized that all organisms are machines and compared human functioning to that of plants, suggesting that the human mind is merely a more sophisticated version of plant behavior (cf. Maher, 2017, p. 9).

explanations. However, it remains unclear, at a substantive level, how much explanatory power this perspective actually possesses (cf. Di Francesco et al., 2017, pp. 102–103).

Nonetheless, overall, these explanatory efforts illustrate the complexity of analysis in philosophy of mind. While most contemporary philosophers agree that mental events and experiences are definitely physical phenomena, many are reluctant to claim that a conscious mental event can be fully explained by descriptions of neural activity (cf. Gazzaniga, 2019, p. 98). The most radical materialist perspectives reject the idea that the mind is merely a result or consequence of physical processes in the brain and instead propose that mental events and experiences are those same biological processes. In this view, minds, as products of eco-evolutionary processes, are configurations and activities of matter and energy. Put differently, brain processes do not merely *cause* thoughts and experiences; they *are* thoughts and experiences.

Even so, as Godfrey-Smith (2021, pp. 26–32) noted, this raises the question of why these particular physical configurations give rise to certain experiences instead of others, which may vary over time. Thus, even as the study of the human mind moves away from speculative or scientifically inconsistent approaches, this dual path of analysis may persist, encapsulating the challenge of the mind studying itself (cf. Bianchi, 2024). Ultimately, research on these issues, stemming, in part, from a broader critique of Cartesian dualism, continues to evolve, consolidating in various forms, whether through physicalist views of the mind or through anti-reductionist positions that emphasize the mental as an inner, subjective experience.

Some examples that, within the scope of this text, cannot be explored in all their strengths and limitations include analytic behaviorism, type identity theory, token identity theory, eliminativism, and, among the anti-reductionist views that diverge from orthodox physicalism, various forms of emergentism. Emergentism posits that emergent properties arise from the physical complexity of the brain or from specific networks of relationships between the brain, the body, and the individual's physical and social environments (cf. Clark, 1997, p. 87). This perspective often employs mathematical tools from complexity theory and dynamical systems theory to explain how system features emerge from the interaction of multiple factors. By contrast, radical eliminativism envisions the complete abandonment of the concept of mind, dissolving mental processes into brain activity or, more broadly, into bodily and environmental processes (cf. Di Francesco et al., 2017, pp. 110–113; 196–197).

A critique of the possibility of transforming the concept of the mind into a scientific, empirically testable notion came, for example, from Keijzer (2021), who argued that only the concept of cognition, which can be analysed at different levels across various life forms in accordance with biological science analyses, can be transformed and adapted to the progress of scientific discoveries. The same does not hold for the concept of the mind, as it is commonly understood. The mind is a key concept that has emerged throughout human cultural history and is often associated with topics that extend beyond the main scientific concerns, having a primarily social and ethical significance, such as the idea of free will or individual and collective responsibility, among others.

However, from a conciliatory perspective, there is also potential for the concept of the mind (as well as the more inclusive, process-oriented, and flexible concept of cognition) to avoid being reduced to an "ontological dustbin." Instead, it could be studied through a reflective balance that negotiates between the constraints of neuroscience and the externalist demands of new behaviorist approaches, while also considering the constraints of certain concepts described in the realm of commonsense psychology (cf. Di Francesco et al., 2017, p. 197).

An aspect previously introduced, which warrants revisiting, is the identification of the mind with consciousness, or more specifically, with self-consciousness. The Cartesian criterion of the mental elevated the mind to a different ontological level from matter to emphasize consciousness as the paradigm of mental phenomena. Thus, to "preserve" the mind, there is a potential risk (even today) of focusing excessively on the inner, conscious life and equating the mind solely with self-awareness (specifically in the context of studies on the human mind). For instance, Galen Strawson argued that the mental realm coincides entirely with the realm of conscious experience (cf. Marraffa & Paternoster, 2013, p. 35). According to some, this may be the best, or perhaps the only, way to define the (human) mind. However, the advancement of the science of mind has shifted considerable focus onto unconscious and subconscious physical processes as integral parts of cognitive functioning, the basis enabling the expression of mental capacities, leading to the realization that self-awareness and the mind are not entirely synonymous, as the mind seems to encompass more than self-consciousness (cf. Perconti, 2017). On the flip side, this realization has also led to a more careful delimitation of underlying processes and unconscious mental states, to avoid categorizing any bodily or neural process as a mental state. This has resulted in confusion between the causal role of processes that enable cognition and cognitive phenomena or mental abilities themselves (cf. Marraffa & Paternoster, 2013, p. 31).

Regardless of the precision or limitations of these perspectives, the more narrowly and in human-specific terms the mind is defined, the less likely it becomes to address the minds of other animals, and even less so those of other life forms, such as plants. The use of introspective methodologies is problematic not only because of the inherent difficulties of retrospection, but also due to their anthropocentric nature (although certain aspects of the introspective approach can still be useful for investigating specific features of the human mind's characteristics and functioning). This approach relies on the ability to provide a linguistic, often verbal, account of a mental state or consciousness, thereby excluding most non-human animals and other organisms. This exclusion conflicts with the naturalistic-Darwinian understanding of the mental capacities (cf. Di Paolo et al., 2017, p. 69).

At a more general level, in reflecting on the various possible explanations of the mind—particularly within the context of classical cognitive science and in relation to what can broadly be characterized as a "disembodied" conception of the mind—one alternative outcome of the original mind-body debate, rooted in Cartesian thought, is the strictly functionalist approach to the study of the mind. According to this view, mental states are identified solely by their functional roles and possess an identity that is independent of their physical substrate (cf. Cellucci, 2008, p. 158).

Specifically, computational functionalism, which aligns with the information-processing model, proposed a view of the mind not strictly grounded in the functioning of biological systems, focusing on architecture rather than material composition (cf. Di Francesco & Piredda, 2012, p. 32). In this framework, the central nervous system plays a critically important role by realizing a specific functional architecture, with mental phenomena described at the level of cognitive functions and structures enabling psychological states (cf. Perconti, 2017, p. 34). In the decades following the rise of the classical model of the mind, functionalism, particularly in its more radical forms, faced increasing criticism, targeting its notions of information processing and representation for neglecting the biological basis of cognitive systems and their interactions with the environment. Consequently, the emergence of "postclassical" cognitive science highlighted the embodied and embedded nature of mental processes within physical and social contexts (cf. Di Francesco & Piredda, 2012, pp. 33–34). As a result, many scientists and philosophers have grown increasingly impatient with the idea that the scientific study of mental states should abstract away from the biological medium and disregard environmental influences. Today, it is widely recognized that biological cognition cannot be understood without acknowledging its bodily, situated, and relational nature (cf. Di Francesco et al., 2017, pp. 148–149), although the importance of information organization processes in various systems cannot be overlooked.

4.5.2 Embodied, Ecological and Processual Conception of the Mind

If, therefore, the study of the (human) mind has often been conducted in relation to aspects of self-conscious inner life, based on various components of subjective experience (e.g., ideas, opinions, beliefs, and feelings), and was initially disconnected from a deeper investigation of bio-ecological processes, it is now possible to recognize that our conception of the mental capacities has changed, and has been integrated in many ways, thanks to cultural evolution and technological advancements, which have enabled the development of increasingly accurate theoretical and methodological approaches. The underlying idea is to examine the physiological and behavioral characteristics (not only of humans but also of other living systems) across their genetic, ontogenetic, and phylogenetic dimensions, in order to understand how different (mental) capacities have emerged in various organisms as a result of evolution, how they change during individual development, and how they can be lost or replaced in inadequate contexts or in the presence of malfunctions and pathologies (cf. Morabito, 2020, p. 122).

In the conception of the embodied mind, mental capacities are deeply reflective of the body's unique characteristics that make them possible. In this view, the body with its mental capacities and environment are intrinsically interconnected, to the extent that studying one inevitably reveals insights about the other. For instance, the physical traits of an organism, whether it has feet instead of fins, eyes instead of sonar, partly shape the development and nature of its mental capacities (cf. Palmiero & Borsellino, 2018, p. 84). Different structural-functional dispositions form the basis for varied behavioral-cognitive strategies. Furthermore, this perspective emphasizes the self-organizing nature of living systems and their orientation toward survival goals. Organisms, in a sense, play a role in shaping their minds through the efforts and actions they take to achieve desired outcomes in particular situations. From this standpoint, the mind emerges through functional coupling and relational interactions with its natural—and, in some cases, cultural—ecological niches (cf. Nucara, 2014, pp. 210–211). This concept is encapsulated in Varela's assertion that the "mind is not in the head." Without overstating the idea of change, having a mind requires active engagement with the world, characterized by constant and varied sensorimotor interaction with relevant environmental elements. "Objects" become visible and usable through the organism's activities, just as organisms themselves co-emerge and co-develop through their relationships with the salient features of their surroundings (cf. Bianchi, 2022).

Furthermore, focusing on the active, relational, and process-oriented aspects of the mind helps to avoid the risk of reifying the concept, just as in the case of the concept of "life." The "mind," after all, is not an independently existing entity; it does not refer to a distinct object or organ but rather to a set of processes, activities, and capacities expressed differently across various forms of life (cf. Mayr, 1982, p. 74). On the other hand, this approach also prevents chaotic outcomes that arise from failing to identify the relevant factors necessary to understand a particular capacity, which might otherwise lead to a reckless extension into numerous unrelated processes. Naturally, acknowledging this does not detract from the significance of the organization of the organism system.

One relevant implication of the bodily conception of the mind is the recognition that perceptual, sensory, and motor capacities—essential for discriminating among elements and acting to achieve survival goals—are present not only in complex organisms but also in simpler ones. This suggests that mental capacities, linked to these bodily activities, are common to most, if not all, living systems, albeit in varying degrees. This understanding, in line with the Darwinian paradigm, places humans in a continuum with other organisms, challenging age-old prejudices of our species' supposed superiority, while acknowledging the differences (cf. Bianchi, 2021). The inclusive approach that incorporates the capacities of diverse organisms is grounded in a bodily, ecological-relational, and process-oriented view of the mind, in which organisms are capable of regulating their behavior in various situations (cf. Bianchi, 2024). Adopting a more demanding or restrictive concept of the mind would make it difficult to include "other minds" in such analyses. This does not mean, however, that the salient features, unique capabilities, and differences cannot be specified or that the analysis must remain overly vague. Rather, this conception of the mind, attentive to species-specific traits, may be better suited to the study of mental capacities across the various forms of life (cf. Perconti, 2017, p. 126).

4.5.3 Beyond the Human Mind

This empirically informed approach to studying the mind is not entirely new; it can be traced back at least to Darwin, who made significant contributions by investigating animal mental activity through an evolutionary lens and emphasizing the adaptive functions of behavior (cf. Bianchi, 2025). Darwin's work also helped position human psychological processes within a naturalized framework, establishing continuity with the mental capacities of other organisms (cf. Di Francesco et al., 2017, p. 68). Building on Darwinian thought, James (1892, p. 4) critiqued Herbert Spencer's (1820–1903) theory of knowledge as a mirroring of reality. James highlighted the dynamic nature of the mind, viewing it as an unfolding product of evolution, shaped through causal variation and environmental selection. He emphasized the reciprocal adaptation between the mind and the world, suggesting that the mind is an integral part of the world itself. In this sense, the organism-mind and its environment are outcomes of a shared history of interdependence. According to Dewey (1884), adopting an interactive concept of the environment makes it difficult to view mental life as something individual or isolated, as environments themselves are the result of the developmental processes of organisms: the framework of analysis is bidirectional. Since Darwin's work, this convergence of philosophical theories and experimental research in psychology and neurophysiology has laid the foundation for a biological, dynamic, historical, and context-dependent understanding of behavior and mental capacities. This perspective holds heuristic potential, especially when applied to brainless organisms like plants, and warrants further investigation in interdisciplinary cognitive science research. Special emphasis should be placed on sensory and perceptual-agentive dimensions to explain the development of cognitive abilities, with a focus on phenotypic plasticity in the case of plants (cf. Morabito, 2020, pp. 57–59).

Nevertheless, if this were the prevailing conception, there would be far less work left to do. The approach outlined above remains a minority position, serving as a promising research program for the future, especially regarding the study of mental capacities in non-animal organisms. As Keijzer (2021) pointed out, there is still no consensus within the contemporary debate on which biological systems truly possess a mind or exhibit mental capacities. This ambiguity arises from the different conceptions of the mind, which depend on the criteria used to identify mental traits and abilities. It also reflects the fact that the concept of the mind was originally formulated to explain human capabilities. Among all living organisms, only humans are universally recognized as having a genuine mind, whereas the attribution of mental capacities to other organisms remains a matter of debate.

On the other hand, even when analysis is restricted to the human mind, there is no consensus on a single approach or criterion for identifying the necessary requirements to define the mind. For some, the human mind is essentially tied to the sensation of one's inner life, often metaphorically described as a "private garden," a secret space accessible only to its owner (cf. Godfrey-Smith, 2021). Others argue that while this aspect is relevant, it cannot solely account for the full complexity of mental phenomena, which encompass many additional elements. Still, some perspectives suggest that to discuss the mind meaningfully, one must move to other levels of analysis. For instance, minds might be inferred from observable actions and responses, manifesting in behaviors. This approach could potentially apply to other animals and, in some cases, even to plant organisms, albeit with variations in focus on distinct behavioral traits and strategies.

When extending the analysis to determine whether other organisms possess a mind or mental capacity, various criteria have been proposed, some quite demanding. Among the most restrictive are theories that insist on the use of concepts or propositional aspects, often excluding non-human entities except for a few select animals (cf. Hurley & Nudds, 2006). Others argue that having properly mental abilities requires the capacity for "offline" processing, where an organism can initiate a behavior even in the absence of a direct sensory stimulus, allowing a detachment from the immediate "here and now" (cf. Godfrey-Smith, 2021, pp. 291–292; 323). Additionally, some perspectives maintain that inferential and/or representational activities are necessary to attribute a mind to an organism. Depending on how representation is defined, these requirements might extend to a broader range of organisms. Less restrictive proposals, however, dissolve the concept of the mind into a series of biological processes occurring at various levels, or they adopt an instrumental perspective, treating the mind as a collection of capabilities without adhering to a rigid set of criteria for what qualifies as "mental." From a functionalist standpoint, having a nervous system is not synonymous with having a mind in the traditional sense. Instead, it can be more accurately described as a behavioral control mechanism primarily aimed at maintaining the organism's integrity. This description can also apply to aneural systems, which consist of networks of interacting cells that transmit chemical and electrical signals and process information across different organizational levels. By adopting a functionalist approach to information processing and organization, whether explaining how minds operate or exploring the biological processes underlying mental activity, one can identify similarities, without erasing differences, between neural and aneural systems. These systems, believed to possess basic mental capacities, are all oriented toward survival goals and capable of modulating bioecological information, albeit at varying speeds and with differing degrees of integration. Dennett (1996), for instance, suggested that even cells or groups of cells could be viewed as tiny, simple agents pursuing their goals. The world is filled with such organized systems, ranging from the molecular to the macroscopic scale. This perspective invites us to consider that, in research, we are the ones who define the boundaries for identifying and categorizing certain processes. The instrumental view of the mind, which avoids ontological commitments, suggests that "minds" do not represent anything real but are instead interpretative attitudes used for pragmatic purposes (ibid.). This perspective, however, has also had a countereffect, sometimes leading to a more selective concept of the mind and emphasizing the idea of a personal mind. Such an approach seeks to prevent an indiscriminate attribution of mind across the natural world and to avoid associating it too broadly with various biological processes, which could lead to the dissolution of the concept of the subject (cf. Di Francesco & Piredda, 2012, pp. 223–224; 231).

Such reflections are valuable for critically evaluating the strengths and limitations of different positions, from the most restrictive to the most inclusive (cf. Bianchi & Castiello, 2023a).

In any case, in the study of animal minds, it is increasingly rare to find scientists who argue that animals, or at least mammals and birds, lack minds, particularly when "mind" refers to the capacity to remember, learn, and plan behavior (even if it does not extend to self-awareness, beliefs, conceptual thinking, or imagining alternative futures). This implies, first and foremost, that recognizing a mind in non-human animals does not mean attributing to them the same type of mind that humans possess (cf. LeDoux, 2020, p. 399). A similar argument can be extended, at an appropriate level of analysis, to other forms of life, including plants. Exploring the unique abilities of organisms to organize and process information, adapt to new or changing environments, and modulate their behaviors to solve various challenges provides insight into widespread capacities that, although expressed differently across species, are fundamental for survival throughout the tree of life (cf. Lyon et al., 2021).

One reason that discussions about the mind or mental capacities in animals were often avoided in the past was the inability of these life forms to articulate their feelings or reasoning through the sophisticated verbal languages unique to our species. However, as we have deepened our understanding of the communication behaviors of various animal species, different forms of nonverbal communication have become valuable tools for assessing animal mental capabilities. The study of species-specific animal communication has opened new paths for understanding the mental capacities of a wide range of animals (cf. Bianchi, 2021). A commonly referenced example is Karl von Frisch's (1886–1982) discovery of the complex signaling system used by bees. This system, capable of conveying detailed information about the type, distance, and location of food sources, overturned the earlier belief that the twirling movements of bees were random. Such discoveries, along with similar findings regarding complex animal communication and navigation systems, have led researchers to view animal behaviors through a new lens, recognizing the sophisticated intentions and adaptive strategies at play. Studying the cognitive capacities of other organisms thus provides an opportunity to engage with alternative modes of relating to the world (cf. Perconti, 2017, pp. 119–123). Continuing the exploration of the intricate "language" of shapes, colors, and scents—primarily based on a sophisticated form of chemical communication sensitive to dosage and context—allows for a deeper understanding of plant behavior as well (cf. Gagliano, 2017).

A bodily, ecological, and process-based approach to understanding the mind, or mental capacities, can be applied across a spectrum: from observing the behaviors of complex cognitive systems to studying the simplest forms of life, such as unicellular organisms. For instance, many fundamental features of animal brains that underlie mental activity—such as sensory integration, memory, learning, decision-making, and behavioral control—are also present in these simple life systems. Despite lacking a nervous system, unicellular organisms can acquire and process environmental information (cf. Trewavas, 2014, p. 155). These abilities can be explained within a framework that emphasizes the bodily and ecological mind, encompassing sensorimotor capabilities and beyond. Both neural and non-neural systems (regardless of differences in sensorimotor organization [cf. Calvo & Keijzer, 2009]) are involved in chemical-electrical intercellular signaling. This enhances response effectiveness, whether through locomotion, movement, or metabolic adjustments, to address survival challenges. From this perspective, the mind, broadly conceived, can be understood as a set of bodily capacities that enable various organisms, based on their structural and functional characteristics, to solve problems related to communication, reproduction, and survival (cf. Cellucci, 2008). These capacities include decision-making, memory, learning, and the ability to adapt behavioral responses to sustain life or improve conditions.[24] Therefore, it can be argued that evolution has favored the development of knowledge and mental capacities not for mere contemplation but for effective action in the world (ibid., p. 167; pp. 196–197; Cellucci, 2017).

Conceiving the concept of mind in these terms also invites reflection on the connection between mind and life, a topic explored within enactive approaches. In *Mind in Life* (2007), Evan Thompson argued that being alive inherently entails having a mind, a claim he extends to all living systems. In this view, having a mind involves opening up to a world of aspects imbued with "meaning"—even if only in an operational sense—that motivate the organism to behave in various ways. This meaning-making arises through the organism's interactions, enabling it to act and achieve its goals within its *Umwelt*, an environment it helps shape through relationships with elements that hold value for it. The capacity to discern what is "significant" for an organism, or sense-making, is the outcome of various eco-evolutionary pathways. Different living self-organizations have developed through evolutionary histories that have uniquely shaped their structures, biological organizations, and corresponding functional capabilities, enabling them to distinguish opportunities from threats within specific ecological-relational contexts. Depending on the approach and author, there are nuanced differences in how the relationship between mind and life is analysed. For example, the idea that all living beings, from the simplest to the most complex, possess some form of mind or mental capacity contrasts with the more cautious view held by some scholars (e.g., Godfrey-Smith, at least as it currently appears defensible), who argue that while everything that has a mind must be alive, not all living organisms—particularly those without a nervous system—necessarily possess a mind. For now, it is worth noting that asserting bacteria or plants, often excluded from discussions of the mental, possess some form of mind, proto-mind, or basic mental ability does not erase the distinctions between different organisms and complexities. Rather, it highlights similarities while respecting differences, acknowledging that every form of life engages in self-preservation and interacts with its environment in species-specific ways and through unique strategies. These "self-care" modes, characterized by varying degrees of complexity

[24] Another way to express this idea is through a "capacitational" approach, which holds that having a mind is evidenced by the possession or acquisition of certain abilities. This perspective emphasizes not what the organism currently does, but what it is capable of doing (cf. Glock, 2019).

and adaptation, distinguish organisms as particular bio-ecological types (cf. Maher, 2017, pp. 69–70; 121–125). The aim is not to trivialize the concept of mind, but to explore its possible nuances.

The groundwork now appears sufficiently fertile to include plants in this. Building on Maher's work in Plant Minds (2017), that investigates whether and in what sense it is possible to talk of a "plant mind" or mental capacities in plants, it becomes apparent that only certain philosophical positions within the philosophy of mind are unable to address plant organisms. To summarize Maher's analysis, the attribution of some form of mind or basic mental capacity to plants is excluded under Cartesian dualism, which posits the existence of an immaterial entity and has traditionally linked mental states to reflective, conceptual, and metacognitive self-awareness. It is also excluded by the mind-brain identity theory, which requires the presence of an equivalent brain state or process, asserting that the mind is identical to certain types of neural activity, thereby excluding systems without nervous system from the cognitive domain. Conversely, the Aristotelian conception, which characterizes the *psyché* as inherently goal-directed, does not explicitly rule out the possibility of proto-minds or basic minds in plants. The mechanistic view, which sees the mind as the product of matter in motion, similarly does not exclude them. The computational-functional approach, which draws parallels between the information-processing capabilities of computers and organic minds, remains open to the idea. The Darwinian perspective, which understands the mind as a set of adaptive abilities shaped by natural selection, also accommodates this possibility. Likewise, the behaviorist view, which explains mental capacities as adaptive, modifiable behavioral dispositions acquired through learning processes, does not exclude plants. Finally, enactivism, emphasizing the capacity of self-organized systems to vary behavior in response to contextual changes, remains compatible with this idea (ibid., pp. 18–19; 115–127).

Of course, the perspectives outlined represent only a subset of the many theoretical approaches to the study of the mind. Although far from exhaustive, this overview provides a starting point for further investigation and future research on these compelling questions.

The approach proposed here is pluralistic, tailored to the specific type of capacity being considered, whether mental, or cognitive-behavioral. For now, within a bodily, ecological, and process-oriented framework of the mental, the emphasis should be on remaining open to the possibility of including plants in cognitive considerations, at least until further evidence suggests otherwise. This openness allows us to explore where our understanding might lead, what concepts can be clarified, and how to express them accurately. Over time, it will become clearer which aspects warrant further investigation and which should be deprioritized or avoided in studies of the mind.

One aspect that Maher (2017) particularly emphasized is his critique of representational approaches to studying cognitive abilities in certain living systems. According to Maher, adopting a representational framework, especially one based on a strict definition of representation, may hinder meaningful discussions of "plant mind" or plant cognition. This raises the question of whether the issue of representation is truly pivotal when it comes to plants. To some extent, it is, partly because it remains a frequently discussed topic. However, much depends on how the concept of "representation" is conceived and understood.

4.5.4 *Representational Cognition: An Open Question*

In this final section, I aim to revisit and further develop the analysis of the concept of representation introduced in the target paper by Bianchi and Castiello (2023a) and subsequently discussed through a series of commentary articles and a response paper (cf. Bianchi & Castiello, 2023b). The overarching goal of these contributions was to spark an interdisciplinary discussion on the explanatory relevance (or lack thereof) of various interpretations of representation in understanding the cognitive behaviors of plants. This debate arises largely because one of the primary controversies surrounding the recognition of plants as cognitive agents centers on the existence, absence, and role of mental representations (cf. Lee, 2023).

This issue is significant because references to representation—its presence, absence, nature, and context of application—often surface in studies of plant activities and contemporary debates on plant cognition. The most contentious point is that without some form of representation, it becomes difficult to attribute genuine cognition to plants. Cognition, in this sense, is associated with real, individual life experiences shaped by specific histories of interaction, encompassing processes like memory, learning, and anticipation. In the absence of representation, plant responses are often perceived as mere reactions to environmental stimuli, lacking the variability characteristic of cognitive processes over an organism's lifespan.

In the fields of philosophy of mind and cognitive science, some scholars have argued that mental representations—such as beliefs, intentions, concepts, or reflective self-conscious states—arise from the processing and interpretation of sensory signals by the nervous system and serve as indicators of cognitive activity (cf. Adams, 2018). Since plants lack a centralized nervous system capable of generating or housing such mental representations (cf. Wang et al., 2021), it has been argued that they should not be classified as organisms capable of cognition. However, this view is not the only perspective on nature and necessity of representations.

The debate surrounding the link between cognition and representations (and the specific level or type of mental activity required) has long been a topic of interest, extending beyond human cognition to that of other organisms. Investigating the physical basis of mechanisms and processes—whether they are conceptualized as representations or not—that enable plants to exhibit (cognitive) behaviors based on variations in sensory perception is a challenging endeavor, both empirically and conceptually. Such an investigation could also provide insights into the cognitive capacities of other "atypical" cognitive systems, such as prokaryotes (cf. Ben-Jacob et al., 2004; Shapiro, 2007; Heras-Escribano & Calvo, 2019).

Currently, most approaches proposed for studying plant cognition are non-representational. However, this does not imply that, at this early stage of research, the available experimental evidence and literature definitively require non-representational methods for understanding plant cognition. It remains plausible to interpret the concept of representation in a significantly revised way compared to classical theories. This could involve adopting a modified notion of representation or employing alternative frameworks to describe how environmental information is structured and retained. Thus, exploring these processes through new or different conceptual lenses is essential. Importantly, one objective is to move away from certain traditional conceptions of representation that may lack empirical adequacy or explanatory utility in the context of plant cognition.

Building on the ideas expressed by Bianchi and Castiello (2023a, 2023b), it is important to clarify from the outset that this reflection does not aim to reintroduce the concept of representation at all costs, as if it were indispensable for explaining certain plant behaviors. Instead, the primary goal is to explore whether a substantially reformed notion of representation could be helpful in explaining how plants engage in ecological interactions, or whether such a concept is unnecessary, or even counterproductive, and should be entirely discarded. Given the complexity of the topic and the ambiguity arising from the various interpretations of representation, this discussion poses questions rather than providing answers, aiming to spark philosophical debate and inform experimental research. A primary aim for future research is to generate new evidence and develop species-specific tests to clarify the different levels of analysis of the phenomena and processes under investigation. For now, the question remains ultimately open.

In any case, the true value of this conceptual exploration lies in the analysis itself.

4.5.5 Varieties of Conceptions of Representation

Before examining the case of plants, it is helpful to briefly revisit the debate surrounding the presence and types of representation within cognitive science. One might first question whether cognition, at all levels of complexity, is necessarily representational, that is, based on representational elements (cf. Rodriguez & Calvo, 2010). The answer often depends on the approach used to study cognition and the level of analysis applied to the cognitive behavior in question. Additionally, when a representational approach is adopted, it is crucial to specify the kind of representation involved, such as whether it is "mental" or "non-mental," and in what specific sense.

Conventionally, to facilitate an analysis of this multifaceted concept, a useful—though merely orientative and not to be taken literally—broad distinction can be made between mental and non-mental representations. The former aligns with the classical conception of the mind as an organ of thought and the locus of conceptual elaboration. This perspective typically draws on traditional representational theories, whether they emphasize propositional content, internalism, offline or detachable content, or self-consciousness as defining features of the mental (a deeper examination of these distinctions lies beyond the scope of this discussion). The latter category, non-mental representations, emerges from a bodily, ecological, and process-oriented understanding of the mind. It seeks to explore forms of representation (or their presumed equivalents) present in both neural and non-neural organisms, focusing on their capacity for environmental interaction and behavior regulation. Some explanatory frameworks possess a hybrid nature. For instance, representations conceived as models or maps may be applied to both neural and non-neural systems (cf. Bianchi & Castiello, 2023a).

Throughout the various approaches to studying mental capacities, a wide range of representations have been identified, including symbolic/subsymbolic, neural, functional, structural, molecular representations, as well as image-based representations, models like maps, and bodily, somatotypic, sensorimotor, and action-oriented representations. Despite this diversity, the term "representation" has historically, and often continues to be, used as shorthand for "mental representation" (cf. Morgan, 2014).

Regarding the history of the concept of representation, it is worth noting its long-standing tradition across various fields of research. At a basic level, representations have often been defined in line with the notion of a sign: something (a representative) that stands in for something else (an aspect of reality, whether concrete or imaginary, i.e., the referent of the representation) (cf. Nannini, 2011, pp. 44–45). For a representation understood in this way, the degree of complexity is irrelevant; what matters is that it serves as a "surrogate" for a state of affairs. This applies in a general sense.

In the evolution of cognitivist mentalism—which marked a decisive departure from behaviorist explanations and contributed to the emergence of classical cognitive science—internal representations of the external world constructed by organisms were increasingly hypothesized to offer a more comprehensive account of animal behavior. Early formulations of the idea of internal models of reality can be traced to the work of Kenneth Craik (1914–1945), who, in *The Nature of Explanation* (1943), proposed that living beings maintain small-scale models of external reality in their minds, along with possible courses of interaction, to navigate and respond adaptively to environmental challenges. This theoretical insight laid important groundwork for the concept of cognitive maps, particularly through the pioneering research of Edward Tolman (1886–1959). Tolman's studies on rats' learning in mazes suggested the existence of internal representational models that guide behavior by mapping the spatiotemporal structure of the external environment (cf. Di Francesco et al., 2017, p. 118).[25]

Following the behaviorist interlude, psychology once again emerged as a science of the mind, shifting its focus to the mechanisms and processes underlying observable behavior. The concept of representation—conceived as an internal map, model,

[25] A renewed emphasis on the concept of cognitive mapping emerged in the 1970s with John O'Keefe's discovery of specific hippocampal neurons, known as place cells, which form part of the brain's memory system. These neurons are thought to underlie the construction of internal representations—or cognitive maps—of the spatial environment, by encoding relationships among various environmental anchor points used for orientation (cf. LeDoux, 2020, pp. 265–266).

or cognitive schema used for spatial orientation and behavior regulation—has even been hypothesized in organisms lacking a nervous system, emerging from dynamic patterns of molecular signaling activation. The ability of an organism to adapt and survive within its environment—even in the absence of conceptual skills or a high degree of abstraction from the "here and now"—is strongly influenced by its capacity to construct and maintain internal maps of the surrounding environment independently of immediate sensory experiences (cf. Michmizos & Hilioti, 2019).

Over time, additional interpretations of representation emerged. In classical cognitive science, mental representations are regarded as the outcomes of neurophysiological mechanisms triggered by perceptual processing. Within this framework, physical signals are initially detected by perceptual systems and transmitted to the brain, where they are transformed into mental representations (e.g., intentions or concepts). These representations are subsequently processed by the mind and ultimately translated into actions by the motor system (cf. Wang et al., 2021). According to representational theory, thoughts, actions, and decisions depend on an internal mediating element, a representation, that stands in for a physically absent state of affairs (cf. Di Francesco et al., 2017, p. 190).

Within information-processing theories of cognition, a mental representation is understood as an entity that must be computed and processed by mental mechanisms to generate behavior (cf. Marraffa & Paternoster, 2011, pp. 14–15). In classical cognitive science (cf. Fodor & Pylyshyn, 1988), this concept found an explanatory metaphor in the comparison of cognitive systems to electronic computers, functioning through symbolic representations governed by transformation rules. Cognitive systems were thus described as physical symbol systems (cf. Newell, 1980). Jerry A. Fodor (1935–2017) systematized this perspective in *The Language of Thought* (1975), integrating representational theory with the computational theory of mind. He proposed that intentional mental processes consist of causal sequences of transformations of representations, guided by rules sensitive to their syntactic form rather than their semantic content (cf. Di Francesco et al., 2017, pp. 128, 142). Building on this approach, Steven A. Pinker argued that mental entities could be understood as information encoded in symbolic configurations—physical states of matter, such as neurons or computer chips—that symbolize real-world entities, activated through sensory organs (cf. Perconti, 2017, p. 31).

With advances in research on the functioning of the mind, embodied theories have gained prominence, particularly within the motor paradigm, which emphasizes a bodily and agentive understanding of cognitive abilities over a purely propositional one. This shift has drawn increasing attention to sensorimotor formats of representation, providing a theoretical framework aimed at overcoming traditional dichotomies, such as the separation between bodily mechanisms and mental content (cf. Morabito, 2020, p. 7). Extending these insights, scholars have explored the notion of action-oriented representations, forms of representation that encode procedural and motor information essential for enabling action (cf. Jeannerod, 1996).

As highlighted in Andy Clark's work, the classical concept of mental representation has been reframed as a continuum of representational types. At one end lie internal substitutes (e.g., concepts, intentions, beliefs) that address "representation-hungry" problems requiring complex reasoning (cf. Clark & Toribio, 1994). At the other end are perceptual mechanisms and processes that handle information through causal correlations or direct sensory control of environmental signals, with action-oriented sensorimotor representations positioned in between. Depending on the degree of coupling or decoupling between agent and environment, cognitive behaviors may require either dynamic, non-representational explanations or representational accounts specifying the type of representation involved (cf. Di Francesco et al., 2017, p. 191). Support for the existence of functional motor representations comes from the work of Michael Graziano (1996), who demonstrated that electrical stimulation of specific cortical sites in animals can elicit complex gestures, such as bringing food to the mouth or climbing. These findings led to the hypothesis that such actions are represented within the stimulated brain areas, encoding the spatiotemporal information necessary for accurate execution. Moreover, the modulation of stimulation effects by contextual factors suggests a relatively dynamic interpretation of motor representations. A major advancement in the study of body representations came with the discovery of two types of neurons, canonical and mirror neurons, located in the premotor region of the neocortex, an area involved in action planning, selection, and execution (cf. Rizzolatti et al., 1996; Gallese et al., 1996). These findings indicate that motor programs not only facilitate object identification but also encode potential interactions with objects, providing pragmatic representations through simulations of possible actions (cf. Palmiero & Borsellino, 2018, pp. 106–107; 143).

As with the concept of mind, the notion of representation has been interpreted not only through various theoretical frameworks for studying cognitive activities but also as a potential explanatory tool or hypothesis, independent of specific references to the material and organizational bases of representations (cf. Marraffa & Paternoster, 2013, pp. 10–12). For instance, Dennett (1987) advocated a methodological or instrumental view, emphasizing its utility in explaining and modeling cognitive processes and behaviors, where representation can serve an instrumental and explanatory role in the interpretive activities of knowing subjects (cf. Zipoli Caiani, 2016, pp. 143–144). However, despite its theoretical convenience and practical usefulness, this approach does not appear entirely conclusive. Describing representation in instrumental or metaphorical terms might function merely as an abbreviation that points toward, or presupposes, underlying factors enabling or guiding behavior, factors that themselves demand further explanation (cf. Maher, 2017, pp. 107–108). Moreover, once representation is established as an explanatory tool, it raises the question of whether it will remain relevant even if the physiological processes underlying the cognitive behaviors of different organisms have been more fully clarified. Alternatively, the notion of representation—as a descriptor of mechanisms underlying cognitive abilities—might ultimately give way to purely relational and causal explanations (cf. Rodriguez & Calvo, 2010).

In short, the continued reliance on some form of representation may ultimately prove to be a temporary requirement. It is plausible that the concept of representation will be gradually phased out as our understanding of the nervous system—and, more broadly, of cognitive processes, including those occurring in aneural organisms—becomes increasingly accurate and detailed (cf. Zipoli Caiani, 2016, p. 144; Bianchi & Castiello, 2023a).

Given the wide range of meanings attributed to the term "representation"—of which only a few have been discussed here (a full exploration lies beyond the scope of this text)—researchers in cognitive studies, such as Ramsey (2007), have questioned what general criteria must be satisfied for something—whether a set of physical structures carrying information, a particular organization, or a process—to count as a representation within a system. Typically, the necessary conditions for employing the concept of representation in explaining cognitive activity include several requirements: for something to function as a representation, it must genuinely represent something; it should exhibit some form of resemblance to its referent, not necessarily "iconic" but at least functional; it must play a causal role and guide behavioral responses in specific contexts; it should be detachable, meaning it can persist and be recalled independently of the stimulus; and it must allow for the possibility of misrepresentation. This set of criteria has been termed "the job description challenge," aimed at preventing the dilution of the concept through overly broad or careless applications. Naturally, questions arise regarding the validity of these criteria: Are there alternative interpretations worth considering? Are these characteristics, or some combination of them and others, the most relevant?

Through detailed and problematizing analyses, it has become clear that multiple conceptions of representation now coexist, each subject to its own critiques, underscoring that the term no longer refers to a singular, unified concept (cf. Morgan, 2014).

4.5.6 Advantages and Disadvantages of Representations

Following the argument outlined by Bianchi and Castiello (2023a), it is possible to highlight both the benefits and drawbacks associated with certain conceptions of representation. On the one hand, there are advantages ("pros") to using representations, aimed at addressing or simplifying specific challenges in understanding cognition. On the other hand, there are disadvantages ("cons"), where the inclusion of representations in explanations has led some approaches to view them as unnecessary or even counterproductive.

Starting with the "pros," an aspect that many cognitive scientists emphasize is that what makes a phenomenon genuinely cognitive is the capacity for offline reasoning. This form of reasoning is difficult to imagine without some representational component. Representations allow organisms to decouple their responses from the stimuli immediately present to perception, enabling more complex cognitive processing. In this context, reliance on internal representations of things or events, especially when the external referent is absent, is what differentiates cognitive processes from simpler biological information processing. Consequently, processes that trigger immediate behavioral responses to sensory stimuli are not considered to be under the same level of cognitive control as those guided by internal representations (cf. LeDoux, 2020, p. 260).

The central question for advocates of representations is whether it is possible to ignore the causal role that representations play in guiding behavior. The idea of representation, whether as a mental state or a bodily perceptual variation in response to a stimulus (which can exist even in the absence of that stimulus), is often introduced for its role as an "intermediary." This intermediary function connects the organism to its environment and links cognitive systems to behavioral possibilities. Without acknowledging some form of representation, one risks reducing behavior to mere stimulus-driven reactions, which lack the flexibility and active modulation seen in behaviors shaped by individual histories of interaction and experience. Such behaviors suggest an organism's capacity to generate specific responses and learn selectively in a context-dependent manner. On the other hand, it could be argued that such an intermediate state amounts to a particular state of the system itself within a context.

Furthermore, some argue that the issue of "meaning," the informational content of mental states, cannot be addressed without invoking some kind of representational system that can be shared among agents (cf. Zipoli Caiani, 2016).

Representations also help explain the potential for error in cognitive processes. From this perspective, an error arises when there is a mismatch between the content of a representation and the actual state of affairs, a discrepancy that can be verified intersubjectively. If reality were accessed directly, without representational mediation, cognitive errors would not occur; instead, only system malfunctions could be observed (cf. Michaels & Carello, 1981).

Among the apparent advantages of introducing representations, one of the strongest criticisms directed at radical antirepresentational approaches is that cognition is a complex phenomenon requiring analysis across multiple organizational levels. Without positing some form of (mental) representation, critics argue, there is a risk of limiting explanations to the "lower" levels of cognitive abilities, that is, to basic functional processes (cf. Di Francesco & Piredda, 2012, pp. 75–78).

One field where non-representational models have been extensively developed is robotics. In the case of Brooks' robots, for example, enactivist scholars skeptical of the concept of representation (cf. Chemero, 2009; Hutto & Myin, 2013) have emphasized the capability of these devices to maintain continuous engagement with their environment without necessarily invoking representational mechanisms. However, it has been otherwise argued that maintaining constant contact with the environment does not automatically exclude the need for representations. A system that interacts continuously with environmental properties (such as light or humidity) might still selectively respond to specific features based on internal needs. In Brooks' devices, for instance, sensors transmit signals with varying intensities, allowing the system to respond differently to objects in its surroundings, a method that could be interpreted as having representational aspects, in the minimal sense of organizing information. Even if explicit representations are not evident in these basic systems, it has been argued that as behaviors become more sophisticated, it becomes increasingly difficult to avoid introducing some form of representation. Approaches to situated cognition in robotics that entirely reject representations risk being confined to solving problems tied to the immediate sensorimotor environment. For instance, navigating from point B to point C after moving from point A to B might require an internal representation of the spatial relationships between these points (cf. Matthen, 2014).

Beyond this level of complexity, behaviors tied to advanced cognitive capacities, often described as "representation-hungry" (cf. Clark & Toribio, 1994), come into play. These include abilities such as imaginative thinking about fictional objects and engaging in simulation or counterfactual reasoning (e.g., considering what would happen if one action were taken instead of another), activities that seem to necessitate representational capabilities (cf. Shapiro, 2021). As a result, extreme forms of anti-representationalism face significant challenges in accounting for more complex cognitive functions, such as human language use, especially when explaining reasoning about states of affairs that are not present to the senses or that concern hypothetical or past scenarios. When dealing with tasks that require a certain level of complexity or abstraction, it thus appears difficult to completely exclude the involvement of some form of representation (cf. Di Francesco & Piredda, 2012, pp. 75–80). Moreover, the existence of various types of representations allows for the adoption of different approaches depending on the specific case, the type of organism or cognitive system, the mental capacity involved, and the level of behavior under investigation (cf. Di Francesco et al., 2017, p. 190).

On the other hand, when considering the "cons" or challenges associated with introducing representations, one of the main hurdles is the ontological-epistemological issue. This problem concerns a particular conception of knowledge and truth, namely, the idea that representations "mirror" a reality existing independently of the organism interacting with it (cf. Bianchi & Castiello, 2023a). This perspective connects to concerns about the rigid encoding of information into representational formats, where external inputs are translated into symbolic codes manipulated by cognitive systems. In this view, the representational vehicle serves as an intermediary within the system, "resembling" or substituting the entity it represents. While acknowledging the existence of information encoding and memory processes, the main difficulty lies in verifying the correspondence between representations and aspects of reality. This difficulty raises the risk of infinite regress, requiring meta-representational structures to account for each encoding, since encoded vehicles lack inherent content (cf. Frazier, 2021). A potentially more effective approach involves focusing on observing agents' contextual behavioral dynamics rather than on verifying the truth value or correspondence between the encoded representation and what is represented (cf. Bianchi & Castiello, 2023a).

A related issue with this argument concerns the tendency to describe representations in terms of their resemblance to what they refer to, rather than in functional terms. The problem does not lie in the notion of resemblance per se, in all its possible forms, since, if the existence of representations is posited, it is reasonable to expect some form of relationship between external situations or stimuli (aspects of reality) and the material bases of representation. The difficulty emerges when an "iconic" conception of similarity is adopted, whereby a representation figuratively resembles some trait of reality. As noted in Craik's work (1943), the focus should instead be placed on dynamic-functional similarities, emphasizing systems that can be functionally exploited without requiring an external interpreter to perceive these similarities (cf. Morgan, 2014). Regardless of one's stance on these views, it remains valid that a structural link between two elements, such as a key and a lock, does not necessarily imply that the elements represent each other. For instance, although a key fits into and opens a lock, it does not follow that the key "represents" the lock, or vice versa (cf. Kirchhoff & Robertson, 2018). Thus, the issue appears to concern a representational interpretation rather than a straightforward functional relationship between elements.

A similar example involves the teleological interpretation of circadian oscillators. These coupled systems, functioning as adaptive mechanisms regulated by feedback loops, are adapted to track the alternation of geophysical periodic cycles. Their role has been interpreted in terms of "representing" the time of day, thus contributing to the teleological coordination of the system's behavior. However, according to Rodriguez and Garzon (2010), such an interpretation may be unnecessary for explaining regulatory adaptations that align with environmental cues, a point that could also extend to certain basic cognitive behaviors. Instead, a less conceptually demanding account, based on causal relationships or signaling dynamics within functional circuits, might be sufficient to explain the adaptive behaviors of biological-cognitive systems. The issue is not the presence of emulators or controller-environment dynamics (cf. Bechtel, 1998, 2009a, 2009b), but the tendency to ascribe representational status to such systems and processes. In such cases, positing the existence of internal representational models could be not only unnecessary but potentially problematic. As Maher (2017, pp. 94–95) points out, explanations framed in terms of feedback loops are often sufficient and do not necessarily imply a strong representational framework. A process that qualifies as a signaling cycle simply requires responsiveness to its effects, adjusting its activity accordingly, rather than representing or "standing for" something else.

On a broader level, it is worth questioning why the explanation of a mechanism, organ, or set of processes should be framed in representational terms. According to Ramsey (2007), this is precisely the type of inquiry that should be pursued when statements about representation are employed in cognitive science explanations (cf. Morgan, 2014). This consideration brings us back to the "job description challenge," which, regardless of whether the proposed criteria are the most appropriate or the only possible ones, emphasizes the need for caution. It is important to avoid extending representational reasoning into contexts where it may be superfluous or counterproductive, unnecessarily complicating explanatory accounts (Bianchi & Castiello, 2023a).

Another issue introduced by representational reasoning—particularly when cognitive reasoning lacks a sufficiently systemic or process-oriented perspective—concerns the problem of locating representations (cf. Bianchi & Castiello, 2023b). This can lead to the mistaken belief that representations reside in a specific place or organ, where they are formed and stored in a fixed format. Such thinking overlooks the dynamic, multi-process, and often distributed nature of cognitive systems. The risk, therefore, is to expect the existence of some concrete "entity"—a scale model, a schema, or a coding mechanism—fixed and structured within organisms, rather than recognizing the fluid and interactive functioning of cognitive processes, despite the material bases of every process (cf. Bianchi, 2021).

Even if all or some of these critical considerations are accepted, the fundamental question remains: if representations are no longer the focus, what exactly are we discussing? Is this merely a matter of terminology used to describe the same processes across different branches of scientific research, or would the exclusion of representations altogether leave an explanatory gap in understanding at least some levels of organismal cognition? Various perspectives exist, ranging from radical eliminative positions to systemic and complex-based approaches to cognitive behavior. By refining the level of analysis, these approaches offer a substantially revised conception of representation that avoids falling into biological reductionism.

If it were considered impossible or unnecessary to rely on any form of representation (cf. Lee, 2023), treating it as a redundant element, this stance would not necessarily entail the exclusion of plants from research on behavior and cognition (cf. Segundo-Ortin & Calvo, 2019).

4.5.7 Representations in Plants?

Attention can now be directed more specifically to the case of plants. Within representational approaches, the prevailing assumption is that the nervous system provides the necessary architecture and functionality for the existence of mental representations, which are considered essential for the manifestation of cognitive phenomena. However, mental representations, as conceived in common-sense psychology or classical cognitive science, represent only one category of representations. Despite being largely modular, and “dissectable” organisms characterized by the autonomy of their various modules (cf. de Kroon et al., 2009), plants nonetheless exhibit a notable capacity to organize signals into complex intercellular signaling networks. This ability to process environmental information supports tasks related to communication, reproduction, and overall survival, supported by physiological coordination mechanisms. As discussed in Chap. 3, plants possess phenotypic plasticity, which enables a continuous spectrum of behaviors in response to environmental cues. As a result, many plant activities do not appear to be strictly determined from a physiological standpoint, that is, they do not follow simple “one-to-one” stimulus-response patterns without room for variation or modulation. Thus, plants can be studied as interactive systems with self-organizing capacities, dynamically coupled to their environment (cf. Gomila & Calvo, 2008).

Many contemporary approaches to plant behavior and cognition have moved away from traditional notions of representation, opting for alternative explanatory frameworks. Historically, the possibility of representational aspects in plants has often been overlooked. Early comparative cognition studies typically excluded plants, classifying them as non-cognitive entities whose physiological activities were believed to depend solely on non-representational perceptual states (cf. Debono & Souza, 2019). This exclusion likely stemmed from strong mentalistic conceptions of representation, according to which representations are rooted in mental states, a process thought to require the presence of a nervous system (cf. Maher, 2017, pp. 105–106). Given the absence of such representations in plants, an important question arises: should the idea of any form of representation in plants be entirely discarded, or might a revised concept of representation be applicable to explain plant cognition? The aim is to determine whether some interpretative framework involving representation could be useful for understanding their behaviors (cf. Bianchi, 2024). Certain embodied, systemic, and action-oriented approaches, since they are not radical, do not entirely dismiss the possibility of representations. Continuing the reflections of Bianchi and Castiello (2023a, 2023b), it is worth questioning whether the long-standing exclusion of plants from the "cognitive club"—justified by some precisely due to the absence of (mental) representations—stems from an inadequate conception of representation for describing the behaviors of these organisms, or whether, regardless of the conception adopted, the use of the notion of representation inevitably and unnecessarily complicates the explanation of plant capacities.

Regarding the offline aspect of cognition—often considered a hallmark of cognitive sophistication and a basis for hypothesizing the presence of representations—in plants, even though our understanding of mechanisms for processing information detached from direct environmental interaction remains limited, some authors have already suggested, without invoking cerebral representational internalism (cf. Hiernaux, 2021), that plants may exhibit a minimal form of offline control. Examples include heliotropic leaf movements (cf. Kreps & Kay, 1997; Garcia Rodriguez & Calvo Garzon, 2010), the ability to choose between competing stimuli, behaviors involving the inhibition of responses to certain stimuli in anticipation of greater future benefits, such as resource acquisition (cf. Novoplansky, 2016), and forms of learning, as demonstrated by *Mimosa pudica*, which can remember the harmlessness of certain stimuli while still responding to others (cf. Gagliano et al., 2014).

In connection with the potential representational aspect, some researchers have suggested that plants may possess internal systems for integrating environmental information (cf. Trewavas, 2003). These would function more as a distributed and dynamic models, continuously tracking environmental conditions across the plant body. The use of terms as "internal map" is thus appropriate only in a metaphorical sense, reflecting the non-centralized and processual nature of plant informational dynamics (cf. Cvrčková et al., 2009). Even so, interpretations of certain information organization schemes in representational terms are not universally accepted. In the case of plants, this is particularly relevant because, despite having internal signaling networks like all other organisms, much of their life is directed toward environmental interactions (cf. Parise & Marder, 2024). Furthermore, when considering the notion of a bodily representation or internal map, a crucial question arises: does such a model or process persist independently of the plant's direct perceptual experiences, and if so, for how long (i.e., how detachable is it)? Additionally, to what extent can it be modified in response to new or changing environmental conditions (cf. Michmizos & Hilioti, 2019)? A related concern is the risk involved when terms like "representation," "map," "scheme," or "model" are employed to explain behavior: such terms may inadvertently reify a set of elements and processes that are, in fact, more diffusely integrated into the organism's overall bodily functioning (cf. Bianchi & Castiello, 2023a).

However, for the sake of explanatory clarity, it is also necessary to employ a single term to encapsulate and describe a set of physically grounded processes that are the focus of investigation.

Aware of these potential pitfalls, it may be advantageous to rely on the explanatory power of the network concept when discussing biological signaling dynamics. Chemical-electrical networks are notable for maintaining a structure that preserves organizational integrity while remaining adaptable and responsive to functional changes driven by contingent demands of the interactional context (cf. Bianchi, 2021, 2022, 2024). Framing the understanding around the functioning of bodily networks offers a flexible perspective to explain information exchanges at the level of intercellular signaling, which can initiate cascades of effects culminating in observable behavioral responses. Biological networks exhibit the capacity to selectively recognize relevant signals, retain information, and modulate it in response to environmental pressures.

Biological diversity is the result of different eco-evolutionary histories, in which functional structures and organizations arise from forms of channeling of information that have a genetic basis (the possibility of building, by repeating memorized patterns, the different phenotypes) in the relationship with environmental influences. Through a complex interplay of multilevel processes—including differentiation, selection, recognition, memory, variation, anticipation, and learning—living systems navigate environmental challenges and adapt to changes while maintaining their functional and agentive identity. This understanding has led to growing discussions about non-neural systems that, despite lacking brains, possess intercellular information signaling mechanisms enabling them to perform essential survival tasks. As Baluška and Levin (2016) observed, neuronal networks have optimized information processing capabilities that were, to some extent, already present within the somatic cellular networks of non-neural tissues, albeit with varying degrees of integration and processing speed. These pre-nervous system chemical-electrical signaling modes were employed by biological networks to coordinate physiological responses. Mustard and Levin (2014) further proposed that non-neural tissues also exhibit forms of plasticity, involving changes in connectivity through electrical signaling and variations in ion channels (cf. Hedrich, 2012; Baluška & Mancuso, 2013). Interpreting these phenomena through the framework of functional network dynamics may help elucidate eco-physiological responses to environmental signals, thereby supporting certain processes across a broader range of organisms.

In this context, Souza et al. (2018) and Debono and Souza (2019) proposed examining the variable organization of plant molecular networks in representational terms. Here, "representation" does not refer to conceptual constructs but rather to a functional map or a flexible procedural model. Such hypothetical representations have been proposed as potentially guiding plant behavior during environmental interactions. Plants are viewed as modular, body-rooted, and enactive systems capable of engaging with their environment through the remodulation of chemical and electrical configurations. Functional molecular patterns emerge from dynamic changes in metabolic networks, shaped by the various arrangements among cellular components corresponding to the states that cells can have under specific local conditions. The question is whether these process configurations can be memorized and later re-expressed in situations similar to those previously encountered, potentially being modified along with future behavioral responses.[26] Perceptual learning in plants is orchestrated by distributed networks and the organism's overall bioelectrical activity, the plant's electrome, which together enable plants to respond to stimuli both broadly and specifically (cf. Debono & Souza, 2019).

Adopting a procedural perspective that emphasizes the temporal variations in signaling network activity challenges the notion of rigid coding and localization of representations, while still acknowledging the physical grounding of these processes. In this context, a key objective is to understand the flexibility of plant responses (cf. Bianchi et al., 2025a). This flexibility may arise from interactions between internal networks and external, ecologically relevant factors. The integration and contextual modulation of information within plant body networks appears to facilitate diverse organizational states, enabling plants to address a wide range of environmental challenges (cf. Trewavas, 2014).

A potential criticism is that hypothesizing functional metabolic or molecular representations linked to signaling networks may appear redundant, as these processes could ostensibly be explained simply as signaling mechanisms, seemingly unrelated to "cognition." However, some of these processes may play crucial roles in guiding plant behaviors involving processes of memory, transmission, organization, anticipation, and learning of environmental information through complex causal chains of effects (cf. Baluška & Levin, 2016).

Even if the hypothesis of molecular or metabolic representation ultimately proves empirically incorrect, exploring this proposal opens valuable avenues for addressing a range of questions in future research, potentially yielding practical implications. For instance, an interesting connection arises concerning how these considerations intersect with systemic approaches, complex systems theory, network theories, and, notably, the field of network neuroscience, but with a focus on organisms without neurons (cf. Bianchi & Castiello, 2023a). In the context of neo-connectionist theories, particularly in the study of neural dynamics, cognitive activity is explained in terms of activation flows across neural networks, with each network processing only fragments of information. These explanations emphasize distributed and subsymbolic functions, which lack explicit semantic content (cf. Di Francesco et al., 2017, p. 167). Accordingly, a potential research objective could be to investigate whether, and in what form, representational aspects in plants might pertain to the coding of configurational patterns, or whether they align more closely with the subsymbolic nature of signaling patterns within bodily networks (cf.

---

[26] The concept of internal "molecular" representation has been proposed in bacteria to account for their ability to acquire environmental information, such as temperature or oxygen levels, and use it to guide behavior and anticipate future conditions, thereby enabling appropriate responses (cf. LeDoux, 2020, p. 71).

Iachini, 2023). Furthermore, hypotheses regarding forms of distributed intelligence in plants could be explored alongside assumptions from network neuroscience, which has identified essential characteristics for intelligence. These include "crystallized" aspects, akin to memory processes involving robust pathways accessing relatively stable network states, and "fluid" aspects, associated with learning processes engaging transient or weaker pathways accessing less stable states (cf. Simard, 2018; Barbey, 2018). Accordingly, one could investigate which plant behaviors or processes—such as chemical synthesis and recognition abilities or the variability of metabolic activities—might exemplify fluid intelligence. A systemic approach, examining the relationship between memory processes in plants and potential representational aspects, could also link to De la Fuente's (2015) proposal to describe the molecularly and electrically interconnected functioning of complex networks aimed at preserving and transforming information at the cellular level, involving two systems: a more conservative one related to genetic memory, and a more flexible and dynamic one linked to metabolic memory (cf. Galviz et al., 2020). This raises the question of whether a hypothetical concept of representation could be framed within the functional configurations of networks encompassing both conservation and modification processes, thus enabling sequences that are partially constrained yet responsive to contextual variations (cf. Bianchi & Castiello, 2023a).

Attempting for a moment to operate within a non-eliminative framework of representation analysis, as noted by Alegiani and Marraffa (2023), if, from a continuist perspective, various cognitive systems possess some form of simple or more complex representations, then simplicity—understood as the possession of certain shared basic capacities—becomes the fundamental trait of cognition. For instance, although plant systems may exhibit lower capacities for future projection or a reduced degree of freedom (i.e., greater constraints) in anticipating and predicting sensory interactions (cf. Calvo & Friston, 2017), this does not imply that simpler models are any less timely or accurate in their agentive responses. A key question, therefore, is whether a minimum threshold of organizational complexity must be exceeded for specific behaviors to emerge in different systems (cf. Bassett & Gazzaniga, 2011). From an emergentist perspective, while in animal nervous systems cognitive abilities appear to arise from a baseline level of neural complexity, plant systems might reflect a different form of functional organization. Alternatively, as proponents of plant cognition and behavior have argued (cf. Trewavas, 2003; Baluška et al., 2006; Calvo & Keijzer, 2011; Calvo, 2016), there may be functional equivalents in the organization of plant body signaling networks, distributed across various tissue types, each with its own structural constraints (cf. Baluška et al., 2021).

A parallel worth further exploration, as discussed in the previous sections, is the one between the features and functions of connectomes and interactomes (cf. Trewavas, 2014, pp. 237–240). This raises the question of which elements these morpho-functional organizations share and which unique, non-translatable characteristics they possess. In particular, regarding the representational issue—namely, whether a representational interpretation of the processes underlying plant behavioral variation is justified—it is worth asking what kinds of aspects might be involved. Are we dealing with models, maps, or procedural schemes that recur under certain circumstances to guide behavior? Or could these, rather than being a form of representation, be better understood as mechanisms and signaling processes that act as causal triggers or transmitters, without the need to label them as "representations"?

At a broader level, regardless of whether a representational interpretation is adopted, a further line of inquiry might concern how open to modification these functional configurations, or networks, are. One of the main risks to be avoided in developing a representational interpretation of plant capacities is the theoretical challenge posed by rigid encoding and mirroring of the external environment. Instead, the various configurations of network components during signal exchange should be understood from a composite and dynamic perspective (cf. Souza et al., 2018). This approach emphasizes a processual view, describing dynamics that are relatively flexible yet constrained.

A critical area for further research concerns the integration of signals within networks, specifically in determining how, if at all, modular organisms can transition from localized organization to a more extensive systemic organization of signaling and corresponding behavioral responses across the entire organism (for instance, systemic and electrical signaling via the phloem). On the methodological side, claiming that this potential representational aspect—which may play a role in guiding behavior by initiating specific responses—is the sole factor necessary to explain these complex processes would be overly reductive. It is likely that numerous other contributing factors exist, some of which we may not yet recognize or fully understand (cf. Bianchi & Castiello, 2023b).

Recognizing both the heuristic value and the potential pitfalls of such perspectives allows for a more critical exploration. In line with the issues raised by Bianchi and Castiello (2023a), it is possible to investigate whether a notion of representation might be equivalent to discussing, in more descriptive terms, plants' capacities for fluid memory[27] and learning. The central question then becomes: What, if any, is the relationship between plants' memory capacities and the use of representational aspects in explaining cognition? Do the memorization processes themselves constitute a form of "representation," or do they merely contribute to or shape representational phenomena? Ongoing research into plant memory mechanisms, learning capacities, and interactomes, may help to clarify, or even resolve, the question of the existence and nature of plant representations. Alternatively, would it be more appropriate and effective to analyze these processes without incorporating any notion of representation?

To summarize the discussion on the concept of representation: it remains an open question whether the notion of representation is still necessary for the study of plant cognition or whether it would be more beneficial to abandon it entirely. The debate centers on whether it is sufficient to render the concept of representation more

[27] In the case of plants, an important role is played by epigenetic memory mechanisms and by the species' "memory" resulting from intergenerational experience, given that a certain capacity for tracking biological information, a form of encoding, is already embedded in the genome as a result of selection processes.

flexible, dynamic, and process-oriented—thus avoiding the issues associated with rigid coding and mirroring—or whether it is preferable to discard this theoretically charged concept altogether. Such a move would open the way for the development of an ecological-relational and multilevel approach to the study of cognition, including that of plants. Another line of inquiry concerns whether the concept of representation should be reframed as part of a complex and relatively flexible system, and whether it can accommodate dynamic characteristics, particularly in light of the explanatory potential of network-based models. Instead of focusing on representations, it may be more productive to differentiate between inherited traits acquired through phylogeny and those learned or developed throughout an individual plant's life. This perspective emphasizes the potential ability to form and organize networks of information in a manner that is physiologically and ecologically influenced but not rigidly determined. Within this framework, the goal would be to better identify and understand, from a relational perspective, the elements, processes, and capacities that enable plants to respond to environmental information sources.

The concept of representation is not essential. Whether plant cognitive phenomena are reformulated, abandoned, or analyzed through new or alternative frameworks, the goal remains the same: to find the most effective means of explaining the capabilities of plants, i.e., the processes and strategies they use to accurately respond to and anticipate environmental challenges. At present, the only certainty is the lack of comprehensive answers to these questions (cf. Bianchi & Castiello, 2023b). One of the primary contributions that research in plant cognition could offer is the development of an ecological-relational conception of the mind (cf. Bianchi, 2021, 2022, 2024), moving beyond traditional notions of (mental) representation and exploring alternative explanatory frameworks.

4.5.8 Cognition and Temporality of Processes

At a general level, within the study of cognitive abilities across humans and other animals, the importance of the relationship between the expression and modulation of individual cognitive capacities and the varying levels of memorization processes and capacity for learning has been emphasized. For instance, Richard F. Thompson (1930–2014) argued that without memory, there can be no mind as we understand it, as what we perceive, think, and feel in any situation is shaped by our past experiences. It could be further argued that, without memory, no individual identity can be sustained. The present is always filtered through the use of information and knowledge derived from previous interactions in order to make inferences, predict outcomes, and adjust expectations, thereby modeling behavior in accordance with potential future scenarios (cf. Friston, 2005; LeDoux, 2020, pp. 353–355).

On the other hand, much like the concept of representation, the concept of "memory" has also been the subject of significant criticism. For example, the ecological psychology approach challenges traditional cognitive science by rejecting the notion of memorization ability as the storage of representations. This criticism arises because, in memory-based recognition models—where memory is conceptualized as a set of stored information about an event, object, or place—the problem lies in determining how an environmental stimulus specifies which representation should be retrieved. If an input accurately specifies the appropriate memory representation, then the entire concept of stored representation becomes redundant. In other words, if the input alone can specify the event, there is no need for the system to contain it in a representational format (cf. Michaels & Carello, 1981, p. 185). This example illustrates how ecological psychology critiques conventional theories of cognitive systems, often regarded as conceptual descendants of classical cognitivism.

In this context, developing a concept of fluid (variably distributed) body memory appears valuable for understanding individual behaviors that activate various cognitive systems, potentially extending even to plants. The question is whether, and what role, an individual form of bodily memory independent of the nervous system plays, and how it differs from an intergenerational form of memory peculiar to the species. It is also important to examine how these forms of memory are connected to mechanisms of anticipation and learning of environmental information. Investigating this connection could be instrumental in explaining the recurrence and variation of plant activities, as well as in linking the in-depth study of the functioning of biological signaling networks with other relevant factors.

The preceding analysis of the concept of representation in plants—regardless of the specific type hypothesized, and even if it may ultimately prove more reasonable to abandon the notion of representation altogether—aims precisely to deepen the understanding of the nature and relevance of the mechanisms and processes that aneural organisms, such as plants, employ and modify in their perception–action dynamics. For instance, in response to Bianchi and Castiello (2023a), Iachini (2023, p. 402) observed: "either we accept that non-mental plant representations constitute a simplified form based only on implicit memory, or we reserve the term representation for more complex phenomena, involving various forms of memory and flexible simulation." Of course, this reasonable option may not be the only available approach. From a non-definitive perspective, the organization of information can be described as involving a form of memory tied to basic learning mechanisms found across all life forms, such as procedural or implicit memory. This type of ability enables action by recalling how to perform tasks without self-conscious or biographical remembrance of past experiences. In the case of plants, it may be worthwhile exploring how environmental information is retained and modified within plant systems, and connecting this phenomenon to a concept associated with procedural memory. A comparison in examining this possibility is Eric Kandel's (2001) research on the molecular mechanisms of information retention and changes in learning processes within animal brains. Investigating how animal nervous systems store and modify information during learning invites a parallel question for plant cellular systems: what changes occur within plant bodies when they learn, and what functional networks are involved in strengthening new informational pathways (cf. Trewavas, 2014)? Furthermore, how are these processes related to plant epigenetic memory? Ongoing studies on intercellular signaling and the organization and storage of information in non-neural tissues are increasingly elucidating these aspects (cf. Baluška & Levin, 2016). Thus, at a general level, this line of research would benefit from a more systemic, relational, and process-based approach to better understand plant abilities. These capacities should be explored as decentralized intra- and intercellular processes that influence responses at various levels of the organism's body organization, ranging from local and circumscribed responses to systemic ones encompassing the entire organism (cf. Morabito, 2023).

Moreover, while maintaining continuity with the points previously discussed, but from a different perspective within the analysis, it is important to emphasize that organisms have developed mechanisms and strategies that enhance their resilience in the face of future perturbations (cf. Gazzaniga, 2019, p. 173). This adaptability stems from the fact that living systems are the result of historical processes that demand their persistence throughout the lifespan and their adjustment to constantly changing interactions. As products of eco-evolutionary processes, organisms are inherently oriented toward meeting future needs. This understanding highlights the importance of the temporal dimension of cognition (cf. Bianchi et al., 2025b). Consequently, a fundamental connection of this analysis lies in contemporary studies on predictive models that explore the anticipatory capacities of perceptual-cognitive systems. From this perspective, even plants, through their behaviors, demonstrate an ability to modify and adapt the form and activities of their bodies. They actively contribute to creating and reshaping their environment, which they "recognize" and "respond to" functionally, to the extent that they can "anticipate" its transformations through adaptive adjustments (cf. Morabito, 2023).

These topics have also been variously connected to the issue of representation. Once again, beyond the different possible options, the value lies in the continuation of the analysis itself.

Modern cognitive science increasingly acknowledges the critical role of anticipation in the functioning of cognitive systems (cf. Sims, 2021a, 2021b). An ongoing debate concerns whether these anticipatory activities necessarily involve representations, and if so, what kind. Proponents often argue that anticipatory behavior presupposes representational activity (cf. Pezzulo, 2008). Scholars such as Gladziejewski (2016) have suggested that predictive coding theory (PCT), with its broad explanatory power, could offer a comprehensive framework for understanding cognition within contemporary cognitive science. One of PCT's core assumptions is that agents' cognitive interactions with the environment are mediated by internal representations, dynamic, probabilistic generative models that guide action, minimizing prediction error. Consequently, many proponents view PCT as a representational theory. However, is this the definitive approach?

Further specifying, Borghi (2023) raised concerns about the possibility of reconciling predictive approaches with non-representational perspectives. In response to Bianchi and Castiello (2023a), who noted that most approaches to plant cognition are either non-representational or employ the concept of representation in a significantly modified form, Borghi questioned whether a predictive approach can indeed be aligned with a non-representational perspective, given the apparent indispensability of the notion of representation in some models, such as predictive coding, particularly when representation involves detachment from the "here and now." These reflections first require clarifying the type of representation involved.

For instance, both Borghi (2023) and Iachini (2023) referred to the concept of simulation, which is generally rooted in sensorimotor experiences and based on offline processes. These simulations imply a detachment from the present moment and involve memory, learning, and anticipation processes (cf. Barsalou, 2008). Simulation is understood as a form of embodied representation, linked to bodily attributes and arising from multimodal integration processes conditioned by the organism's actions and goals (cf. Mannella & Tummolini, 2023). Exploring plant cognition could provide new insights into the application of influential models, such as, among various possibilities, the theory of perceptual symbols (cf. Barsalou, 2008, 2016)[28] or the theory of event coding (TEC) (cf. Hommel, 2021). Both models, at a general level, share the idea of explaining actions through the anticipation of their perceptual outcomes and the multimodal simulation (or representation) of action with predictive capabilities. Within a naturalized and embodied framework of cognitive processes, these and other models appear capable of reconciling the concept of representation with the anticipatory aspects of cognition.[29] Beyond these, of course, there may be other, potentially more promising, models and approaches for investigating the various capabilities of plants (cf. Bianchi & Castiello, 2023b).

Within the framework of predictive processing (PP) theory, Calvo and Friston (2017) described how plants respond to salient features of their ecological niches through the Free Energy Principle (cf. Friston, 2005). According to this perspective, every self-organized biological system can, by its unique organization, tools, and

[28] Barsalou (2008) significantly developed and formalized the grounded cognition approach, which posits that cognition is rooted in modal simulations, bodily states, and situated actions. This framework accounts for both the flexibility and context-dependence of representations, as well as the capacity for abstraction, which is facilitated by multimodal processes and the connection between conceptual reasoning and concrete life practices.

[29] Embodied and grounded perspectives, which emphasize the role of sensorimotor processes and employ the concept of simulation, have investigated the perceptual foundations of representations, particularly in understanding aspects of human language. For instance, hearing a particular word triggers a simulation through the offline reactivation of neural networks initially engaged during the direct perception of the word's referent (cf. Jeannerod, 2006). In this context, Borghi (2023), aligning with Bianchi and Castiello's (2023a) critique of rigid codification in representation, emphasized the significance of communicative context and, consistent with the enactivist framework, focused more on interactional dynamics than on the traditional representer–represented relationship. For example, hearing the word "fir" reactivates a neural pattern similar to that generated when encountering such a tree in person. Just as research on language functioning informs conceptions of representation, studies on representation can, in turn, illuminate how communication systems operate, particularly in relation to behavior initiation. Building on these ideas, it would be valuable to explore ongoing research on plant signaling and biocommunication (cf. Witzany & Baluška, 2012; Bianchi & Castiello, 2023b). Such studies could enhance our understanding of certain cognitive behaviors in plants, which may involve a representational component linked to diverse forms of sign mediation. Specifically, this could include the recognition of chemical compounds, with behavioral responses modulated by the dosage and contextual factors of plant communication (cf. Affifi, 2013; Witzany, 2016; 2018; Bianchi, 2024).

strategies, reduce predictive errors, essentially minimizing surprise over time. Predictive systems attempt to forecast the causes of incoming sensory stimuli, including through action. The hypothesis is that brains, and indeed the organizational structures of aneural systems, function as predictive devices (probabilistic generative models), using perceptual and agentive capacities to minimize the gap between external sensory signals and internally generated sensory predictions. From a continuity perspective, it is plausible that any organism can embody a predictive model and be mathematically described as a system minimizing prediction errors, the disparity between its predictions and incoming sensory inputs. These processes are essential not only for maintaining system integrity but also for preventing functional disintegration. In the case of plants, Calvo and Friston (2017) suggested that, as organisms in continuous interaction with their environments, plants can detect information and adjust their behavior through active inference processes that anticipate the environmental sources of sensory stimulation. The authors argued that plant behavior, with its unique characteristics, can be examined within a predictive framework that incorporates an expanded understanding of perception and action. This approach aligns with certain principles of ecological theories, which are particularly well-suited to the analysis of non-centralized, modular organisms like plants, characterized by repetitive and semi-autonomous structures. Accordingly, it has been proposed that plants utilize ecological information based on the perception of affordances, rather than relying on detailed internal, reconstructive representations. This perspective distances itself from traditional notions of mental representations aimed at accurately mirroring reality, even though inferential aspects—as a form of active inference linked to action possibilities—remain present. Bianchi and Castiello (2023a, 2023b) emphasized the importance of linking predictive models with enactivist theories (cf. Kirchhoff & Robertson, 2018) and ecological psychology approaches. This connection lies at the heart of the debate over whether learning and anticipatory behaviors necessarily involve inferential and/or representational elements, and, if so, what kinds (cf. Calvo et al., 2016; Gallagher, 2017; Souza et al.,2018).

Nevertheless, substantial theoretical and experimental work remains necessary to determine the most appropriate approaches for studying the various aspects of plant abilities. This includes discerning when to adopt principles from ecological psychology, enactivism (in its more or less radical versions), or other embodied, non-representational frameworks. It remains to be clarified when and how the epistemic tools of one approach should be applied independently or integrated with others. Depending on the specific behavioral trait under investigation and the explanatory level required, it may prove useful to adopt multiple approaches. Looking ahead, a pluralistic and multidisciplinary framework—one that draws upon diverse models and theories to address different levels of cognitive processes—could yield more precise analyses. This would require a more explicit integration of the most promising approaches and related analysis techniques and methodologies.

Proceeding with the analysis, Alegiani and Marraffa (2023), in their response to Bianchi and Castiello (2023a), discussed how predictive theories, from a continuist perspective and within the biogenic approach to cognition, have softened the traditional divide between two conceptualizations of representation. On one side are approaches that conceive of representations as internal substitutes with "epistemological" purposes, aimed at providing a truthful image of the environment. On the other side are approaches that regard representations as internal states serving as causal correlations or models for environmental control, without "epistemic" aspirations. By emphasizing bodily aspects, particularly sensorimotor processes and interactions with the environment, the epistemological stance of more "conservative" theories of representation (cf. Hohwy, 2019) has been counterbalanced by "active" or "action-oriented" representations (cf. Clark, 2015), which prioritize the capacity to act effectively and maintain a strong grip on the environment. In this context, the "epistemological" goal is no longer about mirroring the environment but is instead fulfilled through the accuracy of action. If the representation of a target or goal were insufficiently accurate, the body would be unable to act appropriately in the real world. As studies on the perception-action link have shown, the epistemological and active components can indeed coexist (Alegiani & Marraffa, 2023). Similarly, Morabito (2023), also responding to Bianchi and Castiello (2023a), suggested that exploring how plants memorize and learn environmental information could enrich our framework of analysis by promoting research into "non-hermeneutical," "pragmatic," and "action-oriented" forms of cognition, such as task-oriented abilities. In this view, cognition concerns the "where" and "how" rather than a metacognitive "knowing to know" about the "what" (cf. Bianchi, 2021, 2022, 2024). For plants, agency, their capacity to act, manifests through bodily modifications, such as growing in specific directions to acquire resources or avoid obstacles, modulating metabolic activities, or producing and emitting chemical compounds. These behaviors, which are not rigidly determined, can influence and modify the environment, particularly by affecting the behavioral responses of other organisms.

The ongoing question is whether an accurate response is possible without any representational aspect, even in a reformed sense. Alegiani and Marraffa (2023), in their response to Bianchi and Castiello (2023a), emphasized the importance of investigating whether plants possess any form of "attentional" capability to ensure action accuracy (Bianchi & Castiello, 2023b). As discussed in Chap. 3, Parise et al. (2022) hypothesized that plants might attain attentional states in response to different environmental signals or stimuli. They proposed an empirical investigation using electrophysiological techniques to analyze how the overall bioelectrical activity of plants, their electrome, changes according to the type of ecological stimulus (cf. Souza et al., 2017; de Toledo et al., 2019; Parise et al., 2021). In studying plant behaviors, "paying attention" could refer to the ability to select relevant elements (discriminate differences) and respond accurately to a given stimulus in a specific context. One goal, therefore, is to clarify the relationship between anticipatory behaviors and action accuracy, which involve not only the potential for movement but also the synthesis and modulation of chemical compounds. Exploring the link between attention span, memory and learning processes in plants could offer valuable insights (cf. Thellier & Lüttge, 2012; Galviz et al., 2020; Parise et al., 2022).

The cognitive life of an organism is shaped by a set of processes influenced by past events and organized in anticipation of the future, based on predictions of what may occur or recur. From this perspective, cognition can be viewed as a set of processes and capabilities that enable an organism to orient and act (live) in the present, drawing on the past and preparing for the future. In other words, organisms are historical systems, highlighting the inseparable link among these temporal dimensions, a link that allows for both the repetition and variation of behaviors. In this sense, organisms are evolving, dynamical, and historical processes (cf. Bianchi et al., 2025b).

To conclude this exploration of plant characteristics and capacities, particularly in relation to cognitive science and philosophy of mind, it can be observed that the complex issue of representation, along with others, has served as a platform for addressing several interconnected topics. These include the relationship between complexity and levels of emergence in cognitive systems, the emphasis on predictive theories and embodied, action-oriented approaches, and the interrelations among memory, learning, anticipation, and attention capacities. These areas warrant further theoretical and experimental investigation.

Research in this area is still in its early stages, and numerous questions remain unanswered. However, this is precisely the meaning of "research": to continue posing questions and identifying new connections. At this stage, it is crucial to emphasize the need for further evidence and the development of appropriate experimental approaches to better understand the behaviors of plants. The call is to persist in this exciting and challenging interdisciplinary inquiry. Such research is valuable not only for advancing our understanding of plants' interactional behaviors, but also for deepening our knowledge of cognition as a whole, encompassing the simplest, the most complex, and the simply different cognitive systems. Moreover, this research will refine our epistemic approaches and tools, helping to determine what can be understood and the most suitable methods for achieving that understanding.

References

Adams, F. (2018). Cognition wars. *Studies in History and Philosophy of Science Part A, 68*, 20–30.

Adams, F., & Aizawa, K. (2010). The value of cognitivism in thinking about extended cognition. *Phenomenology and the Cognitive Sciences, 9*, 579–603.

Affifi, R. (2013). Learning plants: Semiosis between the parts and the whole. *Biosemiotics, 6*(3), 547–559.

Alegiani, A., & Marraffa, M. (2023). Cognizione vegetale ed elaborazione predittiva. *Giornale Italiano di Psicologia, 50*(2), 381–386.

Anobile, G., Arrighi, R., Castaldi, E., & Burr, D. C. (2021). A sensorimotor numerosity system. *Trends in Cognitive Sciences, 25*(1), 24–36.

Anscombe, F. J. (1963). Sequential medical trials. *Journal of the American Statistical Association, 58*(302), 365–383.
Anten, N. P., & Chen, B. J. (2021). Detect thy family: Mechanisms, ecology and agricultural aspects of kin recognition in plants. *Plant, Cell & Environment, 44*(4), 1059–1071.
Archetti, M., Scheuring, I., Hoffman, M., Frederickson, M. E., Pierce, N. E., & Yu, D. W. (2011). Economic game theory for mutualism and cooperation. *Ecology Letters, 14*(12), 1300–1312.
Austin, C. J. (2016). The ontology of organisms: Mechanistic modules or patterned processes? *Biology & Philosophy, 31*, 639–662.
Austin, C. J. (2020). Organisms, activity, and being: On the substance of process ontology. *European Journal for Philosophy of Science, 10*(2), 13.
Axelrod, R. (1984). *The evolution of cooperation*. Basic Books.
Axelrod, R., & Hamilton, W. D. (1981). The evolution of cooperation. *Science, 211*(4489), 1390–1396.
Baker, E. N. (2017). The intelligence of plants and the problem of language: A Wittgensteinian approach. In M. Gagliano, J. C. Ryan, & P. Vieira (Eds.), *The language of plants: Science, philosophy, literature* (pp. 136–154). University of Minnesota Press.
Baluška, F., & Levin, M. (2016). On having no head: Cognition throughout biological systems. *Frontiers in Psychology, 902*.
Baluška, F., & Mancuso, S. (2013). Ion channels in plants: from bioelectricity, via signaling, to behavioral actions. *Plant signaling & behavior, 8*(1), e23009.
Baluška, F., Miller, W. B., Jr., & Reber, A. S. (2021). Biomolecular basis of cellular consciousness via subcellular nanobrains. *International Journal of Molecular Sciences, 22*(5), 2545.
Baluška, F., Volkmann, D., Hlavacka, A., Mancuso, S., & Barlow, P. W. (2006). Neurobiological view of plants and their body plan. In F. Baluška, S. Mancuso, & D. Volkmann (Eds.), *Communication in plants* (pp. 19–35). Springer.
Barandiaran, X. E., Di Paolo, E., & Rohde, M. (2009). Defining agency: Individuality, normativity, asymmetry, and spatio-temporality in action. *Adaptive Behavior, 17*(5), 367–386.
Barbey, A. K. (2018). Network neuroscience theory of human intelligence. *Trends in Cognitive Sciences, 22*(1), 8–20.
Barsalou, L. W. (1999). Language comprehension: Archival memory or preparation for situated action? *Discourse Processes, 28*(1), 61–80.
Barsalou, L. W. (2008). Grounded cognition. *Annual Review of Psychology, 59*(1), 617–645.
Barsalou, L. W. (2016). On staying grounded and avoiding quixotic dead ends. *Psychonomic Bulletin & Review, 23*(4), 122–1142.
Barsalou, L. W., & Wiemer-Hastings, K. (2005). Situating abstract concepts. In D. Pecher, & R. A. Zwaan (Eds.), *Grounding cognition: The role of perception and action in memory, language, and thought*, (pp. 129–163).
Barthélémy, D., & Caraglio, Y. (2007). Plant architec-ture: A dynamic, multilevel and comprehensive approach to plant form, structure and ontogeny. *Annals of Botany, 99*(3), 375–407.
Bassett, D. S., & Gazzaniga, M. S. (2011). Understanding complexity in the human brain. *Trends in Cognitive Sciences, 15*(5), 200–209.
Bateson, G. (1976). *Verso un'ecologia della mente*, Longo, G., & Trautteur, G. (Trans.). Adelphi (Original work published 1972, *Steps to an ecology of mind: Collected essays in anthropology, psychiatry, evolution, and epistemology*. Chandler Publishing Company).
Bechtel, W. (1998). Representations and cognitive explanations: Assessing the dynamicist's challenge in cognitive science. *Cognitive Science, 22*(3), 295–318.
Bechtel, W. (2009a). Constructing a philosophy of science of cognitive science. *Topics in Cognitive Science, 1*(3), 548–569.
Bechtel, W. (2009b). Generalization and discovery by assuming conserved mechanisms: Cross species research on circadian oscillators. *Philosophy of Science, 76*, 762–773.
Beer, R. D. (2000). Dynamical approaches to cognitive science. *Trends in Cognitive Sciences, 4*(3), 91–99.
Bentley, K., & Chakravartula, S. (2017). The temporal basis of angiogenesis. *Philosophical Transactions of the Royal Society B: Biological Sciences, 372*(1720), 20150522.

Bessonov, N., Morozova, N., & Volpert, V. (2008). Modeling of branching patterns in plants. *Bulletin of Mathematical Biology, 70*(3), 868–893.
Bianchi, M. (2021). *La vita ramificata.* Cognizione e comportamento nelle piante fra scienza e filosofia. Mimesis.
Bianchi, M. (2022). Networks and ramifications: Relational perspectives in plant cognition. *Rivista Internazionale di Filosofia e Psicologia, 13*(2), 157–168.
Bianchi, M. (2024). Plant cognition. Challenges and horizons in philosophical reflection. *Paradigmi, 1*, 21–46.
Bianchi, M. (2025). Oltre il Mondo Animale: i Darwin e la Botanica. In Gagliasso, E., Pollo, S., & Severini, E. (Eds.), *Origine e attualità dell'Origine dell'uomo. Scienza, Etica, Letteratura* (pp. 33–54). Edizioni ETS.
Bianchi, M., Avesani, S., Bonato, B., Dadda, M., Guerra, S., Ravazzolo, L., Simonetti, V., & Castiello, U. (2025a). Plant behavior: Theoretical and technological advances. *Current Opinion in Psychology, 64*, 102026.
Bianchi, M., Guerra, S., Bonato, B., Avesani, S., Ravazzolo, L., Simonetti, V., Dadda, M., & Castiello, U. (2025b). Temporal Dynamics With and Without a Nervous System: Plant Physiology, Communication, and Movement. *Cognitive Science, 49*(6), e70079.
Bianchi, M., & Castiello, U. (2023a). Il concetto di rappresentazione nelle scienze cognitive classiche e post-classiche: la sfida della cognizione vegetale. *Giornale Italiano di Psicologia, 50*(2), 349–385.
Bianchi, M., & Castiello, U. (2023b). Riflessioni corali nell'analisi interdisciplinare sul concetto di rappresentazione nella cognizione vegetale. *Giornale Italiano di Psicologia, 50*(2), 419–452.
Bliss, T. V., & Lømo, T. (1973). Long-lasting potentiation of synaptic transmission in the dentate area of the anaesthetized rabbit following stimulation of the perforant path. *The Journal of Physiology, 232*(2), 331–356.
Boncinelli, E. (2020). *Il principio di Indeterminazione*. Il Mulino.
Borghi, A. M. (2023). Le piante e gli esseri umani, tra rappresentazione e cognizione. *Giornale Italiano di Psicologia, 50*(2), 387–392.
Bray, D. (1995). Protein molecules as computational elements in living cells. *Nature, 376*(6538), 307–312.
Bruni, D., & Fugali, E. (2012). Scienza cognitiva incarnata e modelli evoluzionistici. *Reti Saperi Linguaggi, 4*(2), 4–11.
Buhrmann, T., & Di Paolo, E. (2017). The sense of agency—A phenomenological consequence of enacting sensorimotor schemes. *Phenomenology and the Cognitive Sciences, 16*(2), 207–236.
Burkhardt, R. W. (2005). *Patterns of behavior: Konrad Lorenz, Niko Tinbergen, and the founding of ethology*. University of Chicago Press.
Cabal, C., Martínez-García, R., de Castro Aguilar, A., Valladares, F., & Pacala, S. W. (2020). The exploitative segregation of plant roots. *Science, 370*(6521), 1197–1199.
Calabretta, R. (2007). Genetic interference reduces the evolvability of modular and non-modular visual neural networks. *Philosophical Transactions of The Royal Society B: Biological Sciences, 362*(1479), 403–410.
Calvo, P. (2016). The philosophy of plant neurobiology: A manifesto. *Synthese, 193*, 1323–1343.
Calvo, P. (2017). What is it like to be a plant?. *Journal of Consciousness Studies, 24*(9–10), 205–227.
Calvo, P., & Friston, K. (2017). Predicting green: Really radical (plant) predictive processing. *Journal of the Royal Society Interface, 14*(131), 20170096.
Calvo, P., & Keijzer, F. (2009). Cognition in plants. In *Plant-environment interactions: From sensory plant biology to active plant behavior* (pp. 247–266). Springer.
Calvo, P., & Keijzer, F. (2011). Plants: Adaptive behavior, root brains and minimal cognition. *Adaptive Behavior, 19*(3), 155–171.
Calvo, P., Baluška, F., & Sims, A. (2016). «feature detection» vs. «predictive coding» models of plant behavior. *Frontiers in Psychology, 7*, 1505.

Calvo, P., Gagliano, M., Souza, G. M., & Trewavas, A. (2020). Plants are intelligent, here's how. *Annals of Botany, 125*(1), 11–28.

Calvo, P., Raja, V., & Lee, D. N. (2017). Guidance of circumnutation of climbing bean stems: An ecological exploration. *BioRxiv*, 122358.

Caraglio, Y., & Edelin, C. (1990). Architecture et dynamique de croissance du platane. Platanus hybrida Brot.(Platanaceae){Syn. Platanus acerifolia (Aiton) Willd.}. *Bulletin de la Société Botanique de France. Lettres Botaniques, 137*(4–5), 279–291.

Carello, C., Vaz, D., Blau, J. J., & Petrusz, S. (2012). Unnerving intelligence. *Ecological Psychology, 24*(3), 241–264.

Caruana, F., & Borghi, A. M. (2013). Embodied Cognition: una nuova psicologia. *Giornale Italiano di Psicologia, 40*(1), 23–48.

Caruana, F., & Borghi, A. M. (2016). *Il Cervello in Azione*. Il Mulino.

Casper, B. B., & Jackson, R. B. (1997). Plant competition underground. *Annual Review of Ecology and Systematics, 28*(1), 545–570.

Castiello, U. (2019). *La mente delle piante. Introduzione alla psicologia vegetale*. Il Mulino.

Castiello, U. (2021). (re)claiming plants in comparative psychology. *Journal of Comparative Psychology, 135*(1), 127–141.

Cellucci, C. (2008). *Perché ancora la filosofia*. Laterza.

Cellucci, C. (2013). *Rethinking logic: Logic in relation to mathematics, evolution, and method*. Springer.

Cellucci, C. (2017). *Rethinking knowledge: The heuristic view*. Springer.

Chalmers, D. J. (1995). Facing up to the problem of consciousness. *Journal of Consciousness Studies, 2*(3), 200–219.

Chamovitz, D. (2012). *What a plant knows: A field guide to the senses*. Scientific American/Farrar, Straus and Giroux.

Chemero, A. (2001). Dynamical explanation and mental representations. *Trends in Cognitive Sciences, 5*(4), 141–142.

Chemero, A. (2009). *Radical embodied cognitive science*. MIT Press.

Chemero, A. (2016). Sensorimotor empathy. *Journal of Consciousness Studies, 23*(5–6), 138–152.

Civello, A. (2013). Rete. In F. Michelini & J. Davies (Eds.), *Frontiere della biologia. Prospettive filosofiche sulle scienze della vita*. Mimesis.

Clark, A. (1997). *Being there: Putting brain, body, and world together again*. MIT press.

Clark, A. (2008). *Supersizing the mind: Embodiment, action, and cognitive extension*. Oxford University Press.

Clark, A. (2015). Radical predictive processing. *The Southern Journal of Philosophy, 53*, 3–27.

Clark, A., & Chalmers, D. (1998). The extended mind. *Analysis, 58*(1), 7–19.

Clark, A., & Grush, R. (1999). Towards a cognitive robotics. *Adaptive Behavior, 7*(1), 5–16.

Clark, A., & Toribio, J. (1994). Doing without representing? *Synthese, 101*, 401–431.

Cvrčková, F., Lipavská, H., & Žárský, V. (2009). Plant intelligence, why, why not or where? *Plant Signaling & Behavior, 4*(5), 394–399.

Darwin, C. R. (1859). *Origin of species by means of natural selection, or the preservation of favoured races in the struggle for life*. John Murray.

Darwin, C. R. (1875a). *Insectivorous plants*. John Murray.

Darwin, C. R. (1875b). *The Movements and Habits of Climbing Plants* (2nd ed.). John Murray. (Originally published as an article in the *Journal of the Linnean Society* in 1865).

Davidson, D. (1980). *Essays on actions and events*. Oxford University Press.

Davis, H., & Pérusse, R. (1988). Numerical competence in animals: Definitional issues, current evidence, and a new research agenda. *Behavioral and Brain Sciences, 11*(4), 561–579.

De Kroon, H., Visser, E. J., Huber, H., Mommer, L., & Hutchings, M. J. (2009). A modular concept of plant foraging behaviour: The interplay between local responses and systemic control. *Plant, Cell & Environment, 32*(6), 704–712.

De la Fuente, I. M. (2015). Elements of the cellular metabolic structure. *Frontiers in Molecular Biosciences, 2*, 16.

De Toledo, G. R., Parise, A. G., Simmi, F. Z., Costa, A. V., Senko, L. G., Debono, M. W., & Souza, G. M. (2019). Plant electrome: The electrical dimension of plant life. *Theoretical and Experimental Plant Physiology, 31*, 21–46.

De Waal, F. B. M. (1999). Anthropomorphism and anthropodenial: Consistency in our thinking about humans and other animals. *Philosophical Topics, 27*(1), 255–280.

Debono, M. W., & Souza, G. M. (2019). Plants as electromic plastic interfaces: A mesological approach. *Progress in Biophysics and Molecular Biology, 146*, 123–133.

Dennett, D. C. (1993). *Consciousness explained*. Penguin Books.

Dennett, D. C. (1996). *Kinds of minds*. Basic Books.

Dennett, D. C. (2007). Heterophenomenology reconsidered. *Phenomenology and the Cognitive Sciences, 6*, 247–270.

Dewey, J. (1884). The new psychology. *Andover Review, 2*, 278–289.

Dewey, J. (1896). The reflex arc concept in psychology. *Psychological review, 3*(4), 357.

Di Francesco, M., Marraffa, M., & Tomasetta, A. (2017). *Filosofia della Mente: Corpo, Coscienza, Pensiero*. Carocci.

Di Francesco, M., & Piredda, G. (2012). *La mente estesa. Dove finisce la mente e comincia il resto del mondo?* Mondadori Università.

Di Paolo, E., Buhrmann, T., & Barandiaran, X. (2017). *Sensorimotor life: An enactive proposal*. Oxford University Press.

Dietrich, E., & Markman, A. B. (2001). Dynamical description versus dynamical modeling. *Trends in Cognitive Sciences, 5*(8), 332.

Dudley, S. A. (2015). Plant cooperation. *AoB Plants, 7*, plv113.

Dupré, J. (2020). Life as process. *Epistemology & Philosophy of Science, 57*(2), 96–113.

Dybzinski, R., Farrior, C., Wolf, A., Reich, P. B., & Pacala, S. W. (2011). Evolutionarily stable strategy carbon allocation to foliage, wood, and fine roots in trees competing for light and nitrogen: An analytically tractable, individual-based model and quantitative comparisons to data. *The American Naturalist, 177*(2), 153–166.

Falik, O., Mordoch, Y., Ben-Natan, D., Vanunu, M., Goldstein, O., & Novoplansky, A. (2012). Plant responsiveness to root–root communication of stress cues. *Annals of Botany, 110*(2), 271–280.

Falik, O., Reides, P., Gersani, M., & Novoplansky, A. (2005). Root navigation by self inhibition. *Plant, Cell & Environment, 28*(4), 562–569.

Falster, D. S., & Westoby, M. (2003). Plant height and evolutionary games. *Trends in Ecology & Evolution, 18*(7), 337–343.

Fantl, J., & McGrath, M. (2012). Pragmatic encroachment: It's not just about knowledge. *Episteme, 9*(1), 27–42.

Fenster, C. B., Armbruster, W. S., Wilson, P., Dudash, M. R., & Thomson, J. D. (2004). Pollination syndromes and floral specialization. *Annual Review of Ecology, Evolution, and Systematics, 35*, 375–403.

Fodor, J. A. (1987). *Psychosemantics: The problem of meaning in the philosophy of mind*. MIT Press.

Fodor, J. A., & Pylyshyn, Z. W. (1988). Connectionism and cognitive architecture: A critical analysis. *Cognition, 28*(1–2), 3–71.

Forterre, Y., Skotheim, J. M., Dumais, J., & Mahadevan, L. (2005). How the Venus flytrap snaps. *Nature, 433*(7024), 421–425.

Frazier, P. A. (2021). On the possibility of plant consciousness: A view from ecointeractivism. *Mind and Matter, 19*(2), 229–259.

Frezza, G., & Gagliasso, E. (2014). Fare metafore e fare scienza. *Aisthesis. Pratiche, Linguaggi e Saperi dell'Estetico, 7*(2), 25–42.

Friston, K. J. (2005). A theory of cortical responses. *Philosophical Transactions of the Royal Society B: Biological Sciences, 360*(1456), 815–836.

Gagliano, M. (2017). The mind of plants: Thinking the unthinkable. *Communicative and Integrative Biology, 10*(2), e1288333.

Gagliano, M., Renton, M., Depczynski, M., & Mancuso, S. (2014). Experience teaches plants to learn faster and forget slower in environments where it matters. *Oecologia, 175*(1), 63–72.

Galadí, J. A. (2024). The mind-body problem: An overview of proposed solutions. In M. S. Obaid & M. L. Deliens (Eds.), *The theory of mind under scrutiny: Psychopathology, neuroscience, philosophy of mind and artificial intelligence* (pp. 435–467). Springer.

Gallagher, S. (2017). *Enactivist interventions: Rethinking the mind.* Oxford University Press.

Gallese, V. (2009). Motor abstraction: A neuroscientific account of how action goals and intentions are mapped and understood. *Psychological Research, PRPF, 73*, 486–498.

Gallese, V., Fadiga, L., Fogassi, L., & Rizzolatti, G. (1996). Action recognition in the premotor cortex. *Brain, 119*(2), 593–609.

Gallistel, C. R., & Gelman, R. (1992). Preverbal and verbal counting and computation. *Cognition, 44*(1–2), 43–74.

Galviz, Y. C., Ribeiro, R. V., & Souza, G. M. (2020). Yes, plants do have memory. *Theoretical and Experimental Plant Physiology, 32*(3), 195–202.

Garcia Rodriguez, A., & Calvo Garzon, P. (2010). Is cognition a matter of representations? Emulation, teleology, and time-keeping in biological systems. *Adaptive Behavior, 18*(5), 400–415.

Gazzaniga, M. S. (2019). *La coscienza è un istinto. Il legame misterioso tra il cervello e la mente* (Trans. It.). Cortina Editore (Original work published 2018, *The consciousness instinct: Unraveling the mystery of how the brain makes the mind.* Farrar, Straus and Giroux).

Gersani, M., Brown, J. S., O'Brien, E. E., Maina, G. M., & Abramsky, Z. (2001). Tragedy of the commons as a result of root competition. *Journal of Ecology, 89*(4), 660–669.

Gibson, J. J. (1950). *The perception of the visual world.* Houghton Mifflin.

Gibson, J. J. (1966). *The senses considered as perceptual systems.* Houghton Mifflin.

Gilroy, S., & Trewavas, T. (2023). Agency, teleonomy and signal transduction in plant systems. *Biological Journal of the Linnean Society, 139*(4), 514–529.

Ginet, C. (1975). *Knowledge, perception and memory*, Vol. 5. Springer Science & Business Media.

Giolito, B. (2011). *Epistemologia e Ragionamento.* Mimesis.

Gładziejewski, P. (2016). Predictive coding and representationalism. *Synthese, 193*, 559–582.

Glenberg, A. M. (1997). Mental models, space, and embodied cognition. In T. B. Ward, S. M. Smith, & J. Vaid (Eds.), *Creative thought: An investigation of conceptual structures and processes* (pp. 495–522). American Psychological Association.

Glock, H. J. (2019). Agency, intelligence and reasons in animals. *Philosophy, 94*(4), 645–671.

Godfrey-Smith, P. (2002). Environmental complexity and the evolution of cognition. *The Evolution of Intelligence*, 233–249.

Godfrey-Smith, P. (2021). *Metazoa* (I. C. Blum, Trans.). Adelphi. (Original work published 2020, Metazoa: Animal life and the birth of the mind. Farrar, Straus and Giroux).

Goldman, A., & de Vignemont, F. (2009). Is social cognition embodied? *Trends in Cognitive Sciences, 13*(4), 154–159.

Gomila, A., & Calvo, P. (Eds.). (2008). *Handbook of cognitive science: An embodied approach.* Elsevier.

Grice, H. P. (1975). Logic and conversation. In P. Cole & J. L. Morgan (Eds.), Speech acts (pp. 41–58). Academic Press.

Hallé, F., Oldeman, R. A. A., & Tomlinson, P. B. (1978). *Tropical trees and forests: An architectural analysis.* Springer.

Hannula, S. E., Heinen, R., Huberty, M., Steinauer, K., De Long, J. R., Jongen, R., & Bezemer, T. M. (2021). Persistence of plant-mediated microbial soil legacy effects in soil and inside roots. *Nature Communications, 12*(1), 5686.

Hawkins, R. D., & Kandel, E. R. (1984). Is there a cell-biological alphabet for simple forms of learning? *Psychological Review, 91*(3), 375–391.

Hedrich, R. (2012). Ion channels in plants. *Physiological reviews, 92*(4), 1777–1811.

Heras-Escribano, M. (2016). Embracing the environment: Ecological answers for enactive problems. *Constructivist Foundations, 11*(2), 309–312.
Heras-Escribano, M., & Calvo, P. (2019). The philosophy of plant neurobiology. In S. Robins, J. Symons, & P. Calvo (Eds.), *The Routledge companion of philosophy of psychology* (pp. 529–548). Routledge.
Heisenberg, W. (1958). *Physics and philosophy: The revolution in modern science*. Harper & Brothers.
Heylighen, F. (2023). The meaning and origin of goal-directedness: A dynamical systems perspective. *Biological Journal of the Linnean Society, 139*(4), 370–387.
Hickok, G. (2009). Eight problems for the mirror neuron theory of action understanding in monkeys and humans. *Journal of Cognitive Neuroscience, 21*(7), 1229–1243.
Hiernaux, Q. (2021). Differentiating behaviour, cognition, and consciousness in plants. *Journal of Consciousness Studies, 28*(1–2), 106–135.
Hilborn, R. C. (2004). Sea gulls, butterflies, and grasshoppers: A brief history of the butterfly effect in nonlinear dynamics. *American Journal of Physics, 72*(4), 425–427.
Hodge, A., Berta, G., Doussan, C., Merchan, F., & Crespi, M. (2009). *Plant root growth, architecture and function*, 153–187.
Hoeksema, J. D., Chaudhary, V. B., Gehring, C. A., Johnson, N. C., Karst, J., Koide, R. T., et al. (2010). A meta-analysis of context-dependency in plant response to inoculation with mycorrhizal fungi. *Ecology Letters, 13*(3), 394–407.
Hohwy, J. (2019). Quick'n'lean or slow and rich? Andy Clark on predictive processing and embodied cognition. In M. Colombo, E. Irvine, & M. Stapleton (Eds.), *Andy Clark and his critics* (pp. 191–205). Oxford University Press.
Hommel, B. (2021). The future of embodiment research: Conceptual themes, theoretical tools, and remaining challenges. In *Handbook of embodied psychology* (pp. 597–617). Springer.
Hopfield, J. J. (1982). Neural networks and physical systems with emergent collective computational abilities. *Proceedings of the National Academy of Sciences, 79*(8), 2554–2558.
Howard, S. R., Avarguès-Weber, A., Garcia, J. E., Greentree, A. D., & Dyer, A. G. (2019). Numerical cognition in honeybees enables addition and subtraction. *Science Advances, 5*(2), eaav0961.
Huang, A. C., Jiang, T., Liu, Y. X., Bai, Y. C., Reed, J., Qu, B., et al. (2019). A specialized metabolic network selectively modulates Arabidopsis root microbiota. *Science, 364*(6440), eaau6389.
Huebner, B., & Schulkin, J. (2022). *Biological Cognition*. Cambridge University Press.
Hurley, S. E., & Nudds, M. E. (2006). *Rational animals?* Oxford University Press.
Hurley, S. L. (1998). *Consciousness in action*. Harvard University Press.
Hutto, D. D., & Myin, E. (2013). *Radicalizing Enactivism*. The MIT Press.
Hutto, D. D., & Myin, E. (2017). *Evolving Enactivism: Basic minds meet content*. The MIT press.
Iachini, T. (2023). Rappresentazioni, simulazioni embodied e meccanismi cognitivi semplici: quale etichetta per le piante? *Giornale Italiano di Psicologia, 50*(2), 393–398.
Ippoliti, E., & Cellucci, C. (2016). *Logica*. Egea.
James, W. (1892). *Psychology*. Henry Holt and Company.
Japyassu, H. F., & Laland, K. L. (2017). Extended spider cognition. *Animal Cognition, 20*(3), 375–395.
Jeannerod, M. (1996). Reaching and grasping. Parallel specification of visuomotor channels. In *Handbook of perception and action* (Vol. 2, pp. 405–460). Academic.
Jeannerod, M. (2006). *Motor cognition: What actions tell the self* (Vol. 42). Oxford University Press.
Kandel, E. R. (1976). *Cellular basis of behavior: An introduction to behavioral neurobiology*. W. H. Freeman.
Kandel, E. R. (2001). The molecular biology of memory storage: A dialogue between genes and synapses. *Science, 294*, 1030–1038.
Kauffman, S. A. (1995). *At home in the universe: The search for laws of self-organization and complexity*. Oxford University Press.
Keijzer, F. (2021). Demarcating cognition: The cognitive life sciences. *Synthese, 1*, 137–157.

Keijzer, F. A. (2017). Evolutionary convergence and biologically embodied cognition. *Interface Focus, 7*(3), 20160123.

Kelso, J. S. (1995). *Dynamic patterns: The self-organization of brain and behavior*. MIT press.

Kirchhoff, M. D., & Robertson, I. (2018). Enactivism and predictive processing: A non-representational view. *Philosophical Explorations, 21*(2), 264–281.

Kohler, E., Keysers, C., Umilta, M. A., Fogassi, L., Gallese, V., & Rizzolatti, G. (2002). Hearing sounds, understanding actions: Action representation in mirror neurons. *Science, 297*(5582), 846–848.

Koutroufinis, S. A. (2017). Organism, machine, process. Towards a process ontology for organismic dynamics. Organisms. *Journal of Biological Sciences, 1*(1), 23–44.

Krebs, J. R., Ryan, J. C., & Charnov, E. L. (1974). Hunting by expectation or optimal foraging? A study of patch use by chickadees. *Animal Behaviour, 22*, 953-IN3.

Kreps, J. A., & Kay, S. A. (1997). Coordination of plant metabolism and development by the circadian clock. *The Plant Cell, 9*(7), 1235–1244.

Laland, K. N., Odling-Smee, J., & Feldman, M. W. (2001). Cultural niche construction and human evolution. *Journal of Evolutionary Biology, 14*(1), 22–33.

LeDoux, J. (2020). *Lunga storia di noi stessi: Come il cervello è diventato cosciente* (G. Guerriero, Trans.). Raffaello Cortina Editore (Original work published 2019, *The deep history of ourselves: The four-billion-year story of how we got conscious brains*. Viking).

Leduc, N., Roman, H., Barbier, F., Péron, T., Hutché-thélier, L., Lothier, J., Demotes-Mainard, S., & Sakr, S. (2014). Light signaling in bud outgrowth and branching in plants. *Plants, 3*(2), 223–250.

Lee, D. N. (1976). A theory of visual control of braking based on information about time-to-collision. *Perception, 5*(4), 437–459.

Lee, D. N. (1998). Guiding movement by coupling taus. *Ecological psychology, 10*(3–4), 221–250.

Lee, D. N. (2009). General Tau Theory: evolution to date. *Perception, 38*(6), 837–850.

Lee, J. (2023). What is cognitive about 'plant cognition'? *Biology & Philosophy, 38*(3), 18.

Lehrman, D. S. (1953). A critique of Konrad Lorenz's theory of instinctive behavior. *The Quarterly Review of Biology, 28*(4), 337–363.

Lehmann, L., & Keller, L. (2006). The evolution of cooperation and altruism–a general framework and a classification of models. *Journal of Evolutionary Biology, 19*(5), 1365–1376.

Linson, A., & Calvo, P. (2020). Zoocentrism in the weeds? Cultivating plant models for cognitive yield. *Biology & Philosophy, 35*(5), 1–27.

Longo, G., & Montévil, M. (2014). *Perspectives on organisms: Biological time, symmetries and singularities*. Springer.

Longo, M. R., & Lourenco, S. F. (2007). Space perception and body morphology: Extent of near space scales with arm length. *Experimental Brain Research, 177*, 285–290.

Lorenz, K. Z. (1952). *King Solomon's ring* (M. K. Wilson, Trans.). Methuen.

Lüttge, U. (2012). Modularity and emergence: Biology's challenge in understanding life. *Plant Biology, 14*(6), 865–871.

Lyon, P. (2006). The biogenic approach to cognition. *Cognitive Processing, 7*(1), 11–29.

Lyon, P. (2015). The cognitive cell: Bacterial behavior reconsidered. *Frontiers in Microbiology, 6*, 264.

Lyon, P., Keijzer, F., Arendt, D., & Levin, M. (2021). Reframing cognition: Getting down to biological basics. *Philosophical Transactions of the Royal Society B, 376*(1820), 20190750.

Maher, C. (2017). *Plant minds: A philosophical defense*. Routledge.

Mahon, B. Z., & Caramazza, A. (2008). A critical look at the embodied cognition hypothesis and a new proposal for grounding conceptual content. *Journal of Physiology-Paris, 102*(1–3), 59–70.

Malafouris, L. (2013). *How things shape the mind*. MIT press.

Mannella, F., & Tummolini, L. (2023). Kick-starting concept formation with intrinsically motivated learning: The grounding by competence acquisition hypothesis. *Philosophical Transactions of the Royal Society B, 378*(1870), 20210370.

Marder, M., & Geremia Parise, A. (2024). Extending cognition: A vegetal rejoinder to extensionless thought and to extended cognition. *Plant Signaling & Behavior, 19*(1), 2345984.

Marraffa, M. (2011). L'introspezione. In M. Marraffa & A. Paternoster, Scienze cognitive: Un'introduzione filosofica. Carocci.

Marraffa, M., & Paternoster, A. (2011). Scienze cognitive. *Un'introduzione filosofica*. Carocci.

Marraffa, M., & Paternoster, A. (2013). *Sentirsi esistere: Inconscio, coscienza, autocoscienza*. Laterza.

Martin, T., & Schwartz, D. L. (2005). Physically distributed learning: Adapting and reinterpreting physical environments in the development of fraction concepts. *Cognitive Science, 29*(4), 587–625.

Matthen, M. (2014). Active perception and the representation of space. Perception and its Modalities, 44–72, Oxford University Press.

Maturana, H. R., & Varela, F. J. (1980). *Autopoiesis and cognition: The realization of the living*. D. Reidel Publishing Company.

Maturana, H. R., & Varela, F. J. (1987). *The tree of knowledge: The biological roots of human understanding* (R. Paolucci & F. Varela, Trans.). Shambhala Publications.

Mayr, E. (1969). The biological meaning of species. *Biological Journal of the Linnean Society, 1*(3), 311–320.

Mayr, E. (1982). *The growth of biological thought: Diversity, evolution, and inheritance*. Harvard University Press.

Mazzolai, B. (2021). *Il futuro raccontato dalle piante. Cosa possiamo imparare dal regno vegetale e dal suo percorso sul pianeta*. Longanesi.

McClintock, B. (1984). The significance of responses of the genome to challenge. *Science, 226*(4676), 792–801.

McNickle, G. G., & Dybzinski, R. (2013). Game theory and plant ecology. *Ecology Letters, 16*(4), 545–555.

Michaels, C. F., & Carello, C. (1981). *Direct perception*. Prentice-Hall.

Michmizos, D., & Hilioti, Z. (2019). A roadmap towards a functional paradigm for learning & memory in plants. *Journal of Plant Physiology, 232*, 209–215.

Monod, J. (2017). *Il caso e la necessità: saggio sulla filosofia naturale della biologia contemporanea*, A. Busi (Trans.). Mondadori (original work published 1970, *Le hasard et la nécessité: Essai sur la philosophie naturelle de la biologie moderne*. Éditions du Seuil).

Morabito, C. (2020). *Il motore della mente: il movimento nella storia delle scienze cognitive*. Laterza.

Morabito, C. (2023). Nel mondo dei sistemi complessi: consonanza o cosilience? *Giornale Italiano di Psicologia, 50*(2), 405–412.

Morgan, A. (2014). Representations gone mental. *Synthese, 191*(2), 213–244.

Morin, E. (2007). *L'anno I dell'era ecologica. La terra dipende dall'uomo che dipende dalla terra*, B. Spadolini (Trans.). Armando Editore (original work published 2006, *L'an I de l'ère écologique: La terre dépend de l'homme qui dépend de la terre*).

Murphy, G. P., & Dudley, S. A. (2007). Above-and below-ground competition cues elicit independent responses. *Journal of Ecology, 95*(2), 261–272.

Mustard, J., & Levin, M. (2014). Bioelectrical mechanisms for programming growth and form: Taming physiological networks for soft body robotics. *Soft Robotics, 1*(3), 169–191.

Nagel, T. (1974). The Philosophical Review. *What is it Like to Be a Bat*, 435–450.

Nannini, S. (2011). La naturalizzazione delle rappresentazioni mentali. *Sistemi intelligenti, 23*(1), 41–58.

Nash, J. F. (1950). The bargaining problem. *Econometrica, 18*(2), 155–162.

Neisser, U. (1967). *Cognitive psychology*. Prentice Hall.

Newell, A. (1980). Physical symbol systems. *Cognitive Science, 4*(2), 135–183.

Nicholson, D. J., & Dupré, J. (Eds.). (2018). *Everything flows: Towards a processual philosophy of biology*. Oxford University Press.

Nicolini, E., Chanson, B., & Bonne, F. (2001). Stem growth and epicormic branch formation in understorey beech trees (*Fagus sylvatica* L.). *Annals of Botany, 87*(6), 737–750.

Nieder, A. (2020). The adaptive value of numerical competence. *Trends in Ecology & Evolution, 35*(7), 605–617.

Noë, A. (2004). *Action in perception*. MIT press.

Novoplansky, A. (2009). Picking battles wisely: Plant behaviour under competition. *Plant, Cell & Environment, 32*(6), 726–741.

Novoplansky, A. (2016). Future perception in plants. In R. Dillmann, Y. Nakamura, & D. Vernon (Eds.), *Anticipation across disciplines* (pp. 57–70). Springer.

Nucara, L. (2014). *La filosofia di Humberto Maturana*. Le Lettere.

O'Regan, J. K., & Noe, A. (2001). A sensorimotor account of vision and visual consciousness. *Behavioral and Brain Sciences, 24*(5), 939–973.

Oborny, B. (2019). The plant body as a network of semi-autonomous agents: A review. *Philosophical Transactions of the Royal Society B, 374*(1774), 20180371.

Odling-Smee, F. J., Laland, K. N., & Feldman, M. W. (1996). Niche construction. *The American Naturalist, 147*(4), 641–648.

Odling-Smee, F. J., Laland, K. N., & Feldman, M. W. (2003). *Niche construction. The neglected process in evolution*. Princeton University Press.

Oyama, S. (2000). *The ontogeny of information: Developmental systems and evolution*. Duke University Press.

Palmiero, M., & Borsellino, M. C. (2018). *Embodied cognition, comprendere la mente incarnata*. Aras Edizioni.

Parise, A. G., Gagliano, M., & Souza, G. M. (2020). Extended cognition in plants: Is it possible? *Plant Signaling & Behavior, 15*(2), 1710661.

Parise, A. G., Reissig, G. N., Basso, L. F., Senko, L. G. S., Oliveira, T. F. C., de Toledo, G. R. A., & Souza, G. M. (2021). Detection of different hosts from a distance alters the behaviour and bioelectrical activity of *Cuscuta racemosa*. *Frontiers in Plant Science, 12*, 594195.

Parise, A. G., de Toledo, G. R. A., de Carvalho Oliveira, T. F., Souza, G. M., Castiello, U., Gagliano, M., & Marder, M. (2022). Do plants pay attention? A possible phenomenological-empirical approach. *Progress in Biophysics and Molecular Biology, 173*, 11–23.

Parise, A. G., & Marder, M. (2024). Extended plant cognition: A critical consideration of the concept. *Theoretical and Experimental Plant Physiology, 36*(3), 439–455.

Paternoster, A. (2011). *La percezione*. In Marraffa, M., & Paternoster, A., (Eds.), Scienze cognitive. Un'introduzione filosofica. Carocci.

Pepping, G. J., & Grealy, M. A. (Eds.). (2007). *Closing the gap: The scientific writings of David N*. Psychology Press.

Perconti, P. (2017). *Filosofia della mente*. Il Mulino.

Pezzulo, G. (2008). Coordinating with the future: The anticipatory nature of representation. *Mind & Machines, 18*, 179–220.

Piazza, T. (2017). *Che cos'è la conoscenza*. Carocci.

Ponzio A., & Petrilli S. (2002). *I segni e la vita. La semiotica globale di Thomas A. Sebeok*. Spirali Edizioni.

Prinz, J. J., & Barsalou, L. W. (2014). Steering a course for embodied representation. In *Cognitive dynamics* (pp. 51–77). Psychology Press.

Pulvermüller, F. (2005). Brain mechanisms linking language and action. *Nature Reviews Neuroscience, 6*(7), 576–582.

Qu, H., & Zhu, Q. (2009). Automatic approaches to plant meristem states revelation and branching pattern extraction: A review. *Pattern Recognition*, 419–438.

Raguso, R. A., & Kessler, A. (2017). Speaking in chemical tongues: Decoding the language of plant volatiles. In M. Gagliano, J. C. Ryan, & P. Vieira (Eds.), *The language of plants: Science, philosophy*. University of Minnesota Press.

Ramsey, W. M. (2007). *Representation reconsidered*. Cambridge University Press.

Rapoport, A. (1980). Cross-cultural aspects of environmental design. In I. Altman, A. Rapoport, & J. F. Wohlwill (Eds.), *Environment and culture* (pp. 7–46). Springer.

Rapp, H., Nawrot, M. P., & Stern, M. (2020). Numerical cognition based on precise counting with a single spiking neuron. *IScience, 23*(2), 101283.

Ravasz, E., Somera, A. L., Mongru, D. A., Oltvai, Z. N., & Barabási, A. L. (2002). Hierarchical organization of modularity in metabolic networks. *Science, 297*(5586), 1551–1555.

Rispoli, G. (2012). *Dall'empiriomonismo alla tectologia: organizzazione, complessità e approccio sistemico nel pensiero di Aleksandr Bogdanov*. Aracne.

Rizzolatti, G., & Arbib, M. A. (1998). Language within our grasp. *Trends in Neurosciences, 21*(5), 188–194.

Rizzolatti, G., Fadiga, L., Gallese, V., & Fogassi, L. (1996). Premotor cortex and the recognition of motor actions. *Cognitive Brain Research, 3*(2), 131–141.

Rizzolatti, G., Fogassi, L., & Gallese, V. (2004). Cortical mechanisms subserving object grasping, action understanding, and imitation. In M. S. Gazzaniga (Ed.), *The cognitive neurosciences* (pp. 427–440). Boston Review.

Rizzolatti, G., Riggio, L., Dascola, I., & Umilta, C. (1987). Reorienting attention across the horizontal and vertical meridians: Evidence in favor of a premotor theory of attention. *Neuropsychologia, 25*(1), 31–40.

Rizzolatti, G., & Sinigaglia, C. (2006). *So quel che fai: il cervello che agisce ei neuroni specchio*. Raffaello Cortina.

Rodriguez, G. A., & Calvo, G. P. (2010). Is cognition a matter of representations? Emulation, teleology, and time-keeping in biological systems. *Adaptive Behavior, 18*(5), 400–415.

Rovelli, C. (2014). *La realtà non è come ci appare. La struttura elementare delle cose*. Raffaello Cortina. (English translation published in 2016 by S. Carnell & E. Segre, *Reality is Not What It Seems: The Journey to Quantum Gravity*, Riverhead Books).

Rugani, R., Vallortigara, G., Priftis, K., & Regolin, L. (2015). Number-space mapping in the newborn chick resembles humans' mental number line. *Science, 347*(6221), 534–536.

Rupert, R. D. (2004). Challenges to the hypothesis of extended cognition. *The Journal of Philosophy, 101*(8), 389–428.

Russell, B. (1912). *The problems of philosophy*. Williams & Norgate.

Rumelhart, D. E., Hinton, G. E., & McClelland, J. L. (1986). A general framework for parallel distributed processing. In D. E. Rumelhart & J. L. McClelland (Eds.), *Parallel distributed processing: Explorations in the microstructure of cognition* (Vol. 1, pp. 45–76). MIT Press.

Ryle, G. (1949). *The Concept of Mind*. Hutchinson.

Schlicht, T. (2018). Does separating intentionality from mental representation imply radical enactivism? *Frontiers in Psychology, 9*, 1497.

Schlosser, M. E. (2018). Embodied cognition and temporally extended agency. *Synthese, 195*(5), 2089–2112.

Schlosser, M. (2019). Agency. In *The Stanford Encyclopedia of Philosophy*. https://plato.stanford.edu/archives/win2019/entries/agency/

Schneirla, T. C. (1966). Behavioral development and comparative psychology. *The Quarterly Review of Biology, 41*(3), 283–302.

Schulkin, J., & Sterling, P. (2019). Allostasis: A brain-centered, predictive mode of physiological regulation. *Trends in Neurosciences, 42*(10), 740–752.

Segundo-Ortin, M. (2020). Agency from a radical embodied standpoint: An ecological-enactive proposal. *Frontiers in Psychology, 11*, 1319.

Segundo-Ortin, M., & Calvo, P. (2019). Are plants cognitive? A reply to Adams. *Studies in History and Philosophy of Science A, 73*, 64–71.

Segundo-Ortin, M., & Calvo, P. (2022). Consciousness and cognition in plants. *Wiley Interdisciplinary Reviews: Cognitive Science, 13*(2), e1578.

Shapiro, L. (2007). The embodied cognition research programme. *Philosophy Compass, 2*(2), 338–346.

Shapiro, L. (2019). *Embodied cognition*. Routledge.

Shapiro, L. (2013). Dynamics and cognition. *Minds and Machines, 23*, 353–375.

Shapiro, L. (2021). *Embodied cognition*. In E. N. Zalta (Ed.), *The Stanford Encyclopedia of Philosophy* (Winter 2021 Edition). Stanford University. https://plato.stanford.edu/archives/win2021/entries/embodied-cognition/

Shaw, R. E., & McIntyre, M. (1974). Algoristic foundations to cognitive psychology. In W. B. Weimer & D. S. Palermo (Eds.), *Cognition and the symbolic processes* (pp. 305–362). Lawrence Erlbaum Associates.

Sherrington, C. S. (1906). Observations on the scratch-reflex in the spinal dog. *The Journal of Physiology, 34*(1–2), 1–50.

Shinar, G., & Feinberg, M. (2010). Structural sources of robustness in biochemical reaction networks. *Science, 327*(5971), 1389–1391.

Simard, S. W. (2018). Mycorrhizal networks facilitate tree communication, learning, and memory. In F. Baluška, M. Gagliano, & G. Witzany (Eds.), *Memory and learning in plants* (pp. 91–213). Springer.

Sims, M. (2021a). A continuum of intentionality: Linking the biogenic and anthropogenic approaches to cognition. *Biology & Philosophy, 36*(6), 51.

Sims, M. (2021b). How to count biological minds: Symbiosis, the free energy principle, and reciprocal multiscale integration. *Synthese, 199*(1), 2157–2179.

Smokrović, N. (2018). Informal reasoning and formal logic: Normativity of natural language reasoning. *Croatian Journal of Philosophy, 18*(54), 455–470.

Souza, G. M., Ferreira, A. S., Saraiva, G. F., & Toledo, G. R. (2017). Plant «electrome» can be pushed toward a self-organized critical state by external cues: Evidences from a study with soybean seedlings subject to different environmental conditions. *Plant Signaling & Behavior, 12*(3), e1290040.

Souza, G. M., & Lüttge, U. (2015). Stability as a phenomenon emergent from plasticity–complexity–diversity in eco-physiology. Progress in Botany, 76, 211–239, Springer International Publishing.

Souza, G. M., Ribeiro, R. V., Prado, C. H. B. S., Damineli, D. S. C., Sato, A. M., & Oliveira, M. S. (2009). Using network connectance and autonomy analyses to uncover patterns of photosynthetic responses in tropical woody species. *Ecological Complexity, 6*, 15–26.

Souza, G. M., Toledo, G. R., & Saraiva, G. F. (2018). Towards systemic view for plant learning: Ecophysiological perspective. In F. Baluska, M. Gagliano, & G. Witzany (Eds.), *Memory and learning in plants* (pp. 163–189). Springer.

Spaulding, S. (2011). A critique of embodied simulation. *Review of Philosophy and Psychology, 2*(3), 579–599.

Sperry, R. W. (1952). Neurology and the mind-brain problem. *American Scientist, 40*(2), 291–312.

Stancher, G., Rugani, R., Regolin, L., & Vallortigara, G. (2015). Numerical discrimination by frogs (Bombina orientalis). *Animal Cognition, 18*, 219–229.

Stanley, J., & Williamson, T. (2001). Knowing how. *The Journal of Philosophy, 98*(8), 411–444.

Tarassov, K., Messier, V., Landry, C. R., Radinovic, S., Molina, M. M. S., Shames, I., et al. (2008). An in vivo map of the yeast protein interactome. *Science, 320*(5882), 1465–1470.

Thelen, E., & Smith, L. B. (1994). *A dynamic systems approach to the development of cognition and action*. The MIT Press.

Thompson, E. (2007). *Mind in life: Biology, phenomenology, and the sciences of mind*. Harvard University Press.

Thorndike, E. L. (1898). Animal intelligence: An experimental study of the associative processes in animals. *The Psychological Review: Monograph Supplements, 2*(4), i.

Trewavas, A. (2003). Aspects of plant intelligence. *Annals of Botany, 92*(1), 1–20.

Trewavas, A. (2005). Green plants as intelligent organisms. *Trends in Plant Science, 10*(9), 413–419.

Trewavas, A. (2014). *Plant behaviour and intelligence*. Oxford University Press.

Trewavas, A. (2017). The foundations of plant intelligence. *Interface Focus, 7*(3), 20160098.

Turvey, M. T. (2018). *Lectures on perception: An ecological perspective*. Routledge.

Turvey, M. T., Shaw, R. E., Reed, E. S., & Mace, W. M. (1981). Ecological laws of perceiving and acting: In reply to Fodor and Pylyshyn (1981). *Cognition, 9*(3), 237–304.

Tyree, M. T., & Sperry, J. S. (1989). Vulnerability of xylem to cavitation and embolism. *Annual Review of Plant Biology, 40*(1), 19–36.

Umiltà, M. A., Kohler, E., Gallese, V., Fogassi, L., Fadiga, L., Keysers, C., & Rizzolatti, G. (2001). I know what you are doing: A neurophysiological study. *Neuron, 31*(1), 155–165.

Van Gelder, T., & Port, R. F. (1995). *It's about time: An overview of the dynamical approach to cognition.* Mind as motion: Explorations in the dynamics of cognition, 1, 43.

Varela, F. J., Thompson, E., & Rosch, E. (1991). *The embodied mind: Cognitive science and human experience*. The MIT Press.

Waddington, C. H. (1977). *Tools for thought*. Jonathan Cape.

Walton, D. (2005). *Fundamentals of critical argumentation*. Cambridge University Press.

Wang, Q., Guerra, S., Ceccarini, F., Bonato, B., & Castiello, U. (2021). Sowing the seeds of intentionality: Motor intentions in plants. *Plant Signaling & Behavior, 16*(11), 1949818.

Ward, D., Silverman, D., & Villalobos, M. (2017). Introduction: The varieties of enactivism. *Topoi, 36*, 365–375.

Warwick, K. (2000). *QI: The Quest for Intelligence*. Piatkus Books.

Willis, T. (1978). Cerebri anatome [*The anatomy of the brain and nerves*]. Birmingham, AL: Classics of Medicine Library. (Original work published 1664).

Witzany, G. (2016). The Biosemiotics of Plant Communication. *The American Journal of Semiotics, 24*(1–3), 39–56.

Witzany, G., & Baluška, F. (2012). *Biocommunication of plants,* vol. 14. Springer Science & Business Media.

Yu, H., Braun, P., Yıldırım, M. A., Lemmens, I., Venkatesan, K., Sahalie, J., … & Vidal, M. (2008). High-quality binary protein interaction map of the yeast interactome network. *Science, 322*(5898), 104–110.

Zhu, X., Gerstein, M., & Snyder, M. (2007). Getting connected: Analysis and principles of biological networks. *Genes & Development, 21*(9), 1010–1024.

Zipoli Caiani, S. (2016). *Corporeità e cognizione. La filosofia della mente incorporata*. Mondadori Education.

Index

A
Accuracy, 21, 53, 211
Action, 2, 18, 116, 215
Action-oriented representations, 299, 300
Active inferences, 157, 177, 316
Advantages and disadvantages of representations, 302–306
Affordances, 158, 220, 261, 262, 268, 316
Agency, 8, 19, 37, 63, 88, 225–227, 317
Anesthetics, 56, 177
Animal cognition, 4, 30, 60, 61, 68, 76, 78, 79, 152, 156, 251, 268
Animal minds, 11, 284, 294
Animal studies, 19, 67, 155, 165
Anthropocentrism, 5, 60, 67–76
Anthropodenial, 67–76
Anthropogenic approach, 76
Anthropomorphism, 5, 60, 67–76
Anti-anthropocentrism, 5, 67, 68
Anti-anthropomorphism, 5, 67, 70–73
Anticipatory behaviors, 58, 116, 150, 156–159, 162, 189, 314, 316–318
Anti-intellectualism, 239
Anti-representationalism, 304
Artificial intelligence, 34, 92, 257
Associative learning, 154–156, 257
Attention, 2, 3, 11, 13, 22, 28, 44, 60, 63, 76, 80, 145–165, 172, 179, 181, 213, 241, 263, 279, 283, 284, 300, 317, 318
Autopoiesis, 88
Autotrophs, 118, 212
Awareness, 6, 21, 23, 24, 29–32, 35, 36, 38, 69, 87, 116, 125, 165, 169, 172–179, 189, 230, 231, 238

B
Bacteria, 36, 40, 49, 78, 81, 82, 96, 120, 122, 138, 143, 150, 154, 189, 245, 254, 264, 279, 282, 295
Basal cognition, 16, 187, 188
Basal cognitive capacities, 79, 80
Behavior, 2, 16, 116, 206
Behavioral flexibility, 76, 135, 145, 183, 185, 186, 228, 230, 250, 276
Behaviorism, 8, 65, 153, 232–235, 288
Biases, 27, 29–32, 57, 60, 64, 71, 75, 121, 186, 237
Binomial naming system, 44
Bioacoustics, 54
Biocommunication, 141, 144
Bio-ecological cognition, 11, 96, 281–284
Biogenic approach, 4, 17, 60, 76, 182, 190, 281, 317
Bioinformatics, 92
Biological categories, 78
Biological cognition, 6, 7, 11, 60, 63, 157, 255, 281, 283, 284, 290
Biological complexity, 120
Biological intelligence, 6, 39, 150, 214, 215
Biological networks, 22, 23, 58, 147, 148, 261, 269, 271–273, 308
Biological role of knowledge, 21, 242–243
Biosemiotics, 144
Biosphere, 84, 89, 94, 95, 134
Botanical gardens, 28, 44
Branches, 4, 10, 21, 32, 50, 60, 63, 79, 82, 91, 92, 117, 121, 128, 129, 138, 158, 179, 207, 236, 276, 278, 286, 306
Branching architecture, 276
Branching dynamics, 10, 276, 277

M. Bianchi, *Philosophical Implications of Plant Cognition*, Philosophical Studies Series 158, https://doi.org/10.1007/978-3-032-01328-6

Branching history, 278
Branching processes, 275–277

C

Categorizations, 7, 160, 207–220, 238
Causal explanations, 301
Causality, 86, 87
Chemical compounds, 6, 8, 98, 125, 141, 225, 230, 317, 318
Chemical recognition, 132, 174
Chemical signaling, 66, 122–124, 141, 144, 176, 227, 282, 283
Chemical soil memories, 279
Circadian clock, 129
Circular reasoning, 5, 61
Circumnutation, 29, 46, 129–131, 156
Classical conditioning, 154, 155
Classification methods, 45, 211
Climbing plants, 29, 46, 47, 126, 129–134, 158, 161, 164, 178
Cobolism, 281, 282
Coding, 157, 229, 305, 309, 312, 314, 315
Coexistence, 5, 85, 87, 98
Cognitive ecosystems, 99
Cognitive ethology, 19
Cognitive life sciences, 17
Cognitive ontologies, 7, 215–217
Cognitive plant ethology, 17–19
Cognitive sciences, 1, 2, 4, 9, 11, 12, 17, 20, 22, 29, 34, 36, 37, 39, 42, 47, 56, 59, 61, 63, 76, 78, 82, 99, 100, 146, 190, 215, 216, 222, 223, 234, 255–284, 286, 289, 290, 292, 297–300, 305, 306, 312, 314, 318
Cognitive systems, 3, 6, 9, 11, 12, 17, 37, 42, 74, 82, 88, 157, 216, 218, 238, 243, 259, 262–267, 282, 290, 294, 297, 300, 303–305, 310, 313, 314, 316, 318
Cognitivism, 76, 146, 220, 313
Common-sense psychology, 22, 69, 289, 306
Communications, 4, 10, 11, 17, 21, 32, 39, 40, 54, 58, 66, 75, 79–81, 86, 89, 90, 94, 97, 98, 121, 123, 127, 130, 132, 134, 141–145, 161, 165, 168, 173, 180, 184, 187, 214, 215, 217, 222, 229, 230, 240, 243–245, 248–250, 254–257, 267, 274, 275, 278, 279, 285, 294, 295, 306
Comparative analysis of cognition, 63–67
Comparative psychology, 4, 47, 60, 63–65, 69, 72, 232
Competition, 8, 10, 18, 40, 53, 92, 93, 141, 178, 179, 219, 253, 254, 276, 278, 279
Complexities, 2, 5, 8, 12, 24, 30–32, 38, 43, 67, 75, 76, 79, 81, 83–87, 90, 91, 93–95, 99, 120, 121, 146, 147, 163, 169, 179, 183, 186, 187, 190, 210, 211, 218, 222, 224, 225, 235, 237, 244, 246, 254, 265, 274, 277, 281, 288, 292, 295, 298, 299, 310, 318
Complexity science, 1
Complex networks, 66, 152, 272, 285, 310
Complex systems, 85, 87, 89, 95, 146, 209, 218, 229, 236, 257, 269, 270, 309
Component thinking, 92
Computationalism, 259, 261, 265
Conceptions of mind, 281
Conceptions of representations, 12, 298–302, 307, 315
Concept of network, 10
Concept of ramification, 276
Conceptual analysis, 36, 179
Conceptual memory, 147
Connectionism, 256, 258–260, 266
Connectomes, 10, 273, 274, 310
Consciousness, 37, 51, 55, 63, 116, 165–179, 181, 188, 231, 234, 235, 287, 289
Constraints, 9, 10, 38, 53, 66, 67, 75, 77, 78, 82, 131, 133, 134, 141, 156, 161, 209, 235, 236, 243–246, 262, 274, 276, 278, 282–284, 289, 310
Context of interaction, 22, 129, 133, 176, 187
Cooperation, 18, 40, 92, 93, 142, 184, 250, 252, 253, 276, 279
Counterfactual reasoning, 304
Counter-intuitiveness, 19
Critical evaluation, 24
Cybernetics, 84, 86, 87, 90, 256

D

Darwinian revolution, 47
Data, 24, 25, 27, 31, 33, 59, 61, 75, 92, 116, 236, 237
Decision making, 6, 18, 40, 55, 58, 78–81, 116, 145–165, 181, 185, 189, 252, 276, 294, 295
Defense systems, 18
Dichotomies, 75, 160, 220, 227, 234, 300
Direct and indirect perception, 220, 221, 315
Direct knowledge, 238, 239
Disembodied mind, 11, 284, 287, 289

Distinction of quantity and difference, 250–252
Dualism, 49, 64, 151, 287, 288, 296
Dynamical systems theory, 35, 62, 226, 266, 288

E

Eco-Evo-Devo, 92
Eco-evolutionary, 5, 7, 53, 65, 67, 70, 75, 92, 95, 99, 186, 219, 221, 237, 242–243, 262, 288, 295, 308, 314
Ecological cognition, 5
Ecological interactions, 2, 10–12, 16, 94, 97, 100, 123, 176, 186, 222, 244, 269, 272, 276, 278, 280, 298
Ecological memory, 150
Ecological mind, 295
Ecological needs, 27, 243
Ecological networks, 272, 273
Ecological niches, 42, 67, 81, 91, 95, 99, 121, 150, 176, 177, 180, 185, 186, 191, 212, 267, 291, 315
Ecological psychology, 9, 22, 62, 99, 162, 220, 226, 240, 242, 259–261, 268, 312, 313, 316
Ecological-relational perspective, 62, 141, 168, 250, 259
Ecological relationships, 5, 30, 75, 85, 99, 153, 247, 261
Ecological specializations, 53
Ecology, 5, 48, 59, 65, 75, 76, 83, 92–94, 99, 175
Ecology and cognition, 83–100
Ecology of mind, 99
Ecosemiotics, 100
Ecosystems, 5, 50, 53, 69, 83, 84, 86–88, 93–95, 97, 100, 120, 139, 187, 254, 272, 273
Electrical potentials, 124
Electrical signaling, 29, 46, 58, 75, 81, 123, 124, 137, 149, 150, 164, 186, 272, 274, 308
Electrome, 164, 309, 317
Eliminativism, 288
Embedded cognition, 9, 22, 256, 262, 265, 290
Embodied cognition, 262, 263, 265
Embodied mind, 290
Emergence, 12, 35, 38, 56, 92, 93, 99, 116, 120, 170, 237, 256, 290, 299, 318
Emergentism, 170, 288
Empiricism, 75, 227, 234, 242
Enacted cognition, 256, 259–267, 282
Enactivism, 62, 157, 223, 263, 264, 296, 316
Entropy, 88
Environmental communication, 6
Environmental information, 6, 21, 23, 40, 48, 52, 63, 98, 120, 128, 143, 153, 161, 221, 274, 279, 294, 298, 306, 307, 309, 312, 313, 317
Environmental stimuli, 4, 8, 12, 37, 41, 49, 55, 137, 139, 141, 144, 151, 153, 154, 164, 169, 171, 176, 190, 220, 233, 246, 267, 280, 297, 313
Environmental stressors, 11, 27, 119, 142, 143, 149, 273, 280
Environmental uncertainties, 7, 219, 271, 283
Epigenetic landscape, 28
Epigenetics, 23, 28, 60, 77, 92, 134, 148–150, 152, 277, 278, 313
Episodic memory, 147, 148
Epistemic tools, 4, 34, 244, 266, 267, 316
Epistemological contextualism, 239, 240
Epistemological issues, 8, 235–255
Epistemology, 21, 85, 236, 286
Errors, 22, 27, 32, 33, 35, 57, 80, 157, 173, 185, 213–215, 234, 237, 248, 261, 273, 303, 316
Ethology, 18, 19, 53, 71, 72, 227, 232
Evolutionary cognition, 4, 5, 60, 76–82
Evolutionary tree, 11, 21, 80, 182, 281
Evolutionism, 90
Experiences, 7, 12, 21, 24, 25, 50, 53–55, 69, 70, 72, 80, 89, 146–148, 151, 152, 166, 171–177, 182, 212, 215, 219–221, 227, 229–231, 235, 242, 244, 246, 257, 270, 274, 282, 283, 287–290, 297, 300, 303, 307, 312, 313, 315
Experiential issues, 7, 219–235
Experiential plurality, 7, 219, 230
Experimental applications, 2
Experimental observations, 2, 37, 64
Explanatory circularity, 60
Explanatory pluralism, 62
Explanatory reductionism, 91
Extended cognition, 11, 22, 100, 256, 264, 265, 278

F

Fallacies, 70, 248
Family resemblance, 179, 214
Field research, 30, 32

Flexibility, 7, 8, 11, 21, 23, 50, 51, 59, 82, 97, 98, 119, 136, 137, 145, 147, 152, 165, 168, 176, 181, 185, 187, 225, 227, 230, 250, 255, 274, 280, 303, 309
Foraging, 9, 10, 139, 140, 212, 251, 252, 254, 277–279
Formal logic, 247–249
4E cognition, 9, 22, 62, 187, 226, 256, 262, 267–280
Free Energy Principle (FEP), 157, 315
Functionalism, 257, 258, 265, 283, 290
Functional organizations, 7, 35, 39, 66, 77, 87, 116, 135, 167, 168, 184, 190, 271, 310
Fuzzy logic, 248

G
Game theories, 9, 252–255
Generative models, 157, 177, 314
Genetics, 39, 46, 60, 67, 77, 89, 94, 96, 117, 143, 144, 148, 150, 168, 178, 179, 184, 186, 188, 211, 212, 229, 230, 243, 253, 276, 278, 290, 308, 310
Gestalt psychology, 220
Gnoseology, 56, 236
Goal-directed behavior, 35, 63, 159, 164, 181, 189, 225, 249

H
Habits of thought, 209–210, 212
Habituation, 18, 58, 80, 150, 153–155, 165
Hallmarks of cognition, 215–217
Hearing, 122, 124, 126, 222, 245
Here and now, 172, 293, 300, 315
Heterarchical system, 283
Heterotrophs, 118, 212
Heuristics, 27, 34, 57, 70, 71, 225, 242, 252, 292, 311
Historical perspectives, 34, 93
Human cognition, 70, 72, 76, 181, 235, 238, 247, 281, 297
Human mind, 11, 87, 258, 284, 286, 288, 289, 292–297
Hypotheses, 24, 27, 32, 34, 35, 42, 57, 71, 97, 131, 132, 153, 170, 177, 184, 218, 231, 242–244, 250, 258, 279, 301, 309, 310

I
Imaginative thinking, 304
Immunological memory, 23
Implicit assumptions, 23, 27
Individuality, 6, 36, 165–179, 226
Individual life histories, 168
Inferences, 72, 73, 75, 157, 189, 250, 259, 312
Informal logic, 248
Information processing, 39, 40, 57, 78, 146, 164, 169, 175, 184, 186, 188, 220, 221, 257, 258, 261, 263, 268, 274, 280, 282, 283, 286, 290, 296, 300, 302, 308
Information theory, 85, 91, 208, 256, 257
Innate behaviors, 229
Inner life, 215, 290, 292
Instinct, 171, 229
Integrated information theory (IIT), 175
Integration of information, 6, 184
Intellectualism, 171, 180, 181, 235, 238, 239
Intelligence, 2, 4, 6, 16, 35, 37–41, 45, 47, 48, 59, 63, 69, 90, 116, 134, 137, 167, 168, 175, 179–191, 206, 214, 215, 232, 235, 257, 271, 310
Intentionality, 49, 51, 133, 220, 264
Interaction contexts, 9, 20, 152, 280
Interactomes, 10, 23, 186, 273, 274, 310
Interdisciplinarity, 20, 29, 32–34, 58, 60, 64, 65, 76, 78, 85, 90, 92, 99
Interiority, 183, 215
Internal representations, 40, 41, 61, 175, 216, 258, 299, 302, 304, 314
Internal signaling, 141, 307
Interpretive challenges, 25
Invariants, 10, 151, 261
Invisible, 52–54, 142

K
Kin recognition, 18, 178, 179
Know-how, 165, 239, 278
Knowing, 8, 21, 24, 148, 172, 188, 231, 235, 236, 238–247, 301, 317
Knowing of knowing, 165, 278
Knowledge, 8, 9, 145, 146, 151, 166, 168, 169, 172, 182, 206, 207, 209, 219, 231, 235–246, 248–251, 255, 257, 285, 295, 304, 312, 318
Knowledge and action, 240–241
Knowledge by acquaintance, 238, 239
Know of knowing, 13

L
Laboratory research, 30–32, 129
Languages, 37, 39, 54, 65, 68, 87, 173, 181, 207, 210, 232–234, 244, 249, 251, 258, 283, 287, 294, 304
Latent states of life, 52–54
Learning, 2, 3, 6, 8, 10, 12, 13, 16–18, 21, 35, 37, 40–42, 75, 78–81, 99, 116, 134, 136, 145–165, 180, 187, 189, 190, 221, 228–230, 233, 244, 256, 260, 267, 270, 274, 279, 294–297, 299, 307–313, 315, 316, 318
Learn to learn, 151
Level of explanation, 22, 42, 72
Levels of analysis, 3, 5, 16, 19, 45, 48, 52, 59, 63, 94, 168, 190, 207, 218, 232, 244, 256, 261, 266, 270, 281, 293, 294, 298, 306
Levels of complexity, 6, 9, 10, 120–121, 170, 186, 187, 230, 298, 304
Levels of ecological organization, 94–95
Life contexts, 8, 28, 78, 96, 160, 182
Locomotion, 4, 10, 48–50, 117, 119, 121, 128, 136, 219, 275, 277, 295
Logical positivism, 91, 233, 234
Logics, 156, 235, 247–250
Long-term memory, 150, 155

M
Maps, 41, 160, 273, 299, 300, 307, 308, 310
Materialist perspectives, 288
Mathematics, 86, 252
Measurements, 24–25, 27, 30, 123, 236, 237
Medical botany, 44
Memories, 2, 3, 6, 18, 23, 28, 35, 37, 42, 78–81, 99, 116, 117, 134, 136, 145–166, 172, 180, 187, 188, 190, 220, 230, 233, 257, 265, 267, 279, 294, 295, 297, 304, 308–315, 318
Mental capacities, 11, 12, 22, 39, 65, 72, 281, 284–286, 289–296, 299, 304
Mental contents, 133, 169, 234, 264, 300
Mental representations, 12, 41, 188, 245, 258, 263, 266, 280, 297, 299, 300, 306, 316
Meristems, 117, 272, 276, 277
Meta-analysis, 4, 60, 63
Metabolic representation, 309
Metacognitive, 10, 84, 148, 169, 173, 174, 215, 245, 278, 296, 317
Metaphors, 28, 61, 92, 212, 266, 300
Methodological subdivisions, 210–215
Methodological tools, 19, 27
Microbiome, 96
Microorganism interactions, 96
Mimetic abilities, 140
Mind, 2, 4, 7, 11, 12, 22, 23, 35, 37, 39, 49, 52, 71, 72, 162, 176, 222, 233, 234, 236, 258, 259, 263, 268, 280–282, 284–296, 299–301
Mind-body problem, 286
Mind-life continuity, 282
Mirror neurons, 223, 224, 301
Mirror test, 174
Misrepresentation, 32, 302
Modal and amodal, 262, 263, 315
Modularism, 256
Modularity, 118, 166, 168, 270
Modules, 118, 119, 137, 165, 258, 270, 272, 273, 275, 277, 306
Molecular representations, 299, 309, 310
Morgan's canon, 71, 72
Motor cognition, 223, 241
Motor information, 300
Motor intentions, 133–134, 224
Movement planning, 129
Movements of climbing plants, 6, 116, 129–133
Multidisciplinary framework, 22, 32, 86, 185, 219, 256, 267, 316
Multimodal communication, 144
Multimodality, 144
Multisensoriality, 144
Multisensory perception, 3, 18, 54, 133, 222
Mutualistic associations, 95, 254
Mycorrhizae, 96, 97, 142, 254

N
Nastic movements, 128
Natural logic, 247, 249–250, 255
Natural philosophy, 44
Nervous systems, 1, 4, 7, 10, 12, 16, 22, 34, 36, 38–42, 55–58, 65, 66, 76, 78–81, 87, 121, 132, 136, 142, 146, 147, 151, 157, 163–165, 170, 171, 175, 176, 187, 190, 214, 219, 221, 223, 225, 230, 231, 243, 246, 250, 258–261, 267, 268, 270, 272–275, 280, 281, 283, 284, 290, 293–297, 300, 301, 306, 307, 310, 313
Network neuroscience, 10, 309, 310
Networks, 10, 22, 62, 118, 208
Network theories, 10, 92, 118, 269, 272, 273, 309

Neural cognition, 2, 3
Neural networks, 66, 170, 171, 260, 261, 269, 273, 274, 283, 309
Neurocentrism, 38
Neuroscience, 39, 62, 130, 257, 259, 286, 289
New sciences, 91, 286
Nociception, 54, 55
Nonassociative learning, 58, 153
Nonclassical logics, 248
Non-human organisms, 5, 68, 75, 172, 174, 225, 238, 239
Non-neural cognition, 2, 3, 7
Non-neural learning, 152
Non-random organization and use of information, 39
Non-representational approaches, 12, 23, 285–286
Nonstandard cognition, 17

O

Observation process, 25
Obstacle avoidance, 79, 129, 275
Offline behaviors, 216, 217, 280
Online behaviors, 216
Ontological issues, 7, 207–219
Ontologies, 207, 208, 218
Operant conditioning, 153
Operational closure, 88
Organismomorphism, 73
Organization of information, 5, 82, 313
Organogenesis, 117
Other by difference, 43
Other minds, 215, 291

P

Perception-action, 7, 17, 29, 40, 65, 131, 172, 187, 219–224, 232, 259, 262, 317
Perceptual-cognitive challenges, 48–56
Perceptual inferences, 157, 177
Personal mind, 280, 293
Phenotypic adaptability, 10, 278
Phenotypic plasticity, 6, 59, 67, 119–120, 144, 185, 292, 306
Philosophical research, 2–4, 34–42, 222, 239
Philosophy of action, 226
Philosophy of biology, 7, 91, 216
Philosophy of cognitive science, 1, 2, 4, 11
Philosophy of mind, 1, 2, 11, 20, 47, 61, 284–318
Philosophy of physics, 207
Philosophy of science, 30, 62
Photosynthesis, 118, 119, 125, 138, 143
Physics, 5, 83, 86, 129, 208, 268
Physiological descriptions, 19, 20
Physiological reductionism, 286
Phytosemiotics, 144
Pisum sativum L., 129–131, 159, 162, 164
Plant action, 8
Plant agency, 225
Plant anticipatory behavior, 157
Plant attention processes, 163
Plant awareness, 178, 179
Plant awareness disparity, 48
Plant behavior, 2, 4, 8–11, 16, 18, 19, 23, 29, 34, 36, 41–43, 45–59, 62, 63, 66, 73, 82–84, 98, 118, 130, 133, 134, 136–141, 148, 153, 155, 157, 158, 165, 168, 180, 185–187, 189, 190, 206, 214, 216, 232, 235, 254, 255, 272, 277, 278, 280, 285, 294, 298, 306, 308–310, 316, 317
Plant bioacoustics, 126
Plant blindness, 48
Plant bodies, 4, 5, 10, 23, 39, 56, 57, 118, 119, 126, 150, 159, 164, 190, 246, 269, 273, 274, 277, 279, 284, 307, 309, 310, 313
Plant cells, 28, 82, 117, 123–126, 149, 150
Plant cognition, 1–5, 10–12, 16, 21, 22, 33, 38, 39, 51, 59–62, 67–68, 83, 99, 134, 137, 160, 165, 175, 187, 189, 190, 216, 218, 235, 267–280, 286, 296–298, 307, 310–312, 314, 315
Plant cognitive ecology, 56, 95–100
Plant communication, 6, 54, 77, 98, 116, 122, 141–145, 315
Plant consciousness, 6, 33, 175–177, 231
Plant defenses, 143
Plant ecology, 93, 191, 253
Plant electrophysiology, 46, 123
Plant hormones, 142
Plant individuality, 116
Plant intelligence, 6, 39, 119, 124, 137, 179–186, 214, 215, 218, 271
Plant interactions, 6, 59, 134, 186, 209, 255
Plant learning, 152–156
Plant memory, 23, 147–150, 311
Plant minds, 22, 296
Plant movements, 4, 48, 51, 57, 128–131, 133
Plant neurobiology, 56, 57
Plant organization, 5, 116–121
Plant physiology, 47, 59, 122, 231–232
Plant representations, 311, 313
Plant senses, 124

Plant sensitivity, 56, 125, 177, 252
Plant signaling, 57–59, 98, 141, 144, 186, 250
Plant structures, 63, 277
Pleasure and pain, 54, 55
Pluralism, 282
Pluralistic approaches, 34, 59–63
Pluralistic perspective, 22
Plurality of meanings, 36
Practical applications, 24, 44, 255
Practical inferences, 250
Practical intelligence, 11
Practical knowing, 242
Practical knowledge, 10, 238, 242, 247, 267, 278
Pragmatic encroachment, 239
Pragmatic knowledge, 242
Pragmatic logic, 249
Predictability of phenomena, 77, 164
Predictive models, 187, 190, 314, 316
Predictive processing, 62, 157, 177, 315
Pre-reflective consciousness, 171, 174
Probabilistic logic, 248
Problem solving, 16, 21, 37, 40, 42, 75, 80, 134, 137, 154, 164, 181, 183, 188, 243, 250, 256–258, 267, 280
Procedural knowledge, 238, 239, 241
Procedural memory, 147, 148, 313
Process ontologies, 7, 217–219
Processual perspective, 281–284
Progress in science, 32–33
Propositional knowledge, 238, 239
Proprioception, 125, 132

Q

Quantum mechanics, 91, 208
Quantum physics, 236

R

Ramifications, 255
Rationality, 21, 49, 249, 287
Reflexes, 49, 130, 154, 182, 228, 233
Relational contexts, 74, 252, 276
Relational development, 2, 17
Relational processes, 86, 160, 271
Relativity, 180, 208
Renaissance botanists, 211
Representation, 11, 12, 23, 40–42, 69, 132, 147, 148, 158, 159, 172, 173, 177, 188, 217, 240, 258, 262, 266, 285, 290, 293, 296–315, 317
Representational cognition, 297–298
Representational demarcation challenge, 23
Representational explanations, 264
Representational knowledge, 239, 240
Representational models, 157, 261, 299, 305
Res cogitans, 287
Res extensa, 287
Response selection, 160–162
Responsiveness, 56, 163, 165, 176, 177, 305
Risk sensitivity, 162
Role of chance, 47
Role of the context and other organisms, 95–100
Root behavior, 47, 52, 159, 166, 252, 253
Root-brain hypothesis, 47, 184

S

Scala Naturae, 26
Scientific paradigm, 62
Scientific practices, 24–29, 32, 71, 209, 217, 236, 237
Scientific research, 16, 23–34, 50, 55, 90, 306
Seed dispersal, 53
Seed dormancy, 52, 122
Self-awareness, 22, 173, 174, 189, 245, 270, 289, 294, 296
Self-consciousness, 166, 172–175, 183, 232, 289, 298
Self-organization, 10, 62, 87, 88, 90, 245, 259
Self-recognition, 18, 174, 178, 245
Self-sufficiency, 88, 96
Semantic content, 258, 300, 309
Sensitivity, 4, 6, 55, 56, 58, 82, 116, 122, 124–128, 138, 162, 176, 177, 231, 252
Sensorimotor, 10, 11, 49, 66, 136, 174, 190, 191, 223, 224, 241, 259, 262–264, 275, 278, 291, 295, 299, 300, 303, 315, 317
Sensorimotor representations, 301
Sensory capacities, 7, 125, 133, 219
Sensory dominance, 145
Sensory specialization, 124
Sentience, 55, 175, 176
Sessile organisms, 1, 5, 16, 118, 119, 255, 275
Shade-avoidance, 138
Short-term memory, 150
Signaling networks, 10, 38, 58, 137, 149, 153, 187, 258, 269, 270, 284, 285, 306, 307, 309, 310, 313
Silent languages, 4, 52–54
Similarities and differences, 7, 64, 74, 75, 210

Simulations, 262, 301, 304, 313, 315
Situated and natural practices of knowing, 235–255
Social life, 252–255
Soil legacy, 279
Soil microbiota, 11, 280
Somatotypic representation, 299
Species-specific traits, 53, 255, 291
Speed and accuracy, 131
Subdivisions methods, 210–215
Subjective environments, 99
Subjectivity, 166, 225
Sub-movements, 131
Subsymbolic, 241, 258, 259, 261, 299, 309
Supervenience, 287
Symbiotic relationships, 51–52, 96, 138, 280
Symbolic representations, 258, 260, 261, 300
Sympathy, 54–56
Systemic approaches, 91, 309, 310
Systemic-relational reasoning, 85, 92
Systems thinking, 92
System theory, 84

T

Tacit knowledge, 241
Tactile perception, 50, 125, 126, 174
Taxonomy, 45, 210, 211
Teleonomy, 135, 136
Temporality of cognition, 13, 312–318
Temporality of processes, 312–318
Temporal persistence, 144
Terminological and conceptual analysis, 36–38
Terminology, 37, 57, 72, 175, 273, 306
Theoretical and technological advances, 50–52, 56
Theoretical perspectives, 2, 3, 6, 9, 206, 256
Theory and practice, 25–29
Theory of event coding (TEC), 315
Theory of knowledge, 235, 236, 292
Theory of mind, 258, 300
Theory of perceptual symbols, 315
Thermodynamics, 88, 89, 91, 94, 281
Theromorphic approach, 73
Time-lapse, 51, 131, 156, 268
Top-down and bottom-up process, 35
Tree of life, 79, 121, 182, 243, 251, 294
Trial-and-error-like strategies, 154
Trivialization of cognition, 212, 280–281
Tropisms, 46, 128, 129, 158
True/false, 40

U

Unicellular and multicellular organisms, 147
Uniqueness, 68, 74
Utility and simplification, 209–210

V

Vagile organisms, 10, 119
Value assumptions, 26
Vernalization, 28, 149
View from nowhere, 30, 236
Visual perception, 125, 188, 261, 264
Volatile organic compounds (VOCs), 18, 98, 132, 140–144, 250

W

Wood-wide-web, 97
Working memory, 147

Z

Zoocentrism, 68
Zoomorphism, 60, 76

9783032013279